STO

ACPL ITEM
DISCARDED

8.10.77

Earth, The Living Planet

MICHAEL J. BRADSHAW

Principal Lecturer in Geography and Geology,
College of St. Mark and St. John, Plymouth

A HALSTED PRESS BOOK

JOHN WILEY & SONS
NEW YORK

Library of Congress Cataloging in Publication Data
Bradshaw, Michael J
    Earth, the living planet.
    "A Halsted Press book."
    Bibliography: p.
    Includes index.
    1. Physical geography.   2. Ecology.   I. Title.
GB60.B69   574.5'2   77-946
ISBN 0-470-99107-0

Published in the USA by Halsted Press
a division of John Wiley & Sons Inc. New York

First published by Hodder and Stoughton Educational

First printed 1977

Copyright © 1977 M. J. Bradshaw

All rights reserved. No part of this publication may be reproduced or transmitted in any form or by any means, electronic or mechanical, including photocopy, recording, or any information storage or retrieval system, without permission in writing from the publisher.

Printed and bound in Great Britain

Contents

		PAGE
	Author's Preface and Acknowledgements	4
1	Energy from the Sun	5
2	The ocean-atmosphere system	11
3	Heating the ocean-atmosphere system	19
4	Water in the atmosphere	28
5	Movement within the ocean-atmosphere system	50
6	Weather systems	61
7	Predicting and altering the weather	74
8	Climate	85
9	World climates	91
10	Climates which are wet all year	97
11	Climates which are dry all year	108
12	Climates wet at one season	118
13	Polar and mountain climates	146
14	Climates of the past — and future?	151
15	Life on the Earth	159
16	Life in the atmosphere and ocean	177
17	Life and the soil	186
18	Competition and living together	206
19	Ecosystems	214
20	The distribution of life	228
21	Optimum land biome-types	242
22	Land biome-types with seasonal climates	251
23	Land biome-types with permanently low temperatures or continuous water shortages	267
24	Oceanic biome-types	278
	Bibliography	289
	Index	295
	Colour section	129–144

Preface

Despite man's increasing powers for exploring the universe, he has not yet established the existence of an equivalent variety of living forms on any other planet. The Earth is unique in this respect, and this book examines the surface environments which make life possible.

The mid-twentieth century has seen the growth of studies of the Earth environment. These have been given particular point by the realisation that this planet contains limited resources of materials, has limited capabilities for energy reception from the Sun and possesses limited space for dumping wastes. It is thus vital for man's continued existence that he acquires a greater understanding and awareness of the processes in the atmosphere, oceans and living realm. The Earth is not just an unlimited source of good things, but a balanced and interacting mechanism requiring careful use.

The close linkages of each aspect of the Earth environment — weather, climate, oceans, living organisms — mean that they should be studied together. Meteorology, climatology, oceanography (or oceanology) and ecology all apply a variety of chemical, physical and biological methods and principles to the investigation of natural situations. The interactions are complex and far-reaching. The forces at work are often immense, and the vastness of realms like the oceans and atmosphere have defeated attempts at intensive exploration until the last few years.

Earth, The Living Planet, together with its companion, *The Earth's Changing Surface*, is written as a reader for Sixth Form and College courses in physical geography and environmental science, but it is expected that the general reader will also find interest in its pages. It is hoped that the foundation of basic concepts and themes will lead to field observation where possible, and to the regular reading of the popular scientific press where the increasing number of developments in these fields are reported. It is also hoped that the reader, like the author, will get a thrill from attempting to explain the coincidence of so many factors of the Earth environment. Without such 'coincidences' there could be no life as we know it on this planet. An important question for every human being is: will Earth, the living planet, die? It is an aim of this book to assess the situation realistically, so as to remove the conservation-pollution debate from the merely emotional to a more informed position.

A book of this nature, which reviews such a wide range of investigatory sciences, is inevitably dependent on the work and advice of many others. Sources of diagrammatic material are acknowledged in the text, but it is impossible to refer to every written source which has been consulted, except to include them in the Bibliography so that others may follow ideas to their origin.

It is always invaluable to have criticism before a book is published, and I am deeply in debt to my colleagues Mr A.J. Dunk and Mr A.E. Vines, who have read the manuscript between them and have made many useful suggestions. Mr C.E. Everard has played a very important role as Adviser to the series: his comments are always constructive, and it is a pity that illness has prevented a greater degree of participation by him. Mr R. Stone, my consulting editor, has brought the rigour of a physicist to the basic science and his own expertise in the written word to the text in general. Mr D.E. Pedgley was able to read chapters 4 and 11 in some detail and his ideas have helped immensely.

A number of agencies in the United States of America have been very helpful in producing photographic material, and I must mention Mr Les Gaver of the National Aeronautics and Space Administration (NASA), Mr Bill West of the National Oceanic and Atmospheric Administration (NOAA), Mr Wil Dooley of the US Geological Survey, and Mr Joseph Larsen of the US Department of Agriculture Soil Conservation Service. Photographic sources are acknowledged in the text.

Michael Bradshaw
PLYMOUTH, 1977

1

Energy from the Sun

The Earth is situated approximately 150 million km from the Sun, and intercepts only one part in two thousand millionths of the energy radiated to space by the Sun (which is known as **insolation**). Even this tiny fraction is sufficient to provide light and warmth for the maintenance of a varied life on this planet, plus the heat to drive atmospheric and oceanic movements. There are thus two major systems on the Earth powered by solar energy: the living system and the ocean-atmosphere system, and these form the subject of this book. Both are 'closed' systems, in that energy is imported and exported, but there is virtually no exchange of mass with the surrounding space system. Both systems are dependent on the materials supplied by the solid Earth.

The Earth is unique amongst the planets of the Solar System in possessing a complex variety of living forms: this is made possible by the nature of the living system environment. This is effectively the ocean-atmosphere system and its interactions with the solid Earth system (which is studied in *The Earth's Changing Surface*). The Sun not only maintains its planets in orbit around itself by the centripetal attraction of its own gravitational force, but supplies a continuous flow of radiant energy to the whole Solar System. The planets which are closest to the Sun have extremely high surface temperatures (over 400°C was measured by probes on Venus), and those which are farthest away have extremely low temperatures. Life as we know it on Earth will exist only where water is available (0-100°C temperature range). Further restrictions are imposed by the fact that the complex molecules which make up living protoplasm break down when temperatures exceed 40°C. Mars, which is somewhat farther away from the Sun than the Earth, has areas where temperatures may reach 20-30°C at times, and it is possible that limited forms of life exist on that planet despite the dominance of low temperatures, the rarified atmosphere and its high carbon dioxide content. It seems unlikely that any other Solar System planets contain life.

Interactions between the solar radiation and the ocean-atmosphere system cause water to be circulated through the atmosphere to the continents, and interactions between the ocean-atmosphere system and the solid Earth system lead to the disintegration of rocks and the production of soil materials. Life has been able to spread to almost every part of the Earth because of the circulation of water and distribution of heat by atmospheric movements. The energy amounts involved in these transfers and transformations are too vast for most people to comprehend (Figure 1.1).

Daily solar energy received by Earth	1 (i.e. 1.49×10^{22} joules)
Daily solar energy output	10^{10}
Indian monsoon circulation	10^{-1}
Average midlatitude depression	10^{-3}
Average hurricane	10^{-4}
Average midlatitude summer thunderstorm	10^{-8}
Average local shower	10^{-10}
Average lightning stroke	10^{-13}
Individual gust of wind near Earth's surface	10^{-17}

FIGURE 1:1 *Energy quantities involved in weather phenomena. (After Sellars, 1965.)*

Energy is the capacity for doing work. No movement can take place without the transformation of energy. Energy can be in different forms — mechanical, sound, light, heat, electrical, nuclear — and these can be transformed from one to the other (Figure 1.2). All the various forms of energy can be transformed into heat energy, and most transformations give rise to some heat: an electric light bulb converts electrical energy into a little light and more heat. In this case the heat is unwanted and it can be said that it is 'lost' from the bulb to the atmosphere, but according to the Law of the Conservation of Energy — the First Law of Thermodynamics — energy is neither created nor destroyed (cf. also the somewhat similar Law of Conservation of Matter). Many such energy transformations occur in the ocean-atmosphere and living systems (Figure 1.3).

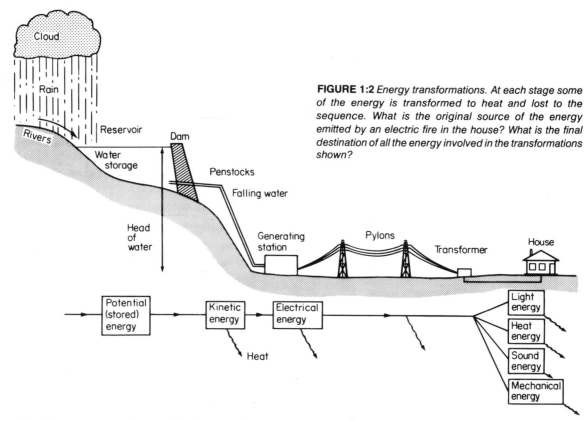

FIGURE 1:2 *Energy transformations. At each stage some of the energy is transformed to heat and lost to the sequence. What is the original source of the energy emitted by an electric fire in the house? What is the final destination of all the energy involved in the transformations shown?*

The Sun's energy is supplied by nuclear sources. The discovery of nuclear energy in the early twentieth century has led to man extending his calculated allowance of the Sun's life: it is now agreed that it will last for at least 5×10^{12} years! The formulation of the laws of thermodynamics in the late nineteenth century suggested that the energy of molten masses like the Sun (and, it was assumed, the original Earth) would be dissipated rapidly to space in a constant manner. A maximum period of 2×10^8 years was allowed for the life of the Earth so far. Such systems were regarded as similar to an oil-fired heating system: the source of energy (oil) would be used up at a constant rate and provide a constant supply of energy until it was finished. The understanding of nuclear energy has shown that some energy sources are not used up in this way. The Law of Conservation of Matter and the Law of Conservation of Energy do not apply individually to nuclear reactions, in which a mass of a radioactive substance is converted into a smaller mass of a 'daughter' substance and vast quantities of energy. Thus mass is being converted into energy, and a new law is necessary, the Law of Conservation of Mass-Energy, which states that the sum of mass and energy are constant. Einstein's equation $E=mc^2$ (energy = mass multiplied by the square of the velocity of light) expresses the relationship for the calculation of the energy released by a given mass and shows how vast this is even for a small mass ($c = 300 \times 10^6$ m sec^{-1}). A radioactive substance will thus produce large quantities of energy from a small mass, and, in addition, will release energy rapidly at first and then at slower and slower rates so that the supply lasts much longer.

In the Sun the nuclear conversion involves masses of hydrogen which are changed into smaller masses of helium plus energy. Gases are involved, rather than the solid mineral substances inside the Earth. The process in the Sun is known as thermonuclear fusion, and this could provide almost limitless supplies of energy for man's use on Earth if the nuclear particles, which must be at temperatures of at least 10×10^{6}°C, could be controlled. The terrifying magnitude of the explosion of a hydrogen bomb gives some measure of both the potentialities and the difficulties of controlling this process, but problems of power station safety and radioactive waste disposal would be fewer than with the fission-based nuclear energy produced today.

Energy is radiated from the Sun as an electromagnetic spectrum of widely varying wave-lengths (Figure 1.4). At the short wave-length end of the scale are high energy waves such as gamma rays, x-rays and ultraviolet rays. These are lethal to organisms since the gamma and x-rays cause the breakup of organic molecules, whilst ultraviolet rays cause sunburn. Fortunately the Earth's atmosphere absorbs most of these rays. The longer wave-lengths include radio waves and infrared waves, which are also absorbed in their passage through the Earth's atmosphere. Light waves dominate the sector of the spectrum reaching the Earth's surface. Two major effects follow from this.

1 The atmosphere is heated in two major zones. In the upper atmosphere a concentration of ozone (O_3) absorbs, and is heated by, the incoming ultraviolet and infrared rays. The lower atmosphere, however, is heated only from below, after the incoming radiation has been absorbed by the Earth's surface and re-radiated as longer infrared waves. These two zones of heating give rise to the particular structure of the atmosphere (Figure 1.5). Heating by re-radiation from the Earth leads to an intense circulation in the lower 10-18 km of the atmosphere and causes surface weather changes together with a range of interactions between the atmosphere and the surface rocks. These include the breaking up of rocks into soil particles and their transport by wind, water and ice. Chapter 3 examines in greater detail the implications of heating the ocean-atmosphere system.

2 Life on the Earth is based on plants which can convert the light energy from the Sun into chemical energy (in the process of photosynthesis). Energy is then passed through the whole food chain via plant-eating and flesh-eating animals and converted into the other types of energy necessary to various life processes. Chapter 15 studies photosynthesis and energy pathways in living systems.

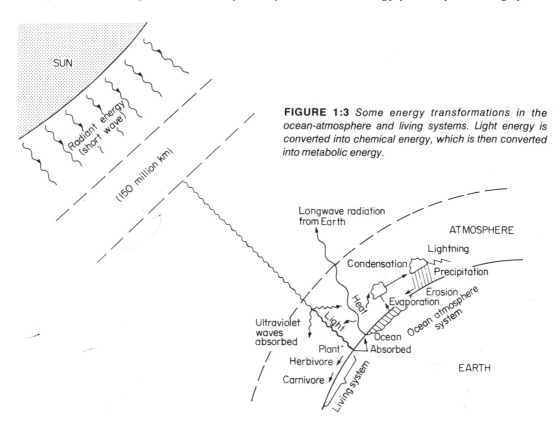

FIGURE 1:3 *Some energy transformations in the ocean-atmosphere and living systems. Light energy is converted into chemical energy, which is then converted into metabolic energy.*

Measurement at the surface of the actual amount of energy received by the Earth from the Sun is not very accurate, since so much is reflected and re-radiated as it passes through the atmosphere. The satellites and space stations being sent up above the atmosphere are enabling more precise estimates to be made. The radiation received from the Sun is measured in terms of the solar constant.

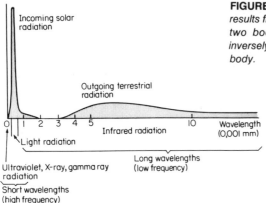

FIGURE 1:4 *Solar and terrestrial radiation. The contrast results from the very different surface temperatures of the two bodies, since the wavelengths of radiation are inversely proportional to the temperature of the radiating body.*

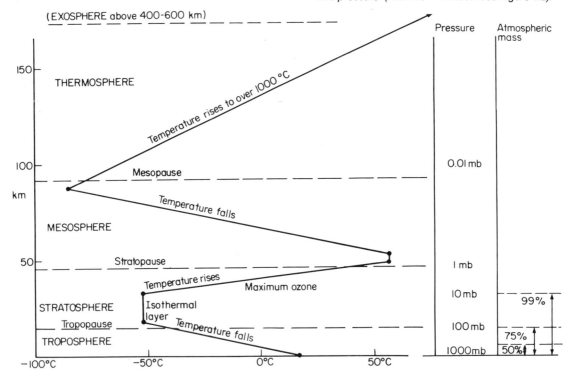

FIGURE 1:5 *The layered structure of the atmosphere. Which are the layers of maximum and minimum temperature? Why do the changes occur? Compare the troposphere and stratosphere in terms of temperature and pressure (N.B. mb = millibar: See Figure 4.2).*

This is defined as the energy in joules received per cm²min on a surface at the top of the atmosphere placed perpendicular to the Sun's rays: it is currently measured as 8.368 joules/cm²min. Another term used in this connection is the langley: 1 langley (ly) equals 4.184 joules/cm². It is interesting to compare the average of 720 ly per day of energy received by the atmosphere from the Sun with the 0.12 ly per day flow of heat from the Earth's interior and the negligible quantity received from other sources in space. The Sun is absolutely dominant in the supply of energy to the Earth's surface.

Variations in the amount of solar radiation, or insolation, received by the different parts of the Earth's surface give rise to differences of heating and to the atmospheric and oceanic circulations. These variations are governed by a number of factors.

1 The Earth is virtually spherical, and it is common experience to observe that more heat is received from the overhead Sun, than when it is shining at a low angle. Thus when the Sun is overhead at the equator (Figure 1.6) it will be at a low angle at the poles. The result is twofold: (a) the Sun's rays pass through a greater depth of atmosphere at the poles than they do at the equator, and (b) equal incoming pencils of rays illuminate a larger area of the Earth's surface at the poles than at the equator, spreading out the incoming energy over a greater area.

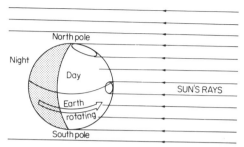
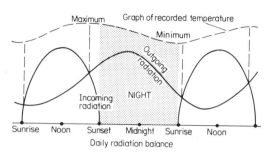

FIGURE 1:6 *Night and day are caused by the rotation of the Earth: only half the planet can be lit by the Sun at one moment. Describe the daily balance of radiation, and its relationship to the graph of recorded temperature. What would be the difference to both situations in (a) the longer nights of midlatitude winters, and (b) the shorter nights of midlatitude summers?*

2 The Earth is rotating on its axis once every 24 hours, so that the half of the planet facing the Sun is in light, and the other half in darkness. Thus each half of the planet is heated and cooled at regular intervals. Compare the situations on other planets with their different periods of rotation (Figure 1.7). Rotation of the Earth is from west to east (i.e. anticlockwise looking down on the North Pole), and the daily experience is of coming out into the light at dawn, followed by the Sun's elevation first increasing and then decreasing until it finally sinks beneath the western horizon and the particular portion of the surface travels on into darkness.

Planet	Mean Distance from sun (Astronomical units: Sun-Earth = 1)	Approximate diameter (Earth = 1)	Mass (Earth = 1)	Average density (10^3 kgm^{-3})	Number of known satellites	Orbital period (Years)	Inclination of orbit to ecliptic (Degrees)	Inclination of equator to orbit (Degrees)	Rotation rate
Mercury	0.38	0.40	0.05	5.1	0	0.24	7.0	(?)	59 days
Venus	0.72	1.00	0.90	5.3	0	0.61	3.4	23	243 days
Earth	1.00	1.00	1.00	5.52	1	1.00	0.0	23	1 day
Mars	1.52	0.50	0.11	3.94	2	1.88	1.9	24	24.5 hours
Jupiter	5.20	11.00	318.00	1.33	12	11.86	1.3	3	10 hours
Saturn	9.54	9.54	95.00	0.69	10	29.46	2.5	27	10 hours
Uranus	19.18	4.00	15.00	1.56	5	84.01	0.8	98	11 hours
Neptune	30.06	4.00	17.00	2.27	2	164.79	1.8	29	16 hours
Pluto	30.44	0.50(?)	0.10(?)	(?)	0	247.69	17.2	(?)	6 days

FIGURE 1:7 *Planets of the Solar System: a comparison of selected characteristics.*
(After Open University Science Foundation Course, 1971.)

3 The Earth's axis is tilted at 66½ degrees to its orbital plane. The amount and direction of tilt do not appear to vary, and so the Sun is overhead at different zones in the course of the Earth's yearly orbit (Figure 1.8). This gives rise to seasonal changes. When the Sun is overhead at the northern tropic (Cancer) in June, the northern hemisphere has its warmest season and so on. Lengths of day and night are also affected: in June the areas inside the Arctic Circle never enter the 'night' zone and for a few days receive 24 hours of daylight, whilst the areas in the Antarctic Circle have 24 hours of darkness. Explain the following facts:

(a) Places near the equator seldom have more or less than 12 hours of daylight and 12 hours of darkness.

(b) British winter days have 8 hours of daylight and 16 hours of darkness, whilst in summer there are 8 hours of darkness and 16 hours of daylight.

Figure 1.9 summarises the distribution of solar radiation over the Earth's surface during the year and emphasises the effect of the dark polar winters and the long summer days in higher latitudes. Just as the 24 hour period of rotation brings areas of the Earth's surface regularly into the Sun's beams, so these seasonal changes lead to a wider distribution of insolation.

4 The ellipse-shaped orbit of the Earth as it revolves around the Sun every 365¼ days brings it closer to the Sun in January than in July. In January the distance between the Earth and the Sun is 146.4 million km and the solar constant is 2.06 ly/min; in July the distance is 151.2 million km and the solar constant is 1.94 ly/min. This increase in solar radiation during January might be expected to affect surface temperatures by up to 4°C, giving the southern hemisphere warmer summers, but other factors, such as the distribution of land and sea, mask the significance of this.

The Earth is thus situated in a good position to become a 'living planet'. Its distance from the Sun leads to neither freezing nor incineration over most of its surface, and its movements enable all parts of the planet to be bathed regularly in sunlight. Differences of insolation exist, but these give rise to vital movements in the ocean-atmosphere system, and to an immense variety of living forms.

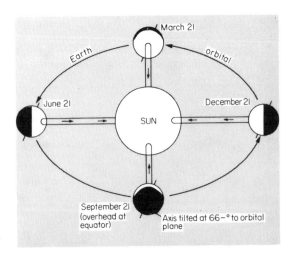

FIGURE 1:8 *The seasons of the year are caused by the tilt of the Earth's axis remaining in the same orientation as the planet revolves around the Sun. How does this affect the position of the overhead Sun and the lengths of day and night in the different seasons (a) at the Equator, (b) in the middle latitudes, and (c) at the poles?*

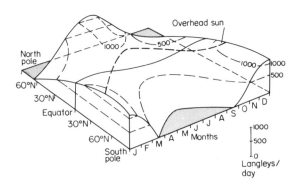

FIGURE 1:9 *The distribution of solar radiation over the Earth during the year. Compare and account for the variations at the poles and the equator.*
(After Davis, in Barry and Chorley, 1968.)

2

The ocean-atmosphere system

The oceans and atmosphere are studied together as a single system because they are linked closely by common interactions and interchanges. The movements within both oceans and atmosphere derive their energy from the Sun: differences in heating lead to the differences in density or pressure which are the basis for wind and ocean current movements. The ocean acts as a store for the heat energy coming from the Sun, being heated to a greater depth than the land areas and releasing this heat to the atmosphere when the waters come into contact with cold air. Ocean currents moving polewards from the tropics accomplish a very important part of the total heat exchange process which gives rise to the weather. Winds blowing over the ocean surface cause waves, and themselves take on the temperature characteristics of the water, together with evaporated moisture. These processes play vital parts in the movement of water around and over the continents and materials are cycled through the combined system. Some of the resultant complex inter-relationships are simplified in Figure 2.1.

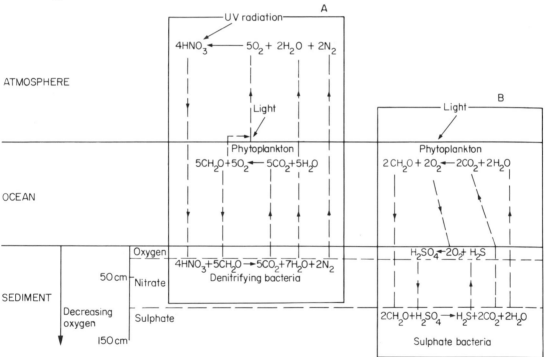

FIGURE 2:1 *Re-cycling materials through the ocean-atmosphere system. Two examples of such cycles are shown (A and B). Note the interactions of atmosphere, water, solar radiation and the bacteria of the marine sediments. Each factor is vital: without the bacteria converting nitrates and sulphates back to water and carbon dioxide, the atmospheric oxygen would be fixed in ocean-floor sediments within 10 million years.*
(After Macintyre, in Moore, 1971.)

Man's activities are affected by the ocean-atmosphere system, and are leading slowly to changes in its characteristics. The weather and atmospheric conditions affect man in many ways (Figure 8.1). Most people prefer a cool, fresh and sunny morning to a hot and sticky afternoon, or a dull, raw winter's day. The athletes involved in the 1968 Olympic Games and the 1970 World Cup experienced difficulties in the rarified air of Mexico City. Extreme heat or cold also lower the efficiency of man — as well as other forms of life. Yet there is much that is not understood about the effect of such factors on man's feelings and emotions. It is clear, however, that activities are often severely restricted by the weather: houses, crops or animals may be damaged or destroyed; transport may be held up by fog, snow, frost and gales; and industrial costs soar when heating or air-conditioning are added. The oceans form zones of difficulty for travel, and the ocean environment as a whole has excluded man as a permanent resident until recently. Experiments in life beneath the sea by man have led to the establishment of small sub-aqua settlements off southern France, California (Sealab) and the Virgin Islands (Tektite). The scientists involved in promoting human life under water would point to the greater economic feasibility of such a move from the overcrowded continents, instead of transportation to other planets!

Atmospheric gases and ocean waters

The gases which make up the **atmosphere** today are dominated by nitrogen and oxygen (Figure 2.2) — all the other constituents together make up less than 1 per cent of the total volume. The proportion of gases (apart from water vapour) is one of the few characteristics of the atmosphere which remains

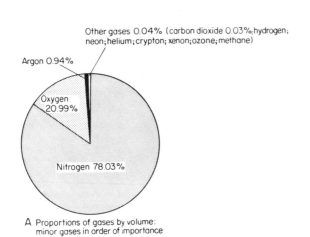

B Gas	Characteristics, functions
Nitrogen	Colourless, odourless, tasteless. Not very active chemically. Dilutant for oxygen. Enters protein molecules via soil, plants.
Oxygen	Colourless, odourless, tasteless. Chemically active: combines readily with other elements. Released by plants in photosynthesis; taken up by animals and plants in respiration.
Carbon dioxide	Absorbs heat radiation from Earth in atmosphere.
Argon, neon, crypton, xenon	Chemically inactive. Present in tiny proportions. Known as the 'noble gases'.
Water vapour	Most variable in proportion and largely concentrated in lowest 10 km. Re-cycled in evaporation-condensation.
Hydrogen, helium	Lightest gases. Present mainly in upper atmosphere.

FIGURE 2:2 *Composition of the atmosphere. The gases in (A) are measured as for dry air (i.e. without water vapour which may constitute 4 per cent by volume locally). Other constituents include pollen and aerosols (i.e. minute particles of dust, salt and smoke).*

constant. Many measurements throughout the world, and at heights up to 80 km, have shown that there is virtually no variation. The main physical changes within the atmosphere come with height:

1 The density of the atmosphere decreases with height. At mean sea level the average density is 1.2 kg m^{-3}; at 100 km above the surface the average density is 8×10^{-7} kg m^{-3}; and at 200 km above the surface it is 1×10^{-10} kg m^{-3}. Ninety-nine per cent of the atmospheric mass is contained in the lowest 30 km (Figure 1.5), and the lightest elements like helium and hydrogen become important only in the upper regions of the Earth's atmosphere.

2 The effect of the high frequency short-wave component of solar radiation is greater at higher altitudes. Some of the ultraviolet radiation is absorbed by the gases oxygen and nitrogen above 80 km, causing them to decompose into free atoms or ionised molecules: this zone is often known as the ionosphere. Below this level, more oxygen molecules are split, but the atomic oxygen recombines with unaffected molecules to give ozone (O_3) in a region called the ozonosphere at 40 km above the surface (Figure 1.5) Nearly all the solar ultraviolet radiation is absorbed, giving rise to heating of the atmosphere at that level, and little reaches the Earth's surface where it would destroy living organisms.

An important division is made in the atmosphere on the basis of these physical characteristics and heating effects (Figure 2.3). The 'lower atmosphere' can be defined as the zone of turbulence, in which the movements causing surface weather largely take place. This zone is known as the troposphere (Figure 2.4) and occupies only the lowest 10-18 km. Above this is the 'upper atmosphere' which has a low density at its base and becomes more and more rarified with increasing height, eventually merging with space at about 600 km. The outermost zone of the atmosphere is a region of neutral atomic and ionised oxygen, nitrogen, hydrogen and helium. The distribution of ions here, in the Van Allen radiation belts of the magnetosphere, is governed by the Earth's magnetic field rather than its gravitational field.

Ground Level	Approx 10-18 km	Atmosphere Space Boundary
Lower atmosphere		Upper atmosphere
Nitrogen and oxygen, together with tiny proportions of other gases		Slightly larger proportions of hydrogen, helium, ozone, ionised gases; increasing with height above surface
Water vapour		Little
Dust particles		Few
Large proportion of atmospheric mass		Extremely rarified

FIGURE 2:3 *Differences between the 'lower atmosphere' and 'upper atmosphere'.*

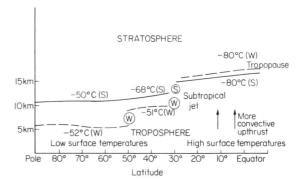

FIGURE 2:4 *The tropopause: seasonal variations in the northern hemisphere. Dashed lines and (W) refer to the winter situation; solid lines and (S) to the summer. Relate the heights of the tropopause at equator and pole to the surface heating pattern. Breaks in the tropopause are the scene of air interchange between the dry, cold Stratosphere and the moister Troposphere. It has been estimated that it takes over two years for a complete interchange in the Stratosphere.*

The **ocean waters** are more variable in composition than the atmosphere, including large proportions of mineral matter as well as water and gases. Dissolved salts, or solutes, are added from the erosion of the Earth's surface rocks and from the eruption of volcanic dust, especially along the oceanic ridges, and are also lost by precipitation to ocean-floor sediments. The concentration of solutes in the water is affected by temporal and regional variations in erosion and precipitation, and also by surface evaporation and the addition of water from rain and rivers. The proportions of salts remain relatively constant in the open oceans (Figure 2.5) due to mixing of the waters, so that the chloride ion content alone is measured to give a rapid indication of the total salinity.

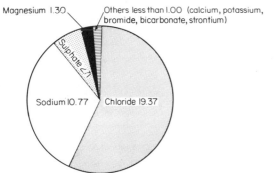

FIGURE 2:5 *The major ions in seawater: figures as parts per thousand (g/kg). The average total concentration, or salinity, is 35 parts per thousand. Sodium chloride is the main compound.*

The salinity of seawater is the mass of dissolved salts in a certain measure, and is expressed as parts per thousand. Thus a 3.5 per cent solution would have a salinity of 35 parts per thousand ($35^o/_{oo}$). The average salinity in the open oceans is 35 parts per thousand, but varies from 32 to nearly 38 (Figure 2.6), and in small confined areas of the ocean margins the waters may be diluted by incoming river water, or may experience an increased salinity because of evaporation (compare the Baltic Sea and the Red Sea).

The sea water also contains very small quantities of materials which are vital for the nourishment and shell structure of marine organisms. Compounds of silicon, carbon, nitrogen and phosphorus together with even smaller quantities (known as 'traces') of elements such as copper, zinc, cobalt, molybdenum, iron and manganese occur in solution in the surface waters where the process of photosynthesis takes place and provides the basis of life by converting mineral salts into organic matter. The food chain continues in the surface waters, but much of the organic matter sinks when organisms die. Bacterial activity returns the nutrients to the ionic form, and the oceanic circulation brings them back to the surface waters (Figure 2.1). In addition, all the atmospheric gases are present in solution in sea water. Thus the oxygen content varies mainly within the range 1-6 ml/l, with high values at the surface and in cold areas. Nitrogen occurs in quantities ranging from 8-14 ml/l, but seems to have little effect on the life in ocean waters.

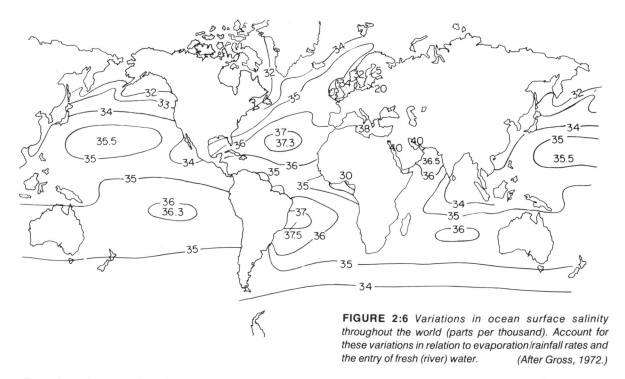

FIGURE 2:6 *Variations in ocean surface salinity throughout the world (parts per thousand). Account for these variations in relation to evaporation/rainfall rates and the entry of fresh (river) water.* (After Gross, 1972.)

Density changes in the ocean waters are associated with differences of salinity and/or temperature: cold, saline water is the most dense form and sinks to the ocean floors, while warm, low salinity water remains at the surface. A well-developed vertical stratification of density is established in ocean waters and strong, deep vertical currents are thought to exist only in high latitudes. Compare the situations in the atmosphere and oceans: the atmosphere is warmed from the base upwards during the day (chapter 3), and warm, less dense air may lie under more dense air, leading to vertical movements which restore the density equilibrium but which give rise to weather changes in the process. The ocean, however, is warmed and cooled at the upper surface and this gives rise to a contrasting situation with three major density zones being recognised.

1 The surface zone, extending to 100 m depth (i.e. only 2 per cent of the ocean volume), experiences seasonal changes where they exist (mostly in the middle latitudes — Figure 2.7), has the lowest density of water in the oceans and is the zone where most marine organisms live since photosynthesis occurs only at this level.

2 The pycnocline zone is a region where water density increases rapidly. In the tropics this is due to marked surface heating throughout the year which keeps the zone above the pycnocline at low density. At higher latitudes salinity is lower at the surface over most of the oceanic area due to rainfall and lower rates of evaporation (Figure 2.7). In both areas the higher density water beneath prohibit vertical circulation and provide a stable layer.

3 The deep zone lies beneath 2 km, coming to the surface in places at high latitudes, and includes 80 per cent by volume of the ocean waters.

The origin of atmosphere and oceans

Whilst much remains unknown about the way in which the waters and gases accumulated around the Earth, they both seem to have originated from the volcanic outgassing which has taken place since the formation of the planet. Some would suggest that an original atmosphere, derived from cosmic gases, was largely lost as the Earth heated up. The basis for this idea is that the atmosphere now has relatively small proportions of neon, argon, crypton and xenon compared with other planets. This event may have led to a new start in which most of the atmospheric constituents came from inside the Earth — water vapour, carbon dioxide, carbon monoxide, nitrogen, chlorine, hydrogen and sulphur

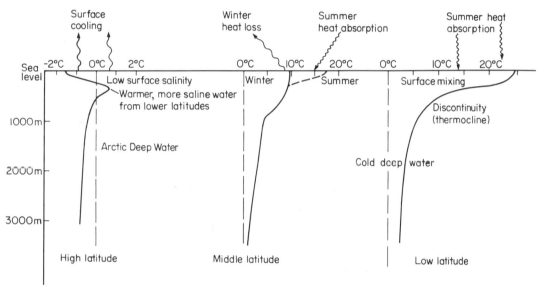

FIGURE 2:7 Some temperature profiles in the deep oceans. How does surface heat absorption and loss affect the profiles? Relate these profiles to the layered structure of the atmosphere (Figure 1.5). (After Tait, 1968.)

dioxide. On cooling, the water vapour would condense into the ocean. Again there is a difference of opinion as to whether the atmosphere and oceans built up suddenly, or whether they have gradually increased their masses through geological time (Figure 2.8). The estimated present volume of the ocean waters is 1370×10^6 km^3. Water is being added at the rate of 0.1 km^3/year, calculated from the quantity of volcanic rock being erupted and the gases released: if this rate has continued since the beginning of the Cambrian period (600 million years ago) a total of only 60×10^6 km^3 will have been added in that time. This would support the idea of an early and rapid build-up, followed by more gradual additions. The free oxygen in both the atmosphere and ocean waters probably did not enter the system until life developed on the planet (chapter 15); the salts dissolved in the ocean waters have accumulated following the erosion of the land areas.

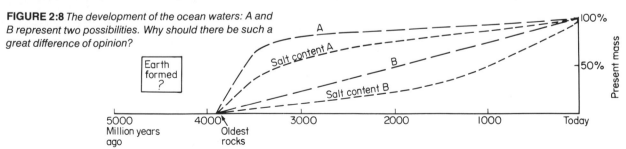

FIGURE 2:8 The development of the ocean waters: A and B represent two possibilities. Why should there be such a great difference of opinion?

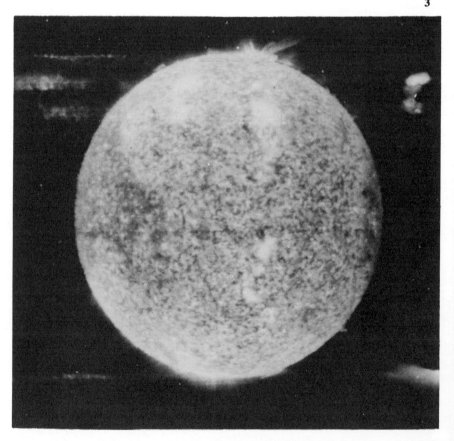

The Sun is an average star, 1.384×10^6 km in diameter, but with 333 000 times the mass of the Earth. It is a gaseous ball, rotating once every 27 days in its middle latitudes, but taking just over 28 days at the poles (i.e. it is not a rigid body). Surface temperatures are approximately 6000°C, but must rise to millions of degrees in the interior, where hydrogen atoms are transformed into helium atoms. The surface region is known as the photosphere, and is a seething mass of hot gases rent by turbulence from below, bubbling to form flares. These flares include the surface sunspots and towering prominences — blasts of hot gas shoot upwards and overcome the gravitational pull of the Sun (280 times that of the Earth). Surrounding the photosphere is the corona, or outer atmosphere, which extends past Mercury and has temperatures rising to millions of degrees.

Plate 1 The Sun, photographed in the hydrogen-alpha wavelength from the NOAA Space Environmental Laboratory at Boulder, Colorado, on 7 August 1972. This just preceded a large flare.

Plate 2 The flare of 7 August 1972.

Plate 3 The solar eruption of 10 June 1973 — a photograph obtained during the first manned Skylab mission by an ultraviolet spectroheliograph: the eruption which can be seen on the left side, extended to one-third of a solar radius. It led to the formation of a coronal bubble. (All NOAA)

Maintaining the environment. Examples of the ways in which man's activities can affect the resources of the Earth's surface. Some activities are more successful because they are related to an understanding of natural processes and the interactions between atmosphere, living organisms and the surface materials.

THE OCEAN-ATMOSPHERE SYSTEM 17

Plate 4 Contour ploughing in Winneshiek County, Iowa. Land is ploughed parallel to the contours to prevent the erosion of topsoil: furrows retain the surface moisture.

Plate 5 The Dust Bowl, Colorado. A farmstead in Baca County, abandoned in the 1930's. Severe wind erosion followed the ploughing of the soils in this semi-arid region. The government then purchased the land, and it was planted with cover crops of sudan cane and broom corn; these will be followed by attempts to replant the natural grasses.

Plate 6 Windbreaks in Waushara County, Wisconsin. This is a dry part of central USA, and a combination of irrigation and planted windbreaks have made it possible to grow wheat without loss of topsoil by wind action.

Plate 7 Lum Hollow, Woodbury County, Iowa. The headward advance of the gully (towards the camera) destroyed the bridge in 1948 and advanced a further 800m by 1956. Increased concentration of runoff waters led to flooding farther down the drainage basin, and the Little Sioux Flood Control Project was initiated. This included a series of small dams to impound the water, and stabilised the situation.

Plate 8 Control of gullying in Texas. The severely eroded area required extensive treatment. The gullies were sloped and grassed, the excess surface water was diverted into special channels, and Bermuda grass (which can now be grazed) was planted. (All US Dept of Agriculture, Soil Conservation Service)

Meteorology and oceanography

The development of scientific studies in the oceans and atmosphere is a recent phenomenon. Whilst man has always realised their effects on his activities, both realms have seemed to be vast, mysterious and even sinister. This attitude was fostered by a lack of knowledge resulting from the limitation of man to one surface of each realm. This limitation has been overcome (Figure 2.9), and a great deal is now known about the full three-dimensional picture, although much more needs to be added. At least it is now known that movements in the upper atmosphere and in the depths of the oceans profoundly affect what happens at the surface.

The meteorologist observes and records weather phenomena including surface pressure, temperature and humidity, wind speed and wind direction, amount of precipitation (hail, rain or snow), duration of bright sunshine and cloud information (i.e. the type of clouds and the proportion of the sky covered). Some of these observations can be carried out at various levels in the atmosphere (Figure 2.9). Each observation is an aspect of the atmospheric state at any one moment and is closely connected with other aspects: pressure is influenced by temperature because heating of part of the atmosphere will cause the air to become less dense and to rise, lowering the pressure at that point; warm air can hold more moisture in the gaseous state than cold air; and winds are related to pressure differences and thus to the heating pattern.

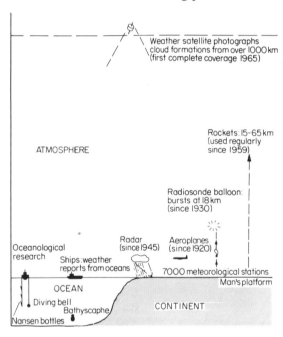

FIGURE 2:9 *Platforms of meteorological and oceanographical observation. Both studies face the problems of investigating the depths of the environment. The Nansen bottles are used to sample seawater and record temperatures at predetermined depths.*

The records of such observations over the past 100 years form the basis of theories concerning movements, reactions and changes taking place in the atmosphere (chapters 3-6). Regular observations taken at meteorological stations and on merchant ships around the world, as well as by the increasing number of weather satellites, provide the raw materials for the weather maps produced by meteorological offices, and the analysis and interpretation of these maps enable weather forecasts to be made (chapter 7).

Oceanography involves the study of 71 per cent of the surface area of the Earth, and includes all aspects of investigation in that realm. The composition of the ocean waters and their movements (chapters 3-6), the living organisms in the oceans (chapters 15-20, 24) and the geological features of the ocean floor (*The Earth's Changing Surface*) all form part of the oceanographer's studies.

Here are two realms of environmental study which have provided a great challenge to man, living on his platform of dry land. It is now seen that the three-dimensional study of these realms is essential, and that they must be treated as a single system.

3

Heating the ocean-atmosphere system

Heat energy from the Sun is the ultimate source of movements in the ocean-atmosphere system. The transfer of heat from the Sun to the ocean and atmosphere takes place in three ways.

1 Radiation enables energy to flow from one body to another in the form of electromagnetic waves. Hot bodies, like the Sun, radiate more intensely and mostly in the short wave-length range, whilst the cooler Earth radiates in a longer wave-length range (Figure 1.4).

2 Conduction takes place when a hot body is in contact with a cooler body: some of the kinetic energy of the molecules in the hot body is transferred to those of the cooler body, so that it heats up. The rate of heating increases with the temperature difference, and is affected by the substances involved: some are good conductors of heat, some poor (Figure 3.1).

3 Convection involves mass movement within a liquid or gas and takes place where the fluid is heated locally, causing one region to become less dense, to rise and to be replaced by surrounding cooler fluid; alternatively it may be cooled, causing sinking (Figure 3.2).

Good conductors	Moderate conductors	Poor conductors
Metals	Rocks	Air
	Soil	Snow
	Water	
	Ice	

FIGURE 3:1 Categories of heat conductors. Notice where snow melts first: roads, grass or earth; compacted or untrodden places? Which is the better conductor, wet soil or dry soil?

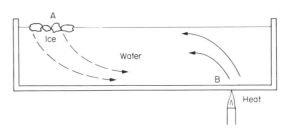

FIGURE 3:2 Convection. At A the water is cooled and becomes more dense, sinking to the bottom of the trough. At B the water is heated, becomes less dense, and rises. A circulation is soon established.

Heat transfer in the ocean-atmosphere system

All of these processes occur in the ocean-atmosphere system. The short-wave radiation from the Sun passes through the atmosphere with little heating effect, although much of the energy is diverted or absorbed on the way (Figure 3.3). Most of the ultraviolet wave-length band is absorbed and transformed to heat. More absorption takes place by the clouds and water vapour lower down. In addition, scattering takes place in the denser, lower atmosphere, where there is more dust and water vapour. The effect of this is greatest at the blue end of the visible spectrum, a fact which accounts for the colour of our clear skies. Some of this scattered energy is deflected Earthwards, but much is lost to space. Pictures of the Earth taken from space show that approximately half of the surface is obscured by clouds at any one moment. In addition to absorbing a small proportion of the incoming energy, they reflect a much larger proportion straight back to space: this reflected energy, lost to the Earth, is known as the **cloud albedo**, and is supplemented by reflection from the Earth's surface to make up the total **planetary albedo**. Such reflection is very high from snow and ice, but minimal when the sun is directly above ocean waters. All these effects vary with latitude (Figure 3.4).

Hardly any heating of the atmosphere has taken place by the time the incoming radiation reaches the Earth's surface: the dominantly short wave insolation does not heat up the air, except in the upper atmosphere, where the ozone layer absorbs the ultraviolet radiation. The Earth's surface absorbs the radiation which penetrates the atmosphere, and re-radiates it as long wave heat. There is also some conduction (but air is a poor conductor), and some air heated near the surface rises convectionally, but radiation is the chief process. The long wave heat energy is absorbed readily by the carbon dioxide and the water vapour concentrated in the lowest levels, and the troposphere is heated from the ground upwards. This is why it gets colder, rather than hotter, towards the top of a high mountain, which may be capped with snow. The process (Figure 3.3) involves not only the radiation from the Earth, but also re-radiation back to the Earth after absorption by the atmosphere and the convectional transfer effect of latent heat through the evaporation of water at the surface (absorbing latent heat) to condensation at higher levels (liberating latent heat).

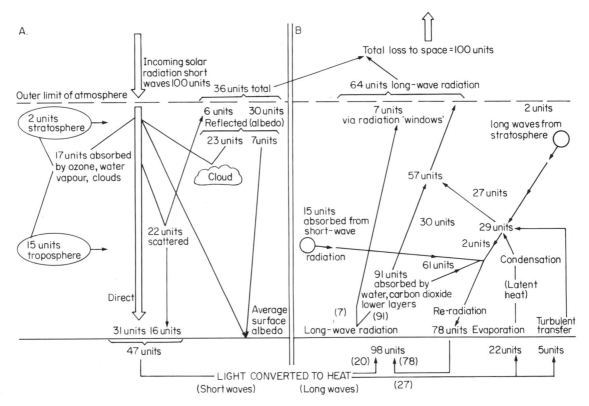

FIGURE 3:3 *The balance of the atmospheric energy budget. Notice the complex system for heating the atmosphere: a combination of radiation and re-radiation, release of radiant heat, turbulent transfer and short-wave absorbed radiation. The items in circles on the right-hand side have been transferred from the left.*
(After Barry and Chorley, 1968.)

The combination of the total planetary albedo, and the loss to space by radiation from the Earth, balance with the incoming solar radiation (Figure 3.4). Thus the ocean-atmosphere system is not becoming hotter or colder. There is a state of balance, in which the continual loss to space is compensated by a continual supply from the Sun. Compare the complex Earth insolation budget with that for the Moon (Figure 3.5).

The oceans play an important role in achieving this balance by storing heat, moving it to cold areas by means of ocean currents and releasing it there. The Sun's rays penetrate deeply into water, especially where the Sun is at an angle of 60 degrees or more and reflection is low: 10 per cent of the radiation may penetrate to over 10 m in such circumstances. Little heat is re-radiated by the water, but is transferred in other ways such as conduction, convection and evaporation.

Temperatures at the Earth's surface

The normal method of observing the effect of solar energy on the weather at a particular place is to measure the temperature of a shady situation (i.e. out of the Sun's direct rays). A map of the world (Figure 3.6) shows January and July temperatures reduced to sea level (in order to mask the effect of altitude) and drawn as isotherms (lines joining all places with the same mean temperature).

A number of factors are responsible for these temperature distribution patterns. It is already clear that the contrasting amounts of solar radiation from the overhead and low-angle Sun account for the fall in temperature from the equator to the poles and that the seasonal differences in temperature are related to the tilt of the Earth's axis and the variations in solar radiation received for this reason. But so far the Earth has been viewed as if it had a uniform surface and no relief, and as if the atmospheric conditions like cloud cover were the same everywhere. This is not so; a number of other factors cause departures from the basic pattern of isotherms running parallel to the lines of latitude.

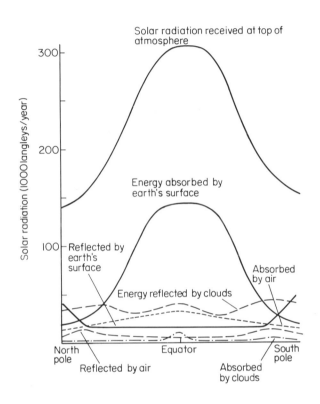

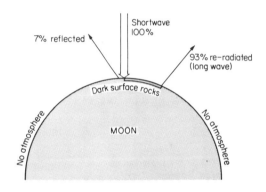

FIGURE 3:4 *Variations in solar radiation received, and in reflection and absorption, at each latitude. Explain the variations shown by the graphs. Why is the solar radiation received greatest at the Equator? Why is the surface reflection greatest at the poles? Why is more energy absorbed by the air and clouds near the Equator?*
(After Sellars, in Barry and Chorley, 1968.)

FIGURE 3:5 *The Moon's energy budget. Compare this with Figure 3.3. What is the effect of the absence of an atmosphere? What would be the difference if the Moon rocks were white?*

The ocean-continent contrast

The most important factor is the difference in heating properties between the two main types of surface: land and sea. This difference is a combination of effects.

1) Whilst the albedo of land surfaces can be high — from 10 per cent over forests to 40 per cent over bare rock desert — it is only 2-3 per cent over the oceans when the Sun is at an angle of 60 degrees or more. Thus, more energy is available to heat the seawater, except in high latitudes, where the Sun is at a lower angle and the oceanic albedo is high, and where both land and sea are often covered by snow and ice reflecting 80-90 per cent of the radiation. The long hours of solar radiation in polar summers have little effect, since the energy is largely used up in melting snow, or is reflected back to space.

2) The specific heat capacity (i.e. the number of joules required to raise the temperature of 1g of a substance by 1°C) of water and soil differ. Water takes more energy to heat it than equivalent masses of most substances, and the comparison with dry soil is 5:1. If equal volumes (which are more significant than equal masses in the present context) are considered, the ratio is still 2:1. Thus, water requires more energy to heat it, and hence the process takes longer. The reverse also holds true: water releases more heat, more slowly. It has been estimated that when the uppermost 1m depth of water cools by only 0.1°C it raises the temperature of the overlying air by 10°C to as much as 30 m above the water surface!

FIGURE 3:6 A World mean temperatures, January: degrees Celsius. B World mean temperatures, July: degrees Celsius. The smaller maps show isanomalies (i.e. lines joining places of equal deviation from the average temperature for the latitude). Notice the positions of the principal anomaly areas for the winter and summer hemispheres. Why are there more temperature anomalies in the northern hemisphere in both seasons?

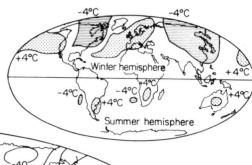

A

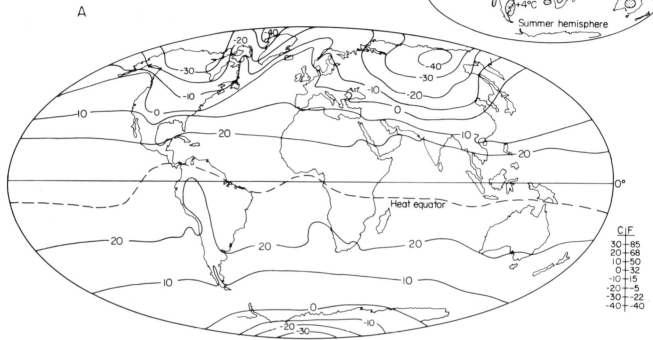

3) Even more important is the fact that radiation penetrates deeper into the transparent ocean waters than into the opaque land surface: 10 per cent of the radiation may reach down as far as 10 metres in tropical waters. At the same time, the restless, turbulent waters of the oceans are continuously on the move due to waves and currents. This leads to an even deeper distribution of the heat energy in the oceans as compared with the land (Figure 3.7). Heat transference in the soil is merely by conduction, and even in wet soils the depth reached is small.

4) Evaporation is generally more important over water surfaces, although the combination of evaporation and transpiration over a tropical forest may be almost as great as evaporation from the ocean. The greatest contrast occurs between the ocean surface, where up to 80 per cent of solar energy may be absorbed in evaporation, not to be released again until the humid air is wafted to high altitude, and a continental desert, where virtually all the energy reaching the surface is available for re-radiation to warm the atmosphere immediately above the ground.

The combination of these factors gives rise to the elements of continentality and marine influence in the temperature variations over the Earth's surface. The land heats and cools more rapidly than the sea. There is thus a shorter time-lag over land between the period of maximum insolation and maximum temperature than over the oceans. Coastal positions will have a smaller annual temperature range than inland places, and, on a larger scale, the greater proportion of land in the northern hemisphere means that the northern summers are warmer than the southern summers, and that the northern winters are cooler.

	Northern hemisphere	Southern hemisphere
Mean summer temperature	22.4°C	17.1°C
Mean winter temperature	8.1°C	9.7°C

What is the major trend of the isotherms? In what direction is there an overall drop in temperature? Where is the 'heat Equator' in January? Where are the world's coldest areas at both seasons? What is the highest temperature along the same line of latitude? Compare the same places in the other season.

Where is the 'heat Equator' in July? Compare the annual temperature ranges over continents and oceans. Compare the seasonal changes in northern hemisphere and southern hemispheres. At which time of the year is there the greatest difference between Equator and poles?

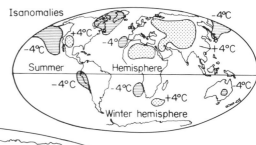

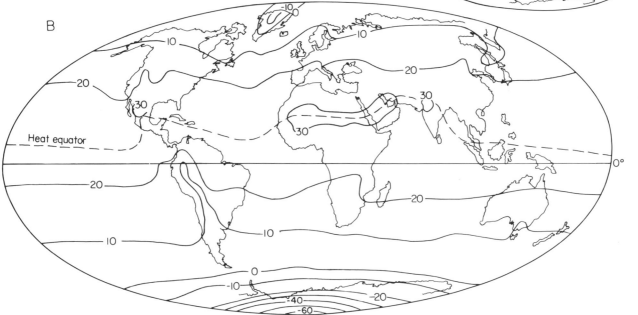

A further result of the contrast in heating properties between land and sea is that the ocean becomes a great reservoir of heat, and at times is able to contribute more energy to the air above it than the Sun. The cold air ($-30°C$) blowing out from the interior of the USA in winter crosses the Gulf Stream current ($10°C$), which transfers up to 1200 ly/day to the air. Even inland lakes, such as Lake Baykal in Siberia, have a similar effect: this lake releases more heat in the winter months than the Sun provides to the area in summer!

Altitude

The effect of falling temperature with increasing height is felt by those who live high in the mountains: there may be snow lying on the ground where mountain ranges cross the equator (e.g. the Andes in Ecuador, and Kilimanjaro in Tanzania). No peak reaches the tropopause, so all mountains are governed by this rule. Although the actual insolation may increase with altitude (average of 720 ly/day at 200 m; 860 ly/day at 2000 m), the temperature is lower because of increased radiation loss through the thinner atmosphere. Sun-facing slopes receive considerably more energy than slopes which spend much of the day in shadow. The north-facing slopes in the Alps remain forested, whilst the south-facing slopes with a much sunnier aspect are farmed and settled.

Cloudiness

The cloud cover varies from one part of the world to another in a consistent pattern (Figure 3.8). Equatorial regions, polar zones and the mid-latitude oceans and coasts are particularly cloudy, whilst the desert areas are almost clear-skied. Clouds have a high albedo and thus reflect a high proportion of the insolation, but they also tend to trap any heat which has reached the lowest regions of the atmosphere. Cloudy regions therefore have smaller daily and annual ranges of temperature than regions where neither insolation nor radiation are interrupted.

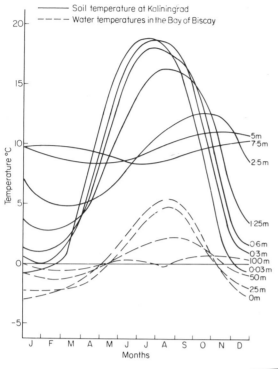

FIGURE 3:7 *Annual variation in temperatures at depth in the soil near Kaliningrad (USSR, 55°N), and in the waters of the Bay of Biscay (47°N). What are the differences in temperature range, times of maximum annual temperature and the depths affected at both places?*
(After Geiger and Sverdrup, in Barry and Chorley, 1968.)

FIGURE 3:8 *Mean annual cloudiness: tenths of sky covered. Use this map to explain the cloud albedo graph in Figure 3.4.*

(After Miller, 1953.)

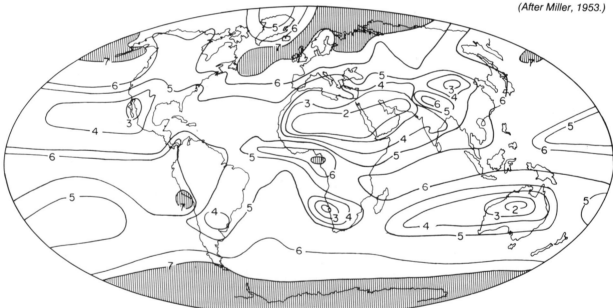

The transfer of energy from the Equator to the poles

The equatorial regions receive over five times as much solar energy as the poles in the course of an average year. The loss of heat by radiation to space varies much less from equator to poles. This means that there is a strong balance in favour of the equatorial regions, and this is reflected in higher surface temperatures. It also means that the regions near the equator have a surplus of energy, whilst the poles have a deficit (Figure 3.9). This could lead to a build-up of heat near the equator, and a sustained loss around the poles: but there is no noticeable change in the temperatures of these regions from year to year. It is concluded, therefore, that a transfer of energy takes place, creating the balance which is observed, by removing the equatorial surplus to the poles. This occurs in several ways, and is responsible for the large-scale movements within the atmosphere and oceans.

	Incoming energy	Outgoing energy	Balance
Equator	9.2	7.5	+1.7
Pole	1.7	5.9	−4.2

FIGURE 3:9 *Energy balance (figures in 10^5 joules/cm year^{-1}). Explain the difference between the amount coming into and going out from each zone.*

Up to 80 per cent of the energy transfer occurs in the atmosphere, this transfer being made by the poleward movements of warm masses of air containing large quantities of water vapour. When the vapour condenses in colder regions it releases latent heat. The oceans also contribute to the transfer process, carrying heated waters away from the tropics to cooler regions and releasing the heat there in contact with cold air (Figure 3.10).

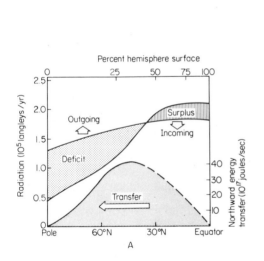
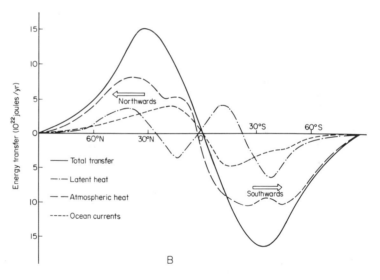

FIGURE 3:10 *Heat exchange. The deficit at the poles, and surplus at the Equator, are balanced by ocean-atmosphere movements, including the transfer of latent heat in evaporation and condensation.*

N.B. In diagram A the surplus is equal to the deficit when the areas involved are compared.
(After Newell, Gabites and Sellars,
in Barry and Chorley, 1968.)

The study of the world temperature maps (Figure 3.6), showed that the difference in temperature between the equator and poles was greatest in winter. This difference is known as the temperature gradient, and can be thought of in similar terms to the gradient of a hill as shown by the relief contours: thus if the isotherms, or contours, are close the gradient is steep and movement across them is rapid. The isotherms are closest in the winter months when the energy transfer is greatest. In addition, the atmospheric circulation is at its most intense in winter. Thus Britain, situated in a part of the world which experiences cyclonic weather systems at the meeting point of tropical and polar air throughout the year, receives the strongest gales in winter.

9

10

11

Platforms of observation. Observations of weather and ocean characteristics are derived from a variety of sources (see colour section).

Plate 9 Radiosondes measure pressure, humidity and temperature in the upper troposphere after release from the ground. The meteorological instruments are carried aloft with a radio transmitter which sends the results back to a ground receiver. Direction-finding receivers are used to determine the path of the radiosonde, and hence the wind direction and strength at various levels. Eventually the balloon bursts, and then a parachute softens the fall of the instruments so that they can be returned and re-used.

Plate 10 Marine data buoy off the coast of North Carolina. Such buoys are 13m across, weigh 100 tonnes, and are moored in up to 3 km of water to relay readings of sea level pressure, air temperatures, precipitation and wind velocity, together with underwater sensors for conductivity (i.e. salinity), water current velocity and water temperatures at five depths. Such buoys continue to transmit during storms and are particularly useful in tracking hurricanes.

Plate 11 Oceanographic vessels combine a wide range of observations, some of which are continuous (water depth, magnetism, gravity and surface temperature measurements), whilst others are taken at intervals (weather and deep water observations) or at slow speeds (seismic reflection profiles, biological tows or trawls). The tethered balloon is used for continuous recordings in the atmosphere such as those made during the summer of 1974 in the Atlantic Tropical Experiment for the Global Atmospheric Research Programme.

12

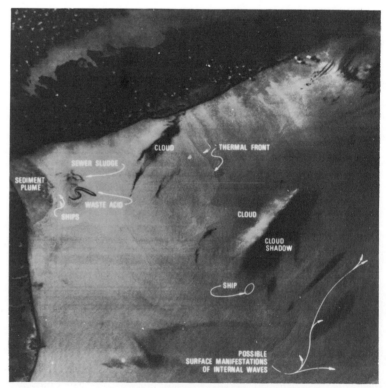
13

Plate 12 A bathythermograph, used to measure temperature changes with depth below the surface whilst the ship is underway. A stylus scratches the record on metal-coated slides, leaving a small graph which can be retrieved.

Plate 13 The LANDSAT satellite has been used to monitor changes in water quality. The area off the mouth of the Hudson river and south of Long Island, New York, shows sediment brought down by the river and the area designated for discharging wastes, some 35 km offshore. Data are gathered at the surface for comparison. Waves and tides maintain particles in suspension near the Long Island coast. This satellite passes over every part of the Earth once every 18 days. (All NOAA)

HEATING THE OCEAN-ATMOSPHERE SYSTEM

Albedo. Reflected radiation from various surfaces is shown up by aerial and satellite photographs. High reflection (and albedo) gives photographs a white appearance; low albedo shows dark. Use the three photographs to compare the albedo of cloud, land, sea, snow, forest, bare rock, blown sand, ploughed land, crops and buildings.

Plate 14 A Skylab 3 photograph of Britain, summer 1973. (NASA)

Plate 15 A Faith 7 photograph of the Himalayas, 13 May 1963. (NASA)

Plate 16 Aerial photograph of Moses Lake area, Washington, U.S.A. (NOAA)

Heating the Earth and Moon. Compare the features of the Earth and Moon which result in the different heat budgets for the bodies.

Plate 17 The edge of the Great Bahama Bank: the darker area is deeper water. Some islands on the edge of the bank can be seen in the centre of the photograph, taken on the Gemini IV mission in June 1965.

Plate 18 The Oceanus Procellarum area of the Moon, photographed by Apollo 15 in June 1971. (Both NASA)

14

15

16

17

18

4

Water in the atmosphere

The presence of water in the atmosphere is one of the most important results of transfers in the ocean-atmosphere system, powered by energy from the Sun. Water is cycled through a series of stages, from evaporation to condensation and precipitation (Figure 4.1), which help to distribute energy in the form of latent heat through the atmosphere. The heating of the Earth by the Sun's radiation, and the transformation of water in the atmosphere between its three phases (Figure 4.2) together largely control the patterns of weather around the world.

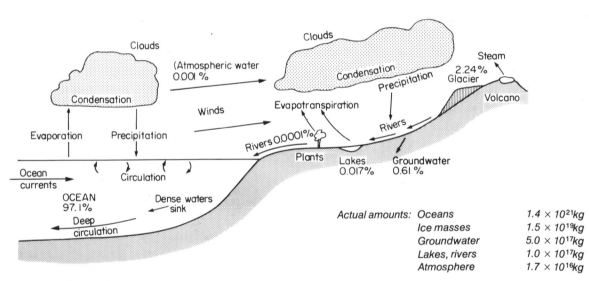

FIGURE 4:1 *The hydrological cycle. What parts do the Earth and Sun play in the circulation of water? What would be the result if either energy source were withdrawn, or the supply decreased? The percentage figures refer to the total volume of surface water on the Earth.*

Atmospheric vapour

How does water get into the atmosphere?

Most water enters the atmosphere by evaporation or transpiration. Evaporation takes place from the surface of a body of water by the transfer of water molecules to vapour form as they move into the air. The process involves a massive use of energy to draw apart the more closely packed molecules in liquid water, where the attractive intermolecular forces play an important role, and to leave them moving freely in the gaseous state where they are too far apart for the forces to be effective. Evaporation thus takes heat from the surrounding air and water, causing a local lowering of temperature, and traps it as latent heat in the vaporised water. The rate at which evaporation takes place depends on the amount of moisture already in the air above the water (measured as the relative humidity), and the movement of air across the surface of the water, which spreads the vapour and brings down unsaturated air into contact with the evaporating surface. The average rate of seawater evaporation varies with latitude (Figure 4.3).

Transpiration is the loss of water vapour by plants after it has been drawn from the soil. This loss is controlled by the stomata in the leaves, which open and close depending on the relative humidity and movement of the air and the concentration of available water in the soil. Transpiration takes place mostly during the day and in summer.

It is often difficult to measure the separate effects of these two processes over the land, and they are usually combined as **potential evapotranspiration** (i.e. that which would occur if plants could always take up soil water — they cannot in time of drought). There are several methods of calculating evapotranspiration in use, but these lead to different estimates for the same location. The annual rate for south-eastern England, for instance, has been estimated variously at amounts ranging from the equivalent of 500 to 600 mm of rainfall. Whichever figure is accepted, a comparison with the average annual rainfall of less than 600 mm/year in this area leads to the surprising conclusion that parts of Britain experience a fine balance between sufficient rainfall for continuous plant growth, and a deficiency. During the 1950's and 1960's, crops such as grass, cereals and potatoes, as well as market garden crops, have been irrigated increasingly until over 10^5 hectares of crops could be watered by equipment in 1973. It is estimated that there could be up to six times as much benefit from irrigation, but problems of water supply are now becoming apparent and prices of water are increasing. This is particularly true in areas, like southern Essex, which have some of the lowest rainfall totals in Britain and where the peak demand from farmers coincides with the lowest state of storage facilities.

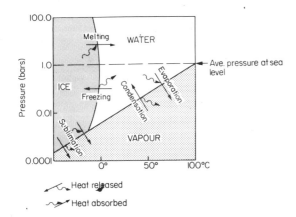

FIGURE 4:2 *Water phases and changes between them. What happens to the vaporizing temperature (i.e. boiling point) of water as the pressure falls?*
N.B. 1 bar = mean atmospheric pressure, sea level

$= 10^3 mb$ (1 mb = 10^3 dynes/cm^2)
$= 10^6$ dynes/cm^2
$= 10$ Newton/cm^2 (1 Newton = 10^5 dynes)

(Diagram after Weyl, 1970.)

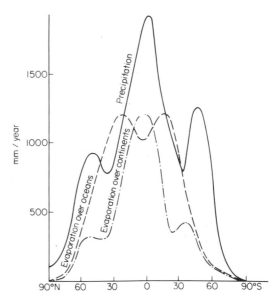

FIGURE 4:3 *Precipitation and evaporation: the average latitudinal distribution. Is evaporation always greatest over the oceans? At which latitudes is it greatest over land or sea? Can you explain the differences? Which regions of the world have an excess of precipitation over evaporation, and which a deficit? (After Sellars, 1965.)*

Measuring atmospheric vapour

The moisture content of the atmosphere is often measured in terms of **vapour pressure** (i.e. that part of the total atmospheric pressure contributed by the water vapour content). This is, of course, much lower than the total atmospheric pressure (average about 1000 mb), and ranges on average from 0.2 mb in a polar winter to 30 mb in a tropical summer. Because both places have high relative humidities, but strongly contrasting temperatures, these figures illustrate the fact that the proportion of water vapour in saturated air increases with temperature. When air contains the maximum water vapour which can remain in equilibrium with liquid water at a particular temperature it is said to be saturated (Figure 4.4). If more water vapour is added, some of the vapour present should condense to water droplets to restore the equilibrium of saturation. If, alternatively, the temperature of the vapour is lowered, some of it again will condense as droplets: the temperature at which this occurs is known as the **dew point.**

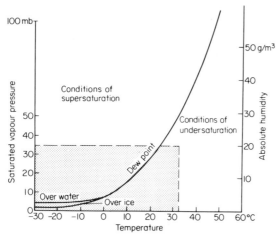

FIGURE 4:4 *Humidity and air temperature. The shaded area covers conditions normally experienced in the atmosphere. Notice the difference in the conditions when ice is present. How much water vapour will there be in saturated air at −20°C, 0°C, 10°C, 20°C and 30°C?*

The **relative humidity** at a particular place is the ratio of the measured vapour pressure to the saturated vapour pressure at the existing temperature (cf. Figure 4.4). It is expressed as a percentage.

$$\text{Relative humidity} = \frac{\text{Measured value of vapour pressure}}{\text{Saturated vapour pressure at air temperature}} \times 100$$

If the measured vapour pressure and saturated vapour pressure values are the same, then the relative humidity is 100 per cent. Humidities in Britain seldom fall below 40 per cent and usually exceed 60 per cent. High relative humidities combined with high temperatures, such as occur in tropical regions like the Red Sea and Persian Gulf areas, can make life intolerable, and the relief provided by sea breezes in lowering the relative humidity on the coast is welcome. Relative humidity is a concept which is in some ways more useful to the meteorologist than that of absolute humidity (i.e. the actual mass of water vapour in the atmosphere, expressed as g/m^3). This is because the relative humidity gives an indication of the likelihood of condensation or further evaporation as well as the general comfort level of the environment. Weather forecasts in the USA may include an assessment of the comfort level: the US Weather Bureau uses a temperature-humidity index (THI), according to which 10 per cent of the population feel uncomfortable when the THI reaches 70, over 50 per cent at 75 and almost all above 80.

Condensation
The process of condensation

Almost all the water vapour is in the lowest few thousand metres of the atmosphere, but it is impossible to talk about moisture in the atmosphere in static terms. The relative humidity is changing constantly due to evaporation, condensation, precipitation and air movement. Condensation is the reverse of evaporation. In condensation, water vapour is released from the gaseous state and becomes droplets of water. This is brought about by cooling the air to below its dew point (Figure 4.5), and leads to the release of latent heat.

Condensation does not take place easily in clean air, which can be cooled below its dew point without condensation occurring; it is then said to be supersaturated (i.e. the relative humidity is over 100 per cent) before water droplets form. Large numbers of tiny particles of dust, smoke and salt, commonly present in the lower atmosphere, play an important part as condensation nuclei. Condensation takes place on these: it may even take place before dew point is reached, and has been recorded at a relative humidity of only 78 per cent in the presence of common salt particles. Condensation nuclei are particularly common over the land (5000 per cm^3), and there is often less than a tenth of this concentration over the open ocean.

Once a droplet has been formed, it continues to grow by further condensation. As its size increases to that of a cloud droplet (less than 0.1 mm diameter) this process becomes extremely slow and virtually ceases. There is a vast difference in size between cloud droplets and rain drops, which often

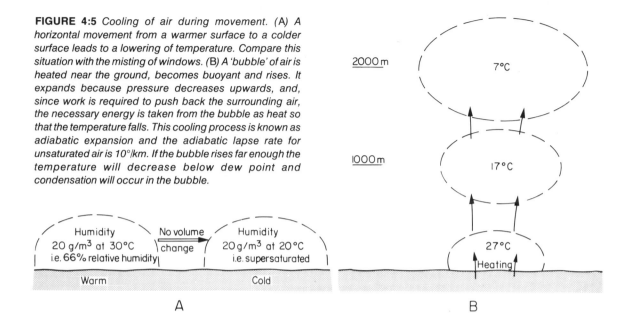

FIGURE 4:5 Cooling of air during movement. (A) A horizontal movement from a warmer surface to a colder surface leads to a lowering of temperature. Compare this situation with the misting of windows. (B) A 'bubble' of air is heated near the ground, becomes buoyant and rises. It expands because pressure decreases upwards, and, since work is required to push back the surrounding air, the necessary energy is taken from the bubble as heat so that the temperature falls. This cooling process is known as adiabatic expansion and the adiabatic lapse rate for unsaturated air is 10°/km. If the bubble rises far enough the temperature will decrease below dew point and condensation will occur in the bubble.

exceed 1 mm in diameter, and for which another special process of accretion is necessary. The process of condensation releases the latent heat of evaporation and leads to a reduction in the rate of cooling due to adiabatic expansion. This is another phase of atmospheric heat transfer.

Condensation forms: (1) fogs

When condensation takes place it produces visible forms in the atmosphere. These are the most obvious features recorded on standard satellite photographs of the atmosphere, and so cloud studies in particular have received a new impetus in the last few years due to the increasing availability and use of satellite information. Clouds and fogs are both produced by the cooling of bodies of air below dew point leading to condensation of some of the atmospheric water vapour.

Fog is condensation in the atmosphere in contact with the ground where the visibility is reduced to under 1 km. Hill fog is really a cloud resting against a hillside. Ground fog has two main causes.

1 Advection fog is caused where moist air flows over a surface whose temperature is below the dew point of the air. This occurs most commonly over the sea where warm air blows over a cold ocean current leading to condensation which is carried upwards by the wind movement. Winds may be up to 20-30 km/hour: at greater wind speeds low cloud is often formed instead unless the rate of cooling is extremely great. The coastal fogs of Newfoundland, central California and northern Chile are typical examples of this process. Coastal fogs in Britain may be caused by warm air from the south-west, or from the warming continent of Europe, being cooled over the seas around the British Isles in early summer.

2 Radiation fog occurs over the land. Prolonged cooling of the ground at night due to continuous radiation through a cloudless sky causes the air in contact with the ground to be cooled, and if there is any slight movement the cooling will be distributed upwards, resulting in condensation to heights between 15 and 100 m (up to 200-300 m in extreme cases). If there is no movement in the air, condensation will lead to the formation of dew, or of hoar frost if freezing temperatures are reached. Condensation above the ground takes place following radiation through the layers above the surface, and from the droplets condensed at lower levels. The process of cooling may take several hours, since the dew point falls as well as the temperature of the air, but once condensation begins it spreads rapidly and visibility decreases suddenly. Autumn and winter in Britain are times when such fogs are particularly liable to form. Days can be warm with considerable evaporation into the lower layers of the atmosphere, but the nights are also long enough for a protracted period of cooling to give rise to the condensation.

Radiation fogs are associated with relatively calm conditions (i.e. winds of less than 10 km/hour) and may persist for days below an **inversion.** An inversion situation occurs when cold air underlies warmer air (Figure 4.6), thus preventing surface air from rising very far. When this happens over a large city, smoke is trapped and may build up to a concentration where it causes danger to transport by reducing visibility below 10 m (clean water fogs seldom cause it to fall below 50 m), and becomes a health hazard. Such smoke-fog, or **smog**, was such a health hazard in British cities that the Clean Air Acts of 1956 established smokeless zones which have been extremely successful in reducing the incidence of fogs in cities and thus increasing the amount of sunshine (chapter 8). Smog is still a problem in urban areas around the world. In Los Angeles, for instance, the inversion situation is common, since air at the surface is cooled in contact with the ocean water, whilst warm air from the desert interior flows over the top. Chemical reactions taking place between oxygen and oxides of nitrogen released in car exhausts produce irritants like ozone and nitric acid: the City Council is very seriously interested in the development of car engines which do not use petrol. Tokyo is a city which has seen rapid industrialisation of its local region especially since 1950, resulting in pollution of the atmosphere in parts of this city so great that pedestrians and traffic police have to wear masks.

Condensation forms: (2) clouds

Clouds occur at varying heights above the Earth's surface. They have always had an important place in weather lore:

> "If woolly fleeces spread the heavenly way,
> be sure no rain disturbs the summer day."
> "When mountains and cliffs in the sky appear
> Some sudden and violent showers are near."
> "Mackerel sky, mackerel sky,
> Never long wet and never long dry."

The first careful classification of clouds is attributed to the French naturalist, Lamarck (1802), but the English pharmacist, Luke Howard produced another in 1803 which became the basis for succeeding work (Figure 4.7). Other cloud form names were added in the nineteenth century, but the major contribution of this period was to divide clouds on the basis of height: low clouds (all water), intermediate clouds (partly frozen) and high clouds (all frozen). The main cloud types are described in the International Cloud Atlas (Figure 4.8 and Figure 4.9).

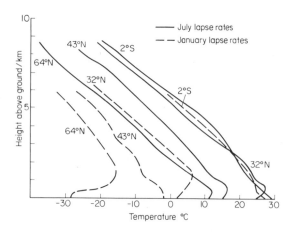

FIGURE 4:6 *Seasonal variations in observed lapse rates. What are the effects of (a) latitude, and (b) season? A reversed lapse rate (i.e. the temperature increases with altitude above the surface for a while before beginning to fall) is known as an inversion. This can be illustrated by listing the temperatures at 0.5 km intervals above the surface for the 64°N January graph. Suggest the changes which might occur in lapse rates between day and night.*

Lamarck (1802)	Meaning	Howard (1803)
Nuages groupés	Grouped, piled clouds; heap clouds	Cumulus
Nuages en voile	Veil, sheet clouds; layer clouds	Stratus
Nuages en balayures	Sweep clouds; hair-like, fibrous clouds	Cirrus

FIGURE 4:7 *The earliest classifications of clouds recognised the three basic forms. It has been suggested that Howard's classification was taken up because it was based on Latin names and thus had international relevance.*

Clouds are masses of water droplets in which there is continuous accretion by condensation and loss by evaporation. Time-lapse films of clouds moving across an area show these processes occurring together. The temperature of a rising body of air is cooled to dew point and condensation takes place, often at a particular level above the Earth's surface which becomes the cloud base. At the same time the air in the margins of a cloud will mix with drier air surrounding it, causing the relative humidity to fall below 100 per cent and leading to the evaporation of the cloud droplets. Alternatively, rain may fall from the cloud and remove a proportion of the cloud water.

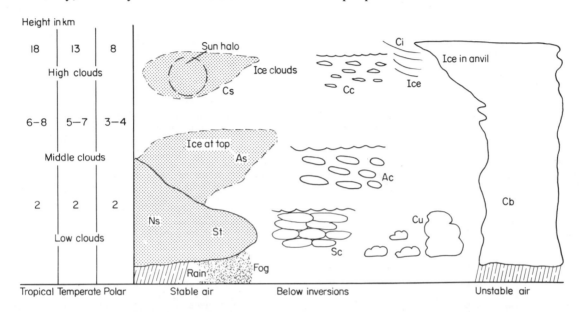

FIGURE 4:8 *A classification of clouds according to shape and height, related to conditions of air stability and composition. (N.B. ice clouds are labelled; the rest are water clouds.) Note the height differences with altitude, and relate these to the height of the tropopause (Figure 2.4). The abbreviations used are explained in Figure 4.9.*

The most important process by which air is cooled to form clouds is **adiabatic expansion** as air rises in widespread masses or in smaller 'bubbles'. The troposphere is characterised by a decrease of temperature with height (Figure 4.6). This is essentially because the atmosphere is heated from below (by contact with the ground) and cooled from above (by radiation to space) during the day. Adiabatic temperature changes within a rising, buoyant bubble of air lead to a lapse rate with a particular value. This is 10°C/km of ascent (Figure 4.5), and a similar lapse rate is produced in air forced to rise over a mountain barrier or over a mass of colder, more dense air. The adiabatic lapse rate in unsaturated air (the dry adiabatic lapse rate, DALR) is greater than that in saturated air (saturated adiabatic lapse rate, SALR) because condensation occurs and latent heat is released.

The relationship between the observed and adiabatic lapse rates, together with the degree of marginal mixing, determines the extent to which the bubble (or larger mass) of air will rise through the atmosphere, and hence the likelihood of it cooling adiabatically below dew point. Although air is a poor conductor of heat, most bubbles do not reach levels that might have been expected from their original buoyancy because cooler, surrounding air is mixed in, reducing buoyancy (Figure 4.10, Figure 4.11, Figure 4.12 and Figure 4.13).

These diagrams suggest what can happen to rising bodies of air as they expand adiabatically, but it must also be remembered that bodies of air may become more dense than their surroundings and sink. The reverse processes apply with adiabatic warming due to compression as air enters lower sections of the atmosphere with higher pressure. In addition the temperature rises, but the dew point does not rise so rapidly, leading to a decrease in relative humidity as the air descends: water droplets which may be present are evaporated. Such downslope movements tend to be slower than the upslope movements (because the latter are aided by the release of latent heat), but they often have an

Name	Symbol	Main Constituent	Level	Over Britain		Appearance	Mode of Formation	Precipitation
				Height of Base (km)	Likely Temperature at Base (°C)			
Cirrus	Ci	Ice	High	5-13	−20 to −60	Detached, delicate white filaments, often fibrous. Larger ice crystals in trails.	Widespread ascent, often frontal	None
Cirrocumulus	Cc	Ice	High	5-13	−20 to −60	White patch or layer with smaller granular or rippled elements.		None
Cirrostratus	Cs	Ice	High	5-13	−20 to −60	Translucent cloud veil. Halo phenomena.		
Altocumulus	Ac	Water (Some Ice)	Middle	2-7	+10 to −30	White-grey layer with shading or broken elements. Sharp outlines.		Little/none
Altostratus	As	Water (Some Ice)	Middle	2-7	+10 to −30	Grey-blueish sheet, uniform appearance. Many have layers of ice, water; may be in rows.		Occasional
Nimbostratus	Ns	Water + Ice	Middle Low	1-3	+10 to −15	Dark grey layer, rendered diffuse by falling rain		Persistent rain or snow
Stratocumulus	Sc	Water	Low	½-2	+15 to −5	Grey-white patchy sheet. May occur in long rows.		Little/none
Stratus	St	Water	Low	0-½	+20 to −5	Grey uniform layer	Cooling widespread at surface	Drizzle or none
Cumulus	Cu	Water	Middle Low	½-2	+15 to −5	Detached clouds: dense with sharp outlines. May grow vertically.	Ground radiation leading to convection and instability	None if small showers
Cumulonimbus	Cb	Water + Ice	High Middle Low	½-2	+15 to −5	Huge towering clouds: dense, dark and mountainous. Top flattened, spreading to anvil.		Heavy showers Thunderstorms

FIGURE 4:9 *A descriptive classification of clouds: ten basic forms of world-wide occurrence. (Data from Pedgley, 1962.)*

important effect on the climates of the lee sides of mountain ranges (Figure 4.14). Thus the chinook winds east of the Canadian Rockies give rise to a dry, barren area, and the föhn winds in the Alps often melt and evaporate the snow cover in one action.

The three main cloud forms recognised by Lamarck and Howard are known now to result from the state of the atmosphere into which the air rises and is cooled. Whilst many other cloud forms have

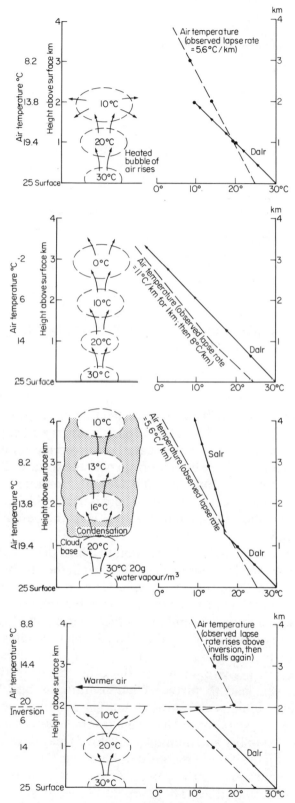

FIGURE 4:10 *Air near the ground is heated and rises. The observed lapse rate is greater than the dry (unsaturated) adiabatic lapse rate (DALR): rising air cools more rapidly than the rate of temperature fall in the surrounding air. When the temperature of the rising bubble falls slightly below that of the surrounding air, ascent will cease. A situation which encourages rising air to cease its ascent is known as a stable atmosphere. This may occur on sunny summer days, although the surface heating often leads to an increased observed lapse rate in the lowest 1000 m. This encourages the bubble to rise at first in the lowest, unstable layer, and then to cease rising in the overlying layer.*

FIGURE 4:11 *Surface heating on this occasion has given rise to an increased observed lapse rate, especially near the ground. The result is a deep unstable layer, through which the bubble will continue to ascend.*

FIGURE 4:12 *Condensation will occur in air when the rising bubble cools below its dew point. This may result in a situation where air, which had been reaching a stable part of the atmosphere, becomes unstable. This happens because latent heat is released by condensation and the adiabatic lapse rate is lowered from 10°C/km (DALR) to 3-5°C/km (SALR). The atmosphere is thus stable for the displacement of a dry bubble, but unstable for a cloudy bubble. Such a situation is known as conditional stability, and condensation may be followed by deep convection.*

FIGURE 4:13 *An inversion. This may occur when warm air rises over cool air, or subsides from higher levels; it also occurs following intense cooling at ground level during the night. Ascending bubbles of air penetrate the warmer level at a temperature which prevents them from rising farther. Inversions commonly give rise to stable layers in the atmosphere by preventing upward movement.*

been named since the early nineteenth century, they can be related to the three main types, and many of them are regarded as species of the ten main genera (Figure 4.9). There are also varieties of these species which cover most eventualities, but it is most important to view clouds as being formed in either stable or in unstable conditions.

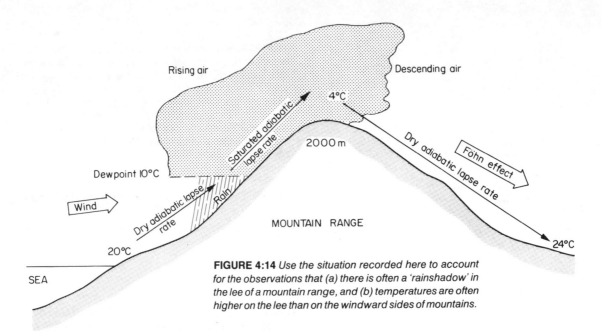

FIGURE 4:14 *Use the situation recorded here to account for the observations that (a) there is often a 'rainshadow' in the lee of a mountain range, and (b) temperatures are often higher on the lee than on the windward sides of mountains.*

1 Cumuliform clouds are formed in an unstable atmosphere and result from convectional lifting of bodies of air, which cool by adiabatic expansion as they rise. Three main types can be distinguished according to their vertical development and the heights of their bases (Figure 4.15).

a) *Cumulus* clouds are formed commonly on warm days. The ground is heated irregularly by the Sun, depending on whether the slopes are facing towards or away from its rays, and whether the ground surface is bare soil, roads and buildings, or woodland. The irregular heating of the ground leads to irregular heating of the air immediately above with sections of less dense, warmer air known as 'bubbles' or 'thermals' (Figure 4.16). The heating also leads to increased observed lapse rates near the ground and this unstable situation assists in lifting off the bubbles from near the surface. The height of the cloud base is determined by the humidity of the bubble as it leaves the ground: lower humidities require a greater cooling and have a higher cloud base. The cloud base is usually higher in summer than in winter, and rises during the day to a mid-afternoon maximum. Thus the base of a cloud may rise from 500 m to over 1200 m in one day.

The actual shape and size of the cloud depend on several factors. The extent to which the bubbles will continue to rise is determined by whether or not a stable situation is established in the air above the cloud base. A stable situation is always reached at some stage with *cumulus* clouds, when the observed lapse rate is greater than the SALR (Figure 4.15). A second factor is the relationship between the rising bubble of air and its environment in terms of humidity: if the environment has a low relative humidity, the effect of mixing will be to evaporate rapidly the cloud droplets, leading to the cooling of the cloud 'skin' and the sinking of the margins. This acts as a brake on upward movement for air rising through the cloud. In a moist environment, on the other hand, mixing involves smaller losses in the cloud, and may even lead to further condensation, as bubbles containing cloud droplets rise into air of high relative humidity above the cloud. The combination of stability and mixing means that with a slight degree of instability and a dry environment only small, flat clouds will form (if any), whilst with increased instability and humidity tall clouds may develop. In addition the cloud shape will be affected by the winds: if there is considerable shear (i.e. a sharp increase in wind speed with height) the top of the cloud will be pushed over. At times the winds will also cause the *cumulus* clouds to be arranged in 'streets' or lines. These are developed best where there is a stable layer above the cloud base and where there is sufficient wind shear to prevent the clouds spreading out to form *stratocumulus*. Streets are often spaced at two or three times the depth of the convection layer and thermals become concentrated along the lines of clouds and a sinking motion is established between them.

Name		Shortened form	Characteristics
Genus	Species		
Cumulus	fractus	Cu fra	Small, ragged *Cumulus*, just formed
Cumulus	humilis	Cu hum	Shallow *Cumulus*: stable layer just above condensation level
Cumulus	mediocris	Cu med	Increasing *Cumulus*: up to 1000m deep
Cumulus	congestus	Cu con	Towering *Cumulus*: 2000-3000m deep
Cumulus	pileus	Cu pil	A cap above *Cumulus* due to local lifting of environment. Shortlived: *Cumulus* pushes through or it subsides and evaporates.
Cumulus	radiatus	Cu rad	'Streets' of *Cumulus* cloud
Stratocumulus / Altocumulus	cumulogenitus	Sc / Ac cugen	*Cumulus* spreads sideways at stable layer giving lumpy clouds decaying from lower surface
Cumulonimbus	capillatus	Cb cap	*Cumulonimbus* with fibrous, cirri at top
Cumulonimbus	cap. var. incus	Cb inc	Anvil cloud
Cumulonimbus	calvus	Cb cal	Transition to *Cumulonimbus* from *Cumulus*: cloud top becomes smoother and diffuse (Cu con→Cb cal→Cb cap)
Altostratus / Nimbostratus	cumulonimbogenitus	As / Ns cbgen	Layers formed by spreading of *Cumulonimbus* tops
Cirrus spissatus	cumulonimbogenitus	Ci spi cbgen	Dense cirriform streak clouds formed by glaciation from *Cumulonimbus* fall streaks
Altocumulus	castellatus	Ac cas	Row of turreted clouds with well-defined base
Altocumulus	floccus	Ac flo	Irregular tufts with cumuliform tops and ragged bases
Cirrus	castellatus	Ci cas	As above, but higher and with finer detail
Cirrus	floccus	Ci flo	

FIGURE 4:15 *The cumuliform association of clouds. (Data from Pedgley, 1962.)*

Cumulus clouds disperse when the bubble supply is cut off. This occurs when winds move the cloud away from the source, or when surface heating falls off. Each cloud has a lifetime of 5-30 minutes.

b) *Cumulonimbus* clouds develop from large *cumulus* (Figure 4.17). A major factor in their makeup is the upward extension of the cloud to a point where the water droplets turn to ice. When the cloud extends through the level where the temperature is $-20°C$, most of the water above that level turns to ice. Ice particles may begin to form below this level, but the process is slow there. In order to reach such low temperatures the clouds have to reach immense heights. Thus in warm weather (British summers or generally in the tropics) the $-20°C$ level is reached at 4-6 km above the ground, and the tops of *cumulonimbus* clouds may reach 12 km in Britain (18 km near the equator) if the entire troposphere is unstable. In colder weather the $-20°C$ level is reached sooner, often at less than 3 km, and *cumulonimbus* clouds may be shallower. The great vertical extent of these clouds means that internal vertical currents of over 50 km/hour may be experienced, causing problems for aircraft passing through.

As the ice top develops, the sharply defined, bulbous top of the cumulus cloud becomes diffuse with fibrous margins. When it reaches the upper limit of convection at a stable layer (which may be the tropopause) there will be a spreading out of the ice cloud to form an anvil shape or thick masses of *altostratus* or *nimbostratus* clouds. Decay of *cumulonimbus* clouds occurs when convection ceases, and this may result in a complex breakdown into clouds at several levels, including high level dense *cirrus*.

19

20

Fog. Explain the ways in which fog has formed on these occasions.

Plate 19 Wymondley, Hertfordshire, England.

Plate 20 At Port Erin, Isle of Man. (Both Aerofilms)

Cumuliform clouds. Describe the varied weather conditions you might expect from this series of photographs.

Plate 21 Cumulus humilis. (C E Wallington)

Plate 22 Cumulonimbus calvus. (E R Trendell)

Plate 23 Cumulus mediocris, Ochill Hills, central Scotland. (Aerofilms)

Plate 24 Cumulonimbus capillus. (G Nicholson)

21

22

23

24

WATER IN THE ATMOSPHERE 39

Stratiform clouds

Plate 25 Stratus bank over the sea in the Orkney Islands.

Plate 26 Low stratus at Gibralter.
(25 & 26 Crown Copyright)

Plate 27 Altrostratus translucidus. (Royal Meteorological Society)

Cirriform clouds. How are they formed, and with which weather patterns are they associated?

Plate 28 & Plate 29 (Royal Meteorological Society)

Plate 30 (E R Trendell)

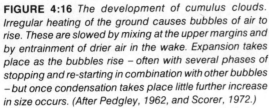
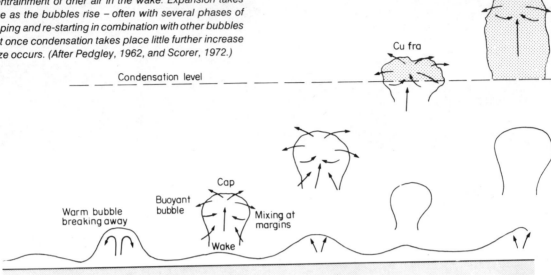

FIGURE 4:16 *The development of cumulus clouds. Irregular heating of the ground causes bubbles of air to rise. These are slowed by mixing at the upper margins and by entrainment of drier air in the wake. Expansion takes place as the bubbles rise – often with several phases of stopping and re-starting in combination with other bubbles – but once condensation takes place little further increase in size occurs. (After Pedgley, 1962, and Scorer, 1972.)*

c) *Cumuliform* clouds at high levels may develop due to convergence of air near a depression, or ascent caused by a mountain range. This is more common at medium than at the highest levels and instability in the middle troposphere may lead to the upward spread of *altocumulus* clouds. Such clouds are often a useful indication to the ground-based observer of instability in the upper layers of the troposphere, and hence of the likelihood of the development of *cumulonimbus* clouds in the vicinity.

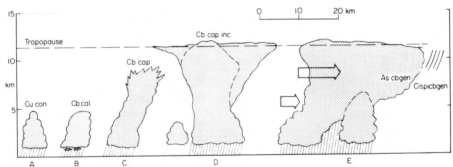

FIGURE 4:17 *The development of cumulonimbus clouds. The key to the symbols used is given in Figure 4.15. (A)-(C) shows the sequence of cumulonimbus development. In (D) the central portion represents a new cell, which has grown from a small cloud like that at the base, and is replacing an older cell which has spread out in anvil form. (E) shows the effect of wind shear, causing the upper part of the cloud to spread out: the lengths of the arrows reflect wind speeds. (After Pedgley, 1962.)*

2 Stratiform clouds are associated with a stable atmosphere, and with widespread cooling due to uplift over a large area, or to the passage of warm, moist air over a cool surface (Figure 4.18). The cloud forms produced by these means contrast markedly with the cumuliform varieties. They have a wide horizontal extent compared with vertical depth, and are usually associated with comparatively feeble air currents. Upward growth is prevented by stable conditions in the overlying atmosphere. There are three main groups of stratiform clouds.

a) *Stratus* clouds are often associated with ground fogs and may be formed in similar ways, but with winds lifting the masses of condensed water droplets off the ground. The cloud base is thus often low, and seldom above 500 m. The layer of cloud covering the sky leads to poor visibility and dull

Name		Shortened form	Characteristics
Genus	Species		
Stratus	nebulosus	St neb	Base shows little or no detailed structure
Stratus	fractus	St fra	Broken cloud layers: just forming or dispersing
Stratus	pannus	St pan	Low *Stratus* beneath *Nimbostratus* or *Altostratus*
Cirrocumulus Altocumulus Stratocumulus	stratiformis	Cc Ac str Sc	Extensive horizontal sheets of these genera *var. perlucidus (pe)*: broken into cloudlets due to radiation from cloud sheet top and shallow convection; giving thickening in upcurrents and dispersal in down currents; thickness depends on depth of convection — shallow at high levels ('mackerel' sky) *var. translucidus (tr)*: cloud shallow enough for Sun or Moon to shine through *var. lacunosus (la)*: rare, netlike pattern with spaces clear due to cellular convection *var. undulatus (un)*: equally-spaced, parallel bands *var. opacus (op)*: Sun obscured
Cirrocumulus Altocumulus Startocumulus	lenticularis	Cc Ac len Sc	Lenticular clouds

FIGURE 4:18 *Varieties of stratiform clouds. (Data from Pedgley, 1962.)*

weather. Most *stratus* cloud is formed by advection (Figure 4.19) and is commonly experienced in Britain when warm, moist air comes from the south-west or east across cool seas in early summer. *Stratus* cloud is dispersed by heating, which leads to evaporation.

b) Thin stratiform clouds may have some degree of structure associated with them. There are varieties of *cirrocumulus*, *altocumulus* and *stratocumulus* clouds of the species *stratiformis*, for

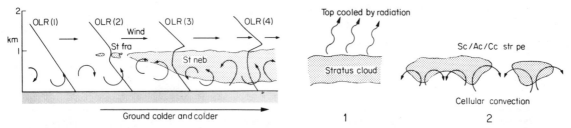

FIGURE 4:19 *The formation of stratus cloud by advection. Notice the changes in the observed lapse rates (OLR, (1)-(4)) and the progression towards the establishment of a marked inversion. How is the thickness of the cloud layer related to the temperature difference between ground and air? (After Pedgley, 1962.)*

FIGURE 4:20 *The formation of stratiformis clouds, resulting in 'mackerel sky'. (After Pedgley, 1962.)*

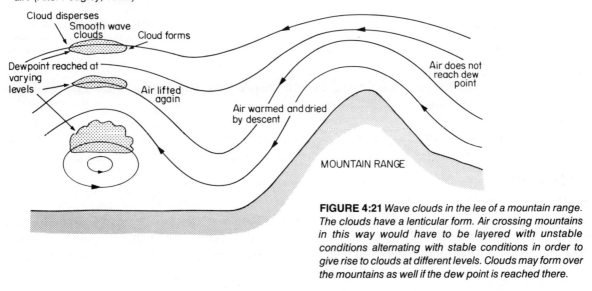

FIGURE 4:21 *Wave clouds in the lee of a mountain range. The clouds have a lenticular form. Air crossing mountains in this way would have to be layered with unstable conditions alternating with stable conditions in order to give rise to clouds at different levels. Clouds may form over the mountains as well if the dew point is reached there.*

instance (Figure 4.18). Of these the variety *per lucidus* is familiar in the high level 'mackerel sky' form, and occurs when shallow convection develops due to radiation from the top of the cloud sheet (Figure 4.20). The variety *undulatus* occurs when the cloud is broken into parallel bands, sometimes known as 'billow clouds') near the jet stream flow.

Another major occurrence of stratiform clouds is the species *lenticularis*, which is in the form of lens-like clouds formed in the wave-like turbulence produced as air flows across high ground (Figure 4.21). The clouds may appear to be motionless, but air is moving through them with condensation at the windward end and evaporation on the lee. They may develop at several levels at the same time, depending on the humidity stratification of the air as it approaches the barrier, and the altitude of the induced waves.

c) Thick stratiform clouds and multilayered clouds are associated with fronts and may give widespread and persistent rain. These may include unstructured sheets of *cirrostratus*, *altostratus* or *nimbostratus* and thick layers of *altocumulus* or *stratocumulus stratiformis* of the variety *opacus*. As the front advances high *cirrostratus* will spread in, formed of ice crystals, with the Sun shining through a halo. The cloud thickens to 1000 m with *altostratus* formed of water and snowflakes through which the Sun may just (var. *translucidus*) or may not (var. *opacus*) shine. *Nimbostratus* has a much lower base, and is much thicker (several km) than these varieties, and gives rise to continuous precipitation. Many fronts are associated with a more complex pattern of layered clouds. One common grouping is of *cirrostratus* and *altostratus* above, with *altocumulus* below giving slight, intermittent precipitation. Figure 6.6 shows how these cloud types occur in frontal developments.

3 Cirriform clouds are extremely high level clouds, formed entirely of ice. The base height is at least 8-11 km, and 10 per cent of bases are over 13 km; most are in the top 2 km of the troposphere. The low temperatures at these levels prevent more than tenuous clouds from forming, and many are formed of ice particle trails descending from the level of ice formation. The crystals form and fall until they sublime at a level saturated for ice. The fall streaks produced may be several km long. Most cirriform clouds (Figure 4.22) are formed in air which is rising slowly ahead of fronts associated with developing depressions (Figure 6.6).

Name		Shortened form	Characteristics
Genus	Species		
Cirrus	*uncinus*	Ci unc	Distinct head and tail
Cirrus	*fibratus*	Ci fib	Diffuse fibrous tails only
Cirrus	*spissatus*	Ci spi	Denser forms

Particle	Mean radius (mm)	Terminal velocity (m/sec)
Cloud droplet	0.001 - 0.05	0.0001 - 0.25
Drizzle drop	0.1 - 0.25	0.70 - 2.0
Rain drop	0.5 - 2.5	4.0 - 9.1
Snowflake	0.5 - 4.0	0.7 - 1.7

FIGURE 4:22 *Varieties of cirriform clouds. (Data from Pedgley, 1962.)*

FIGURE 4:23 *Condensed water and forms of precipitation. N.B. terminal velocity = the constant speed attained when an object falls in relation to its environment after initial acceleration. (Data from Pedgley, 1962.)*

Thus the forms of clouds are closely tied up with the processes producing them. The cooling of air is mostly by ascent — widespread along a front or over a mountain range, or by local convection — but may also be by advection over a cool surface or by radiation at the surface giving a fog, which may later be lifted off the ground. The atmospheric environment, in terms of observed lapse rates and relative humidities, will determine the upward extent of the clouds and the formation of rain.

Precipitation
What causes rain?
Cloud droplets must grow to ten or a hundred times their size before they can fall as rain (Figure 4.23). The terminal velocity of a falling drop increases with its size. The terminal velocity is extremely slow for cloud droplets and the slightest atmospheric motion is sufficient to hold them aloft.

Condensation does not account for growth beyond the size of cloud droplets, since, although initial growth of the droplets is rapid, further growth by condensation from 0.02 mm to 0.2 mm (still far short of rain drop size) takes about an hour, by which time many *cumulus* and even *cumulonimbus* clouds have dissipated.

The heaviest rainfall occurs where mountain ranges face rain-bearing winds (Figure 4.24). Such airstreams are forced to ascend and the intensity and frequency of precipitation (falling rain, snow, hail) increases to 3000m in humid temperate areas. Above this height the effect falls off and

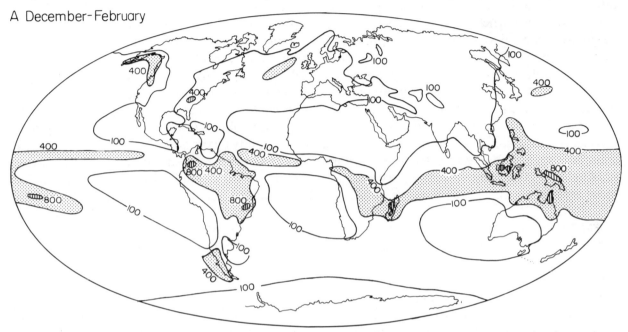

FIGURE 4:24 *World rainfall patterns for three-monthly periods. The figures are in mm. Which regions have (a) most, and (b) least at each of these seasons? Where is there (c) a summer maximum, (d) a winter maximum, (e) little throughout the year, and (f) a lot of rain at all seasons? Draw a world map showing (c) to (f).*

Compare the rainfall patterns with the temperature distribution (Figure 3.6). In what ways is the rainfall distribution more difficult to explain? Note the positions of high mountains, on-shore winds bringing moisture to coasts, and zones where airstreams converge. (After Möller, in Barry and Chorley, 1968.)

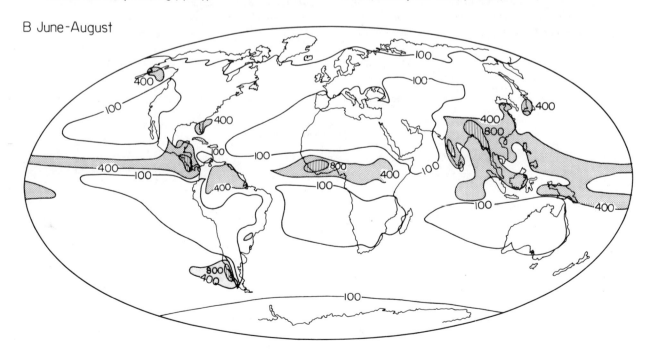

mountain climates become drier. The mountains along the west coast of Britain scarcely reach 1500m in height, and maximum rainfall totals occur at some distance inland: e.g. at the foot of Snowdon in north Wales (15 km from the sea), at Seathwaite in the central Lake District, and around Fort William in north-west Scotland. Other major areas of rainfall occur in equatorial zones, where air-flows converge at the ground causing the ascent of very humid air on a large scale.

One major process of rain formation was first discovered by the Swedish meteorologist, Bergeron, in 1933. His ideas have been modified in the light of further study. The theory is based on the difference in saturated vapour pressure over ice and water surfaces at temperatures below 0°C. In a cloud where ice particles and supercooled water droplets are both present, the vapour pressure will be greater than the saturated vapour pressure for ice and less than that for water. The ice crystals grow by taking up water vapour, which thereby becomes undersaturated for a water surface (Figure 4.4), so the droplets shrink by evaporation. According to Bergeron the ice crystals are the vital ingredient for the production of large-scale rainfall. The ice crystals form around special freezing nuclei, which are less common than the condensation nuclei. Insoluble soil particles are the main source of these nuclei which encourage early freezing, and kaolinite (a clay mineral) is known to be especially important. Once formed, the ice crystals grow rapidly by the addition of water vapour, and splinter to give rise to new nuclei. The freezing of supercooled droplets does not become extensive until below $-20°C$, but takes place very quickly below about $-40°C$. Falling ice crystals combine into loose aggregates which are known as snowflakes; these melt to rain drops when they fall below the 0°C level.

The Bergeron process does not explain all occurrences of rain. Much of the drizzle experienced in Britain comes from clouds in which the lowest temperatures have not reached 0°C, and in the tropics heavy showers come from clouds which do not have an ice phase. The coalescence process is based on the fact that clouds contain a variety of sizes of cloud droplets. The larger droplets fall through the cloud more rapidly (Figure 4.24), and collide with smaller droplets: they coalesce with some, while others may bounce off and will become large enough to fall as rain. Drops may become so large that they break up in very intense falls. Although a variety of factors are important in this mechanism, it seems that the range of droplet sizes needed to give rise to as many collisions as possible is most significant (Figure 4.25).

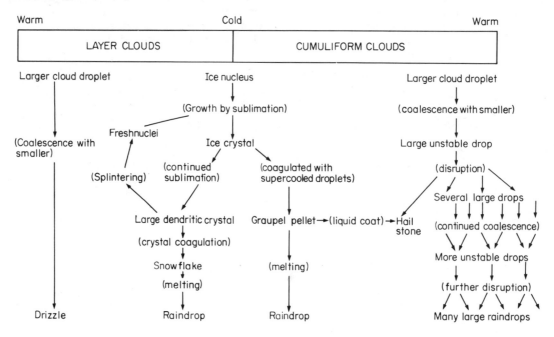

FIGURE 4:25 *Theories of rain formation. (After Mason.)*

Thunderstorms and tornadoes

Thunderstorms and tornadoes constitute some of the most dramatic evidence of atmospheric activity in the middle latitudes, and a large proportion of the rain falling in tropical areas is derived from thunderstorms. Thunderstorms involve an immense amount of energy and electrical effects become important. Severe hailstorms are associated with some thunderstorms, but these seem to be limited to the middle latitude continental interiors: whereas central Europe and the interior Mississippi basin experience up to 10 devastating hailstorms a year, none occur in the 90 average thunderstorm days of Florida, or in tropical areas.

The typical thunderstorm is an agglomeration of convective cells, forming a series of vigorous chimneys of rising air each up to 1-5 km across and persisting only for ½-1 hour. The entire cluster of evolving cells in the storm may persist for up to 12 hours (Figure 4.26). A storm progresses by

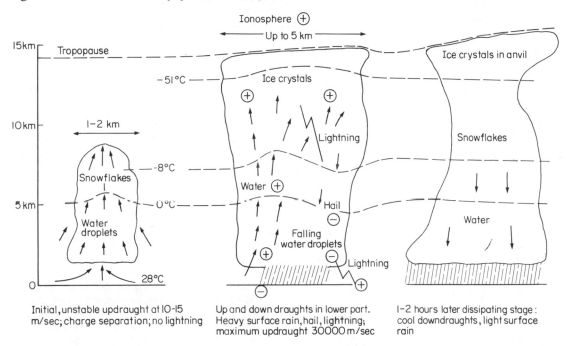

FIGURE 4:26 A thunderstorm cell. The diagrams are simplified with only one cell's air currents: in nature there will be several updraught cells, arranged in clusters and separated by downdraught areas.

replacing rained-out cells with new, marginal cells, and this affects the movement of the storm (Figure 4.27). As the storm moves across an area it will sweep up the low level concentration of water vapour in the atmosphere, and precipitate a proportion back to the ground. This may be as small as 20 per cent, since much evaporates from the cloud droplets or from rain drops whilst falling.

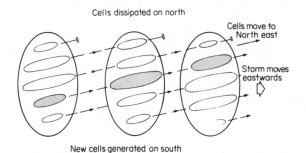

FIGURE 4:27 The development and movement of a thunderstorm. This is a well-organised (and therefore untypical) example, but demonstrates a number of features associated with thunderstorm development.

Hail varies from very small grains, which are more common, to hard masses of ice the size of tennis balls. The largest recorded fell on Coffeyville, Kansas, on 3 September 1970, measuring 19 cm in diameter and weighing 770g. There is a considerable amount of variation in composition, but many hailstones are banded, with layers of finely crystalline, opaque ice separated by coarser-grained, clear ice layers. The two types depend on the rate of accretion of supercooled water droplets on the growing stone. The type and size of hailstone falling is determined by the amount of time it spends in the cloud, and how many times it is carried up and down before final precipitation (Figure 4.28).

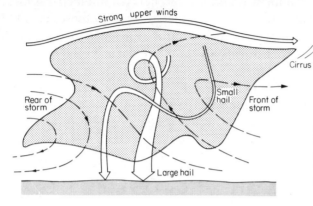

FIGURE 4:28 *A hailstorm. A thundercloud grows in a deep, moist lower layer of the atmosphere, and its top is tilted forward by upper winds. The size and structure of the hailstones is determined by the time spent in the cloud, and the temperatures of the layers through which they descend.*

Thunderstorms are notable for their lightning and thunder effects. It is thought that this may be the principal mechanism by which the large electrical charge of the Earth is maintained in spite of the fact that much is continuously 'leaked' into the atmosphere on average at 1800 coulombs/sec (Figure 4.29). Confirmation of this theory is coming from a number of sources, and it seems that between 3000 and 6000 thunderstorm cells are needed over the Earth at any one time to fulfil such a function: there are probably at least this number. Thunderstorm activity and the Earth's electrical charge both reach a maximum at approximately 19.00 hours each day all round the world.

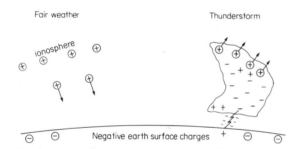

FIGURE 4:29 *Maintaining the Earth's electrical charge. During fair weather, positive ions are drawn downwards and the field is diminished by neutralization. In a thunderstorm wild fluctuations of the field occur, and it may be reversed, renewing the Earth's charge.*

Thunderclouds have regions which are electrically charged (Figure 4.29): so much can be derived from observations of lightning flashes within and between clouds, or from balloons sent into the clouds. It has been suggested that the charge separation is caused by heavier particles (e.g. large raindrops or hail) with negative charges dropping to the base of the cloud, leaving smaller particles (ice) with positive charges at the top. Two mechanisms are thought to be important in this separation: both occur in clouds containing ice particles, and it has not been demonstrated convincingly that lightning can take place in clouds without ice.

1) When two solid ice particles collide they take on opposite charges. That with the higher conductivity (due to a higher temperature or the presence of impurities) takes a negative charge, and the other a positive charge. In a cloud with larger solid precipitation particles, solid cloud particles and supercooled water droplets, the droplets gather on the larger ice particles, freezing and releasing latent heat. This warms the particle and when it collides with cloud ice it continues downwards with a negative charge. This process is not sufficient to account for the entire charge in a cloud, which has to be built up to 1000 coulombs at 1 coulomb/km^{3}min^{-1}.

2) A second mechanism may account for the remaining charge. When a supercooled cloud droplet freezes suddenly in collision with a precipitation particle, the outside of the droplet freezes first, takes on a positive charge (being cooler than the inside of the particle) and then splinters apart as the inner portion of the droplet freezes and expands. This causes small positively-charged particles to

move away from the precipitation particle which falls with a negative charge to the base of the cloud.

Once the charge accumulates in the cloud, lightning occurs. At times it may reach the ground, but normally it does not. Lightning follows a tortuous path and consists of a series of strokes following each other every few hundredths of a second, as Figure 4.30 shows. If it does strike the ground, trees and tall buildings are most likely to be hit because there is often a large difference in electrical field potential at these sharp points: this may cause electrons to be removed from gas atoms, ionising the air and providing an easy path for the lightning flash.

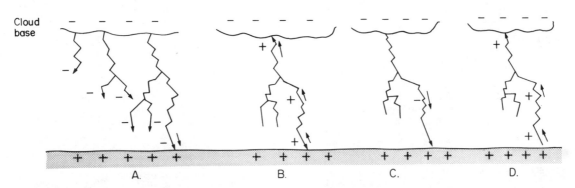

FIGURE 4:30 *A sequence of stages in a lightning flash: these follow extremely rapidly.*
(A) Stepped leader strokes build steadily downwards from a cloud towards the ground at 15 km/sec. At this stage the lightning is only faintly luminous.
(B) A return stroke from the ground to the cloud, lasting about 40 microseconds; intensely luminous.
(C) A dart leader stroke uses the same path after an interval of 0.1 second.
(D) The second return stroke. Dart and return strokes may be repeated, discharging different parts of the cloud.
(After Pedgley, 1962.)

The thunderclap is caused by an explosion resulting from the rapidity of the lightning strikes, which heat up the air column to over $15\text{-}20 \times 10^{3\circ}$C. A pressure wave moves outwards at the speed of sound and gives rise to the noise heard on the ground up to 10 km away. Subsidiary rumbles are possibly associated with zones of sound wave reflection and refraction inside and outside the cloud, or may be due merely to sound intensity reaching the observer in a variable manner depending on the direction and time.

In addition, many scientists have noticed the correlation of a particularly heavy fall of rain with the overhead position of a lightning flash, and it is suggested that the electrification may trigger off a massive coalescence of cloud droplets by charging them with opposite charges. This account serves to emphasise the weakness of our present knowledge. Much work needs to be done on cloud physics before a full understanding of the complex processes can be reached.

Tornadoes are small, but extremely violent storms, and occur particularly in central USA. They consist of a cylindrical or cone-like funnel reaching down from the base of a *cumulonimbus* cloud. The funnel is seldom more than 1 km across at the surface, and is short-lived, moving at 50-60 km/hr across country for up to 15 km, although some have travelled up to 200 km. The pressure in the central area drops to 60-80 per cent of normal, and this is sufficient to lift off house roofs. The surrounding whirl of wind has velocities up to 350 km/hr, and this leads to queer effects, such as the piece of straw found embedded in a wooden plank after the passing of a tornado.

The situation in which these storms develop is one of extreme instability and convection. The instability is built up by a variety of processes so that violent vertical movements are released. This seems to occur where the warm sector of a depression is lifted off the surface by the advance of a cold front over warmed ground. A situation of instability near the ground extends upwards with extreme conditional instability, sucking up moist air from the lower layers. Rotation, associated with **horizontal wind shears** at the advance of a squall line with towering *cumulonimbus* clouds, becomes concentrated beneath the upward currents and leads to the vigorous tornado circulation. Each tornado lasts for a few minutes, and may be replaced by another from the base of the same cloud.

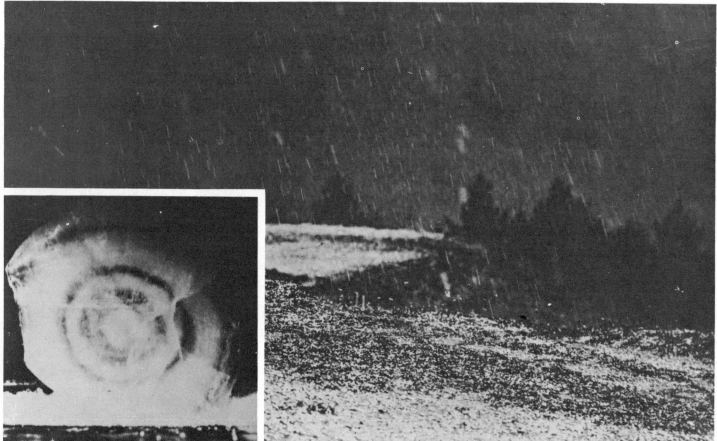

Thunderstorms and lightning. Describe the events depicted in these photographs. That at bottom left is a cross section through a hailstone.

Plate 31 (NOAA)

A

B

Development of a tornado, as seen at Enid, Oklahoma. At first a long, thin, light-coloured funnel touches down to the ground surface; then it darkens as debris is drawn up, thickening and coarsening until it reaches a ropelike mature stage.

Plate 32 (NOAA)

C

D

5

Movement within the ocean-atmosphere system

The ocean-atmosphere system is a realm of continuous movement. The resultant phenomena are often terrifying to human beings: waters whipped up to 20 m waves, or the swirling hurricane with wind speeds of over 150 km/hr. Studies of these phenomena, largely in the twentieth century, have led to a general understanding of the causes behind the movements which take place.

Movements of the ocean waters

The major large-scale movements of ocean waters are the steady transfers of water in relatively narrow zones, known as **ocean currents.** The world map of ocean currents (Figure 5.1) demonstrates a clear pattern of approximately circular cells, or **gyres**. The Atlantic and Pacific Oceans each have major gyres north and south of the equator, centred about 30°N and 30°S; the Indian Ocean has one only south of the equator. There are also smaller subpolar gyres, particularly in the North Pacific Ocean and around Antarctica: these circulate in an opposite direction to the subtropical gyres of their hemisphere.

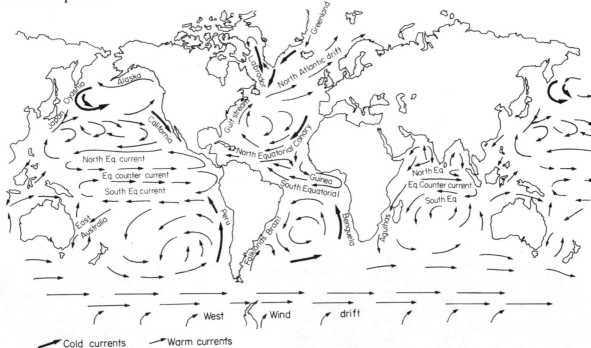

FIGURE 5:1 *World distribution of ocean surface currents: a simplified pattern. Some areas experience seasonal variations due to wind shifts (e.g. northern Indian Ocean), and there are greater complexities due to the eddying of waters near coasts.*

Most ocean currents have an east-west (or west-east) flow, related to major easterly and westerly wind belts (cf. Figure 5.17), but are diverted into north-south movements as they reach the continents, and split into currents flowing along the coasts. Such north-south (or south-north) currents are known as **boundary currents**, and those on the western sides of oceans are particularly well-defined, deep-reaching and rapidly-moving. Thus the Gulf Stream in the western North Atlantic Ocean flows outside the continental shelf of eastern North America at speeds of 100-300 cm/sec in the central 50-75 km of the current. The surface waters have a temperature of 20°C and a salinity of 36 parts per thousand (i.e. both are high relative to the surrounding waters). A definite

MOVEMENT WITHIN THE OCEAN-ATMOSPHERE SYSTEM 51

Current type	General features	Speed (km/day)	Transport ($10^6 \times$ m³/sec)	Special features
Eastern boundary	Broad (approx. 1000 km) Shallow (less than 500 m)	Slow (tens km/day)	Small (typically 10-15)	Diffuse boundaries separating from coastal currents Coastal upwelling common Waters derived from west wind drifts and thus cool
Western boundary	Narrow (less than 100 km) Deep (substantial transport to 2km)	Swift (hundreds km/day)	Large (usually over 50)	Sharp boundaries with coastal circulation system Little or no coastal upwelling — waters depleted in nutrients, unproductive Waters derived from trade wind belts and thus warm

FIGURE 5:2 Boundary currents: general characteristics. (After Gross, 1972.)

boundary, or ocean front, is established between the contrasting waters of the current and the continental shelf, marked by changes in colour as well as by temperature and salinity contrasts. This current is also remarkably deep: flows of 1-10 cm/sec have been recorded from depths of 1500-2000 m. Eastern boundary currents are less vigorous, have poorly defined boundaries and flow over the continental shelves, being less than 100 m deep (Figure 5.2).

Boundary currents are particularly important in the transfer of heat from the equatorial to polar regions. The north-moving currents, like the Gulf Stream, transfer large quantities of heat to cooler regions, whilst the 'cold', south-moving waters of the northern hemisphere can be seen as water returning to the equator for re-heating.

The surface map of ocean currents, however, presents only a partial view of oceanic circulation. There are also some vertical movements within the oceans, and although these are restricted (cf. chapter 2) they lead to deepwater currents (Figure 5.3).

Major movements of ocean waters are set off by two main factors: the interaction of winds blowing across the ocean surface, and density differences within the oceans.

Movements resulting from these forces are deflected by the shapes of the ocean basins and the

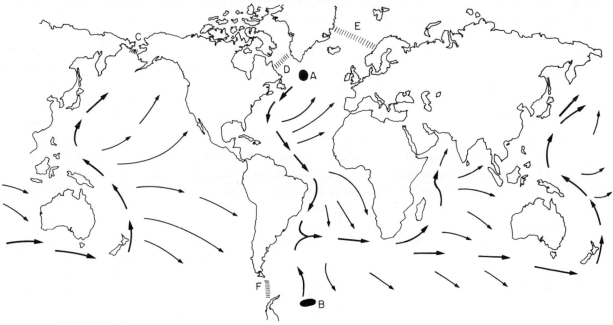

FIGURE 5:3 Deep ocean circulation. The main source regions (A and B) give rise to broad currents which spread through the oceans. Barriers (C-F) prevent the passage of such water at depth. The pattern depicted here is still tentative, having been compiled from few observations. (After Stommel and Stewart, in Moore, 1971.)

rotation of the Earth. The deflection caused by the Earth's rotation is an effect resulting from the Coriolis Force, named after the French scientist, G.G. Coriolis.

1 Wind-driven currents are related to the prevailing wind systems. They include the easterly currents on either side of the equator, associated with the trade winds, and the westerly 'drifts' between 40 and 55 degrees north and south of the equator. Such surface currents flow at approximately 2 per cent of the wind speed causing them: thus a windspeed of 10 m/sec will give rise to a surface water current flowing at 20 cm/sec. Winds blowing across the ocean surface also cause smaller scale vertical movements within the waters, extending as much as 10 m deep and mixing the heat and gases supplied at the surface. This type of vertical circulation is known as Langmuir circulation and occurs when wind speeds exceed a few km/hr. It may also give rise to deeper penetration of surface conditions as water masses sink beneath the surface convergences set up by the circulation.

2 Density differences in ocean waters are controlled by the temperature and salinity of the waters: high temperatures and low salinities lower the density, and vice versa. Such differences are particularly important in controlling water movements at depth, where the circulation is known as thermohaline. Dense water acquires its characteristics at the surface by cooling, or by evaporation (leading to high salinity) and it will sink until it reaches a level where its density is equivalent to that around it: it will be slightly less dense than the water below, and more dense than the water above. At this point it will spread out to form a shallow layer, and this results in the temperature layering in oceans mentioned in chapter 2.

The most dense water originates in the polar regions and descends to ocean floors, where topographic features control the circulation. The main areas for the formation of this dense water are off the eastern coast of Greenland, where Gulf Stream waters with a relatively high salinity are cooled to $-1.4°C$, and in the Weddell Sea of Antarctica, where water at $-1.9°C$ has a slightly lower salinity than that from Greenland. The Arctic Sea waters do not escape at depth because of ridges north of Iceland and across the Bering Sea. The very dense waters in the open oceans sink, warming a little after mixing, and move slowly along the ocean floors at 1-2 cm/sec, although more rapid speeds occur round the western ocean margins. Thus the southward-moving counter current beneath the Gulf Stream moves at speeds up to 10 cm/sec. New bottom waters of this type form in large quantities in these rather restricted locations, and flow beneath the older waters, gradually forcing them up to the surface at a rate of approximately 4 m/year. This means that such cold water stays at depth for periods of 500-1000 years resulting in an extremely slow vertical circulation of ocean waters.

3 The shapes of the ocean basins affect the movements of waters by splitting the surface currents into boundary currents, and by determining the details of the deep water circulation. Thus the Arctic Ocean waters are excluded from the main ocean deep circulation.

4 The rotation of the Earth has the effect of deflecting particles moving across its surface into curved paths. The general result is that particles are deflected to the right of their wind or density current induced courses in the northern hemisphere, and to the left in the southern hemisphere (Figure 5.4). The deflective movements affect both winds and ocean water movements. Thus, density movements in the oceans are deflected and the resultant currents are known as geostrophic currents. When wind blows over the surface of the ocean it exerts a drag on the surface layer, which moves at 2 per cent of the wind speed; the surface layer in turn affects the next layer of water beneath it, and so on with friction cutting down the speed of movement (Figure 5.5). The effect of wind blowing across the ocean surface is known as the **Ekman spiral** and the net water movement is perpendicular to the wind direction: it is known as Ekman transport. Ekman transport causes upwelling where winds blow parallel to the shore (Figure 5.5(C)), bringing up cold, nutrient rich waters from 100-200 m below the surface. This is an important factor in the climates of coastal areas like Chile-Peru and California.

The net transport of surface waters leads to the piling up of water in zones of convergence such as the subtropical centres of the major oceanic gyres. Divergence occurs around Antarctica.

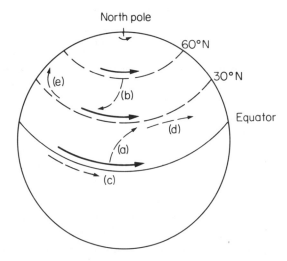

FIGURE 5:4 *Coriolis Force. The Earth rotates on its own axis, and the ocean and atmosphere move with it at the same speed. Rotational speeds (broad arrows) vary from 1600 km/hour at the Equator to 800 km/hour at 60° latitude (i.e. half the length of the Equator) and nil at the poles. If there is no relative motion of water or air with respect to the Earth motion, then no deflection will occur: it is only when water or air is forced to move more rapidly than the Earth, or is slowed down, or moves across a line of latitude, that the Coriolis Force has its effect.*

(a) Particles moving north from the equator have the eastwards rotational component of 1600 km/hour and retain this. When they reach higher latitudes they are moving eastwards faster than the ground beneath, and their course is thus deflected eastwards (or to the right in the northern hemisphere).

(b) Particles moving southwards towards the Equator begin with a slower eastwards rotational component (e.g. 800 km/hour at 60°N) and move into zones of more rapid eastward component. They are thus deflected to the west (or, again, to their right in the northern hemisphere).

(c) Particles moving eastwards or westwards along the Equator are not deflected: this is a neutral effect.

(d) Particles moving eastwards along lines of latitude north of the Equator (e.g. a westerly wind) are in effect moving at speeds greater than the rotational speed, and the centripetal force of the rotating sphere deflects their path towards the Equator – i.e. into zones of higher rotational speeds.

(e) Particles moving westwards along lines of latitude are in effect moving more slowly than the rotational speed and are deflected to the north. Deflection is always to the right of the original path in the northern hemisphere, and to the left in the southern hemisphere.

Differences in the height of the ocean surface result, and, although there are only 2 m between the highest and lowest parts of the ocean surface, the water movements can be seen as running downhill from the highest parts to the lowest; deflection by the Coriolis Force leads to resultant movement at right angles to the slope. The Florida Current (the southern part of the Gulf Stream) has some of the steepest sea surface slopes (10 cm over 100 km) and consequently rapid current speeds (150 cm/sec). Such a view of ocean currents is new, and charts are now being prepared with this information, so that current speeds can be predicted from a knowledge of ocean topography. It is only in the last few years that the technology has been available to measure such small differences in ocean height.

Another idea which has been important in the study of ocean water movements is the concept of water masses. A water mass is a volume of water which has recognisable temperature and salinity characteristics. Distinctly different layers of water exist at particular depths in the oceans so that sampling at one point may locate several strata with peculiar temperature or salinity conditions.

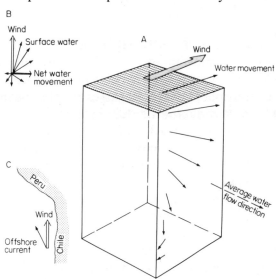

FIGURE 5:5 *(A) The Ekman spiral effect. The wind drives the surface water layer, and the movement is transferred downwards, with loss of energy, by friction. At each level the flow is deflected (to the right in the northern hemisphere). (After Stewart, in Moore, 1971.).*
(B) The average water flow is at right angles to the wind direction.
(C) The effect of winds blowing parallel to the coast causes the Peruvian Current to flow away from the shore (i.e. deflection is to the left in the southern hemisphere).

Water mass	Temperature (°C)	Salinity (parts per thousand)
Antarctic bottom water	−0.4	34.66
North Atlantic deep water	3-4	34.9-35.0
North Atlantic central water	4-17	35.1-36.2
Mediterranean water	6-10	35.3-36.4
North polar water	−1-10	34.9
South Atlantic central water	5-16	34.3-35.6

Water mass	Temperature (°C)	Salinity (parts per thousand)
Subantarctic water	3-9	33.8-34.6
Antarctic intermediate water	3-5	34.1-34.6
Indian equatorial water	4-16	34.8-35.2
Indian central water	6-15	34.5-35.4
Red sea water	9	35.5

FIGURE 5:6 *Some major ocean water masses. (After Defant, in Gross, 1972.)*

These water masses can be traced to formation in certain areas (Figure 5.6): their nature is determined at the surface and they may sink with transport to other areas whilst mixing is taking place. Mixing in deep oceans is slow, so such characteristics persist and can be identified far from the source areas.

The three-dimensional circulation of the Atlantic Ocean can be analysed in terms of surface wind drifts, density-controlled movements, the effect of ocean basin shape and deflective movements of the waters (Figure 5.7).

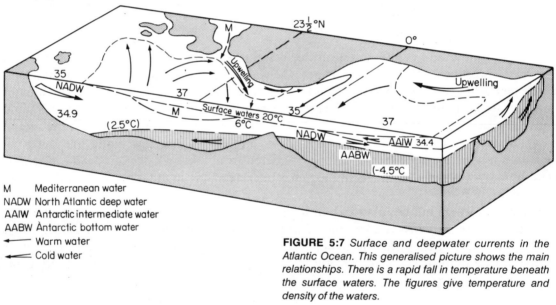

FIGURE 5:7 *Surface and deepwater currents in the Atlantic Ocean. This generalised picture shows the main relationships. There is a rapid fall in temperature beneath the surface waters. The figures give temperature and density of the waters.*

Movements in the atmosphere

Atmospheric movements — **winds** — are subject to many of the forces which affect the oceans: heating, rotation, gravity, friction and pressure gradient. A consideration of small-scale movements in the atmosphere enables an assessment to be made of these complex factors one by one.

Small-scale movements: local winds

It can be cool on a beach on a bright summer's day when a strong breeze blows in from the sea and moves loose dry sand across the surface. Clouds tend to form over the land by day and out to sea at night. The basic cause of such movements is that the air over the land is heated much more rapidly, and cools more rapidly, than the air over the sea (Figure 5.8). Daily wind direction changes also occur in mountainous areas (Figure 5.9). An understanding of this type of situation is basic to an appreciation of the wind-producing forces. Heating by the Sun creates differences of pressure in the atmosphere and these lead to winds, which are movements of air from high pressure zones to low pressure zones so that the differences may be eliminated.

Such movements are most intense when the differences in pressure are greatest. Differences in pressure can be measured and plotted on a map to show their distribution. Isobars (i.e. lines joining

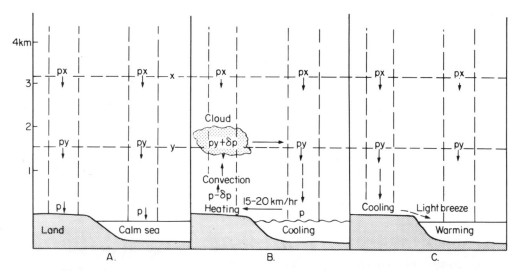

FIGURE 5:8 Land and sea breezes are caused by differences in the heating of land and sea areas.
(A) Early morning: pressure in columns of air over land and sea is approximately in a state of equilibrium.
(B) Late morning/early afternoon: heating of the land and the air above it causes a disturbance in the system and results in flow within a shallow (up to 2 km) cell. The surface density contrast leads to a movement of cooler air from the sea towards the land and the less dense air over the land is rising. The system readjusts to the disturbance and a new equilibrium is established.
(C) Heating of the land and convection over the land ceases at night and the circulating cell breaks down. A reversal of flow may result if the sea temperature is warm relative to the land.

places of equal atmospheric pressure) can be drawn, delimiting areas of high and low pressure (Figure 5.10). Winds blow across the isobars from areas of high pressure towards regions of low pressure. The closeness of the isobars gives a measure of the pressure gradient: gentle if they are far apart, steep if close together (cf. isotherms). Wind speed is related to the pressure gradient, being greatest with steep gradients. The 'mistral' wind of the Rhône-Saône valley is an example of a pressure-induced wind (Figure 5.11). It is important to remember that winds are named after the point from which they blow: sea breezes blow from the sea, westerly winds from the west, etc.

If the Earth was stationary there might be a simple convectional circulation from the warmed equatorial regions to the cooled polar regions along north-south lines, with a movement of heated air to the Poles at high levels and a return of cold air at the surface, like the sea breeze situation. The Earth's rotation, however, has a most vital effect on the way in which this basic convectional pattern is broken down into smaller units.

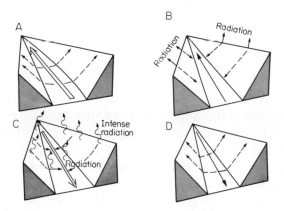

FIGURE 5:9 Valley winds. Why are coffee trees and fruit orchards sited on the slopes rather than on hilltops or valley floors?
(A) Early afternoon: air heated above the valley floor expands and rises up the valley slopes (anabatic winds).
(B) Evening: up-valley winds die out; downslope winds become important as hilltops cool and the cold air drains downwards (katabatic winds).
(C) Night: general downward drainage of dense, cold air. Fog may form in valley bottom.
(D) Sunrise: cold air still drains down from valley head; first upslope winds begin. (After Defant, in Barry and Chorley, 1968.)

Circulation in the upper troposphere

The simplest pattern of wind circulation on a world scale occurs near the top of the troposphere. Increasing interest has been shown in this sector of the atmosphere since jet aircraft began flying there, and since it was realised that conditions at that level affect profoundly the surface weather. Weather observations at these heights are still inadequate, and they are concentrated over the richer

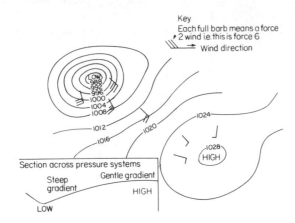

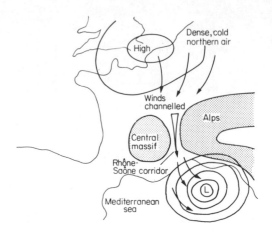

FIGURE 5:10 *Isobars and pressure gradient. Isobars are close together around the low pressure area, giving a steep pressure gradient. How are the wind strengths related to the low and high pressure centres, and the pressure gradients?*

The isobars were drawn from a weather map from 31 October 1969 in the western North Atlantic Ocean. The winds did not in fact blow in the way shown here. Why? What causes ocean currents to be deflected from their density-controlled or wind-blown patterns? (cf. Figure 5.4).

FIGURE 5:11 *The mistral wind. What is the relationship of the winds to the pressure conditions and the mountain areas?*

FIGURE 5:12 *Pressure in the upper atmosphere. The mean contours (intervals of 100 feet or approximately 30 m) of the 700 mb pressure surface over the northern hemisphere are shown. What are the seasonal differences in the pressure gradients, and how would you expect these to affect the winds aloft? (After O'Conner, in Barry and Chorley, 1968.)*

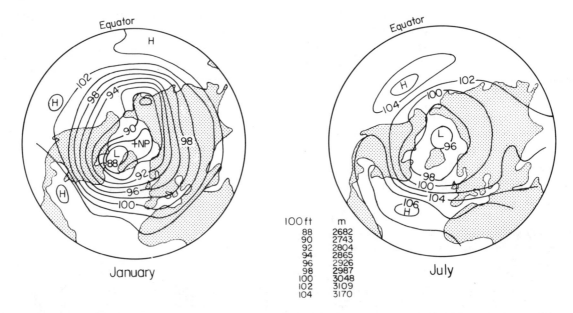

countries of the world which can afford the necessary equipment, but a general pattern of events has emerged from the results and satellites now provide a world wide coverage.

It is important to realise that pressure maps drawn at various levels in the atmosphere have a different basis from those drawn relative to sea level. Instead of plotting pressure differences at a certain height, the reverse process is adopted: the height of a particular pressure surface is plotted and the lines on the map are relief contours. The 700 mb pressure 'surface' is often used (Figure 5.12): low pressure areas occur where this is nearest to the surface, and high pressure where it is highest. This convention does not affect the relationship of the winds to the lines drawn.

The first thing to be noticed is that the winds blow parallel to the isobars instead of across them (cf. Figure 5.10). The reason for the difference between this situation and that involving the sea breeze is the Earth's rotation, which deflects winds as it does the ocean currents (Figure 5.4).

In the upper troposphere the Coriolis Force is often sufficient to balance the effect of the pressure gradient, and this leads to the winds blowing almost parallel to the isobars, when they are known as geostrophic winds (Figure 5.13: cf. the geostrophic currents in the oceans.) Interruptions of this situation, particularly in the form of divergence or convergence of isobars, lead to the transfer of heat across the isobars, a process which is all-important to atmospheric circulation and exchange between high and low levels of the troposphere.

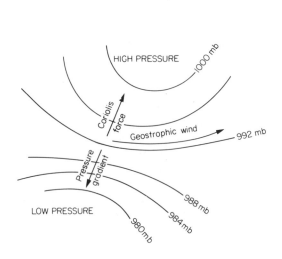

FIGURE 5:13 *The geostrophic wind. What is the relationship of its direction to (a) the pressure gradient, and (b) the isobars? Under what conditions does it occur? In which hemisphere would it blow in this direction?*

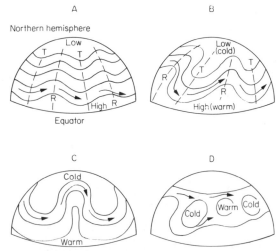

FIGURE 5:14 *The jet stream cycle.*
(A) Small amplitude waves develop in the upper air winds.
(B) These waves become larger. (T = trough; R = ridge.) Jet stream winds are very strong, but allow little heat exchange across their path.
(C) Waves become extremely sinuous.
(D) The cycle breaks up, isolating cold and warm air masses, so carrying out heat exchange. This is the northern hemisphere pattern. Why is this better understood than that of the southern hemisphere? (After Namias, in Barry and Chorley, 1968.)

The strongest of the upper air winds are known as **jet streams**, and they blow with speeds of 80-240 km/hr at heights of 10-18 km on the poleward sides of the surface subtropical high pressure belts. Winter maxima of 450 km/hr have been recorded. At times there may be two jet streams in the winter hemisphere, although the second, in middle latitudes, is less persistent. These jet streams develop wave forms (Figure 5.14) which cause a breakdown in the tight-celled geostrophic pattern. This leads to an important transfer of heat (Figure 5.14(D)). An understanding of jet streams has become important in the world of aviation, and a plane can make a useful saving in fuel on a west-east flight, although pilots avoid the turbulence of the central zones. Even more important is the fact that these movements control the surface circulation, particularly in the middle latitudes. The wave forms, having an amplitude of 1000-10000 km, move from west to east near the top of the troposphere: between three and six waves may occur in each hemisphere at any one time, and the winter movements are most intense. More rapid, smaller amplitude waves move through the larger waves. Exchange of air takes place at definite points:

a) Air ascends to the front of the large wave troughs where the upper air flow diverges (Figure 5.15(D)). The flow aloft here is south-westerly in the northern hemisphere and the diverging winds draw up air from the lower troposphere. There is also some ascent associated with a further zone of divergence to the poleward side of the main jet and at the front of a ridge. The ascending air is replaced by horizontal inflow at the surface. Warm and cold air are brought together in this way, and fronts and cyclonic wind circulations are formed in surface zones of convergence.

b) Air descends where convergence takes place aloft, near the most rapid sections of the jet: the most important area is in the rear of a trough, where the winds are north-westerly. Such areas of descending air lead to the warming of the air and are usually devoid of clouds if they reach down to the surface (where the winds blow outwards). Thus, convergence aloft involves surface divergence, whilst divergence aloft involves surface convergence.

The surface circulation

The study of surface winds adds a further factor to those influencing movements in the atmosphere: the frictional effects of contact with the land and sea (Figure 5.16). At the surface, the combined effects of pressure differences and the Coriolis Force are countered by friction, since the slowing down of the winds lessens the deflective effect. This happens to a greater extent over land, where winds cross the isobars at angles of 25-35 degrees as compared to 10-20 degrees over the sea. The world surface isobar/wind patterns reflect this relationship (Figure 5.17).

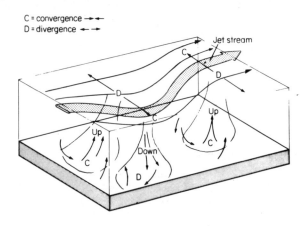

FIGURE 5:15 *The jet stream and its relationship to air circulation in the troposphere. The jet blows almost parallel to the upper air isobars, but its sinuous course leads to the convergence and divergence of air and to an exchange with air at lower levels. Relate the zones of convergence and divergence aloft and at ground level.*

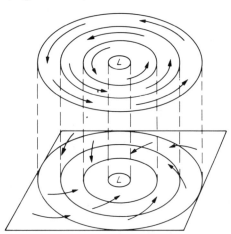

FIGURE 5:16 *Flow of air around a low pressure centre at ground level and at 1 km above the surface. Which winds are geostrophic? What differences in wind speed would be expected at the two levels? In which hemisphere would this pattern be recorded?*

As with the upper air situation, knowledge of what happens at the surface is rather variable for different parts of the world. In particular, little is known about conditions in very high latitudes near the poles. It seems that there are sometimes areas of high pressure, but that low pressure is dominant: even when pressure at the surface is high this is usually only a shallow feature.

The main surface wind systems are related closely to the strong, stable, high pressure areas centred just outside the tropics in each hemisphere (the 'subtropical highs'). On the equatorial side there are mainly easterly winds (known as the *north-east trades* and the *south-east trades*), which are the most extensive and constant winds at the Earth's surface, affecting half the globe in tropical latitudes. They are strongest in the winter season.

The easterlies to the north and south of the equator blow towards the equatorial low pressure trough known as the Inter-Tropical Convergence Zone (ITCZ), which is at its most effective over the oceans. Surface conditions are characterised by light winds and low pressures, air being forced to rise on a large scale. Compare the seasonal positions of the ITCZ (from Figure 5.17) with the thermal equator (Figure 3.6), which is the result of direct heating from the Sun.

On the poleward margins of the subtropical high pressure systems the winds are mostly westerly in direction, but are much more variable than the tropical easterlies because they are in a zone dominated by the passage of cells of low and high pressure with circulating winds. In addition, the larger land masses of the northern hemisphere also affect the pattern, due to differences of surface heating and cooling. Thus, whilst the Scilly Isles (50°North) have 46 per cent of their winds from the westerly quarter (SW-W-NW), the Kerguelen Islands (49°South) receive 81 per cent from these

A January

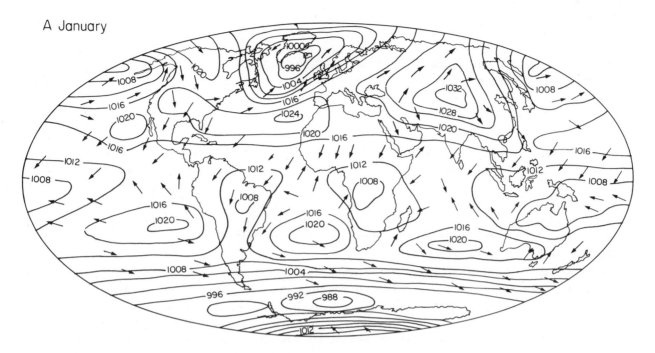

FIGURE 5:17 *Average barometric pressure and surface winds for January (A) and July (B). Measurements of pressure are given in millibars, reduced to sea level. Mark the following on an outline map of the world:*
(a) areas which have high pressure in both January and July ('H');
(b) areas which have high pressure in summer (Hs) or winter (Hw) only;
(c) areas of low pressure in winter or summer (Ls/Lw). Do land areas, or ocean areas, experience most summer-winter changes? Why? Using the information plotted, attempt to draw the surface wind directions, relating these to the pressure patterns and ocean currents. (Maps after Strahler, 1969.)

B July

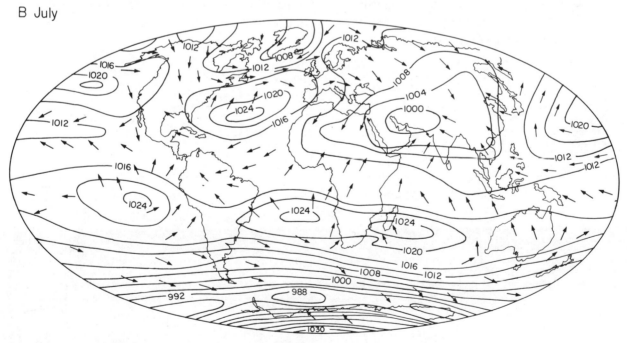

directions. Patterns of variable, but westerly, winds continue dominant to the poles, although there are some local easterly winds in very high latitudes.

Thus the surface pattern is divided approximately between easterly and westerly winds, in contrast to the upper air pattern where the westerly circulation is overwhelmingly important.

A pattern of atmospheric circulation

The atmospheric winds have been studied so far in terms of two-dimensional distributions aloft and at the surface, but of course these two levels are linked closely as was suggested when jet steams were discussed. For many years the most acceptable interpretation of the connection was in terms of a vertical convectional circulation based on air rising following heating at the equator and descent at the subtropical high pressure areas (Figure 5.18). This model was named after the meteorologist Hadley, who put forward the idea of vertical convection cells in the eighteenth century. The pattern was extended polewards on theoretical grounds, as the diagram illustrates.

All sorts of difficulties faced this rather simple-looking model. The equatorial heat source is a complex band encircling the Earth, and there is some evidence to suggest that cross equatorial flows of air take place. Moreover, the trade winds do not form a continuous belt around the globe, and the subtropical high pressure areas are cells separated from each other by areas of low pressure. At the other end of the system it is now realised that the poles are not the centre of high pressure surface conditions as Hadley envisaged.

The increasing number of upper air observations has confirmed that horizontal, rather than vertical, heat transfer is the most important mechanism, especially away from the tropical areas where most heating by solar energy takes place. Transfer in the middle latitudes is carried out largely by the travelling high and low pressure cells at the surface, and by the jet streams in the upper atmosphere. The tropical circulation is more complex, and the high pressure systems are obviously so important that the Hadley cell type of system is still thought to be effective there. This view resulted in a revised model (Figure 5.18(B)). It provides a suitable mechanism for the anomaly (chapter 3) whereby the equatorial regions receive an excess of solar radiation, but do not get hotter, and the polar regions suffer a deficit, but do not get cooler. The balancing factor is this transfer of heat energy from low latitudes to high, involving both the atmosphere and the ocean.

Thus, the atmospheric and oceanic circulations are due to a combination of factors. Heat from the Sun is the main source of energy driving the movements of air and water. The resultant patterns are modified by the Earth's rotation, together with the relationships between the rotating Earth and movements in the atmosphere-ocean system rotating with it. The picture derived from increasing observation becomes more complex!

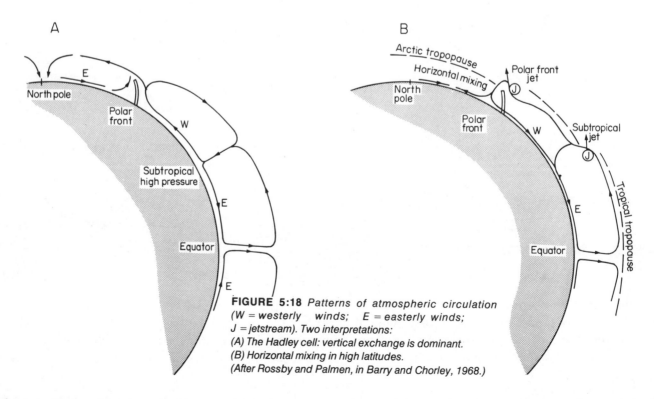

FIGURE 5:18 *Patterns of atmospheric circulation (W = westerly winds; E = easterly winds; J = jetstream). Two interpretations:*
(A) The Hadley cell: vertical exchange is dominant.
(B) Horizontal mixing in high latitudes.
(After Rossby and Palmen, in Barry and Chorley, 1968.)

6
Weather systems

Atmospheric temperature and humidity, sunlight, cloud cover, rainfall, wind speed and direction, pressure, and conditions in the upper atmosphere are all elements in the weather which can be observed, measured and recorded. Knowledge of such elements provides a basic pattern for understanding the working of the atmospheric machine. Generalisations based on experience of the commonly repeated sequences of weather changes enhance an understanding of what occurs, and provide the weather forecaster with basic models which he can use to suggest ways in which a certain progression of events might continue. It is almost needless to say that nature seldom repeats itself exactly, and that scientists are continually bringing to light new facts which modify the models so that they approximate more closely to the truth.

Size (km)	10^4	5000	10^3	500	10^2	50 10	5	1	0.5
Scale	Planetary		Synoptic		Mesoscale		Convective, microscale		Microscale, molecular
Middle latitudes	Long waves		Extratropical depressions		Fronts		Cumulonimbus cloud		Boundary-layer eddies
	Subtropical anticyclones		Anticyclones		Lee waves		Showers		
					Squall lines		Tornadoes		
Tropics	Intertropical convergence zone		Cloud clusters		Mesoscale convective elements		Convective cells		Boundary-layer eddies
			Easterly waves		Tropical cyclones				

FIGURE 6:1 *Types of weather system related to scale of size and latitude.* *(After Mason, in Corby, 1970.)*

A first step in the study of weather systems is to define the scale of the events and patterns under study (Figure 6.1). Many of the items included in the microscale groups (e.g. thunderclouds, tornadoes, convective cells) have been studied already, so that emphasis will now be placed on the mesoscale, synoptic and planetary groups. The world distribution of weather stations is related most closely to the synoptic scale (the scale of system which has been found to be most useful for weather forecasting). The following systems have been identified.

Middle latitude systems:
(1) fronts and air masses;
(2) extra-tropical depressions and long waves in the upper atmosphere;
(3) anticyclones and subtropical anticyclones.

Tropical systems:
(1) cloud clusters;
(2) easterly waves;
(3) tropical cyclones, including hurricanes.

Middle latitude weather systems
Most meteorological research has been carried out from the richer middle latitude countries such as the U.S.A., the U.S.S.R., and the west European countries. This has resulted in a more thorough knowledge of the processes and systems affecting these areas. Although increasing attention is now

being paid to tropical situations, since it is realised that an understanding of the whole atmospheric system is necessary, the state of knowledge lags behind. This is not helped by the wide spacing and relatively recent origin of meteorological stations in the tropics. As more is discovered, particularly in connection with the upper troposphere circulation, it is found that the succession of weather in the tropics is associated with systems like those which are more clearly defined at ground level in temperate latitudes.

Fronts and air masses

As meteorological stations became more widely distributed during the first quarter of this century, it was realised that on the continental scale there were large bodies of air in which conditions of temperature and humidity are remarkably uniform over distances of about 1000 km. It was also recognised that the zones, along which these air masses meet (known as fronts), are important in explaining weather changes: they are zones of little mixing, across which the differences are most important. The study of air masses prompted an extension to the oceanic realm, where water masses have now been recognised (Figure 5.6).

If air remains over a part of the Earth's surface with little movement for 3-5 days it takes on characteristics related to the radiation (temperature) and evaporation (humidity) rates in the area. These source regions for air masses are associated closely with the large, slow-moving anticyclonic (high pressure) belts of the subtropics and high latitudes. Thus polar air (P) and tropical air (T) are distinguished, and sometimes Arctic air (A) is also recognised, although its features are similar to polar air. Continental (c) and maritime (m) influences affect both P and T areas. They control the humidity of the air mass, just as the P or T element controls the temperatures. There are thus four basic source regions (Figure 6.2).

Air masses move out from the high pressure source regions towards the low pressure zones where they converge with other air masses. Changes take place slowly as the air passes over warmer or colder surfaces. These are particularly important in determining the observed lapse rate of the air: if air is warmed at the surface the observed lapse rate will increase, encouraging instability, and vice versa. The conditions in the upper troposphere are important in assisting or restricting instability.

The air masses affecting Britain will be taken as an example, since they include all the main varieties. Britain stands in a region where polar and tropical air are brought together continually, and at the same point air flows meet with marine and continental characteristics (Figure 6.3).

Continental polar air (cP) formed over northern Scandinavia and Siberia in winter brings intensely cold weather as it moves towards Britain. It is usually very dry, since it comes from a land area, and stable, because of surface cooling which lowers the normal lapse rate. When it crosses the North Sea it picks up moisture and is warmed at the surface, arriving with cloudy and drizzly weather. Northern

FIGURE 6:2 Air mass characteristics: the main types affecting Britain.

Air mass types	Abbreviation	Source regions	Properties
Maritime polar (or Arctic)	mP (mA)	Oceans north of 50°N Oceans south of 50°S (Arctic Ocean in summer)	Cold, damp, unstable
Continental polar (or Arctic)	cP (cA)	Northern Canada, NE Siberia, Scandinavia, Antarctica (Arctic Ocean in winter)	Cold, dry, stable
Maritime tropical	mT	Subtropical high pressure belts over oceans	Moist warm, variable stability (more stable on eastern sides of oceans)
Continental tropical	cT	Tropical deserts	Hot, dry, unstable

Canada is another source region for *cP* air, but by the time such air reaches Britain across the North Atlantic it has lost most of its original characteristics.

Maritime Polar air (mP) is modified by extensive travel across the North Atlantic. When it reaches Britain from a northerly direction it brings heavy clouds and rain. At times, however, it comes from the south-west, surface-cooled and stable over the sea with fogginess in early summer, but generally giving light cloud as it moves over the heated land.

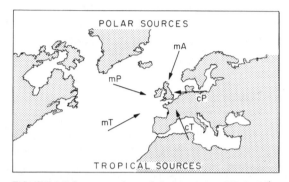

FIGURE 6:3 Source regions of air masses, and their paths towards the British Isles, which stand at the meeting place of polar, tropical, continental and oceanic air.
(After Belasco, in Barry and Chorley, 1968.)

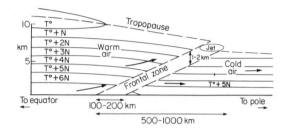

FIGURE 6:4 A frontal zone between two air masses. Why does it slope in the manner shown? It is important to realise that the gradient of the frontal zone is probably only 1 in 100: it is exaggerated in this diagram (cf vertical and horizontal scales). Arrows on the diagram show air movement.
(After Barry and Chorley, 1968.)

Continental Tropical (cT) air, originating in the Sahara desert, rarely reaches Britain without modification, since the high temperature and low relative humidity encourage evaporation into it as soon as it crosses water. Only occasionally do the extremely hot, dry draughts of southerly air which may give exceptionally warm days, penetrate to the British Isles.

Maritime Tropical (mT) air is the major type Britain receives from the tropics, since it originates in the oceanic subtropical high pressure belts and penetrates into the middle latitudes. The air leaves its source with high temperature and humidity, but with clear skies, and often reaches Britain in this form. Surface cooling may lead to low *stratus* cloud or sea fog on the western coasts. Cloud will thicken if instability is encouraged aloft, and, in summer, surface heating of this warm moist air, plus high level instability, leads to intense thunderstorms.

Air masses meet along a transitional zone, or front (Figure 6.4), where there are relatively sharp discontinuities in pressure and temperature patterns. Fronts can thus be plotted on isobaric maps by drawing lines through changes in direction of the isobars (Figure 6.5): they are known as warm or cold fronts according to the relative temperature of the air behind the front as it travels. At one time

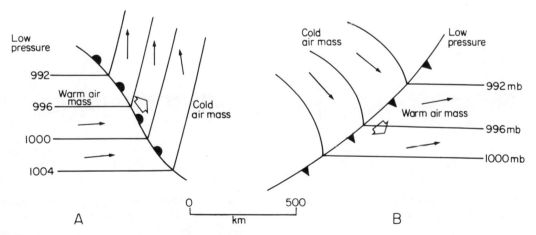

FIGURE 6:5 Fronts on maps. Conventional symbols used to depict warm fronts (A) and cold fronts (B). The large arrows indicate the direction of frontal movement; the small arrows represent wind directions. What happens to wind and pressure conditions as each type of front passes across an area?
(After Pedgley, 1962.)

it was assumed that there was an almost continuous frontal zone in the middle latitudes between the tropical and polar air masses (the 'Polar Front'), and that cyclonic disturbances began when waves developed along this zone. It is realised now that such is not the case, and that the surface cyclonic and anticyclonic systems are connected more closely with movements in the upper troposphere. Surface frontal zones are discontinuous and temporary. The winds in the frontal zone transition, between warm and cold air, increase in velocity with height until they merge with the jet stream.

Fronts are thus essentially zones where dense (colder) and less dense (warmer) air meet (Figure 6.6). As they move the warmer air is forced to flow upwards, so that there is a vertical component to the winds. Where the warm air is rising, the fronts are known as ana-fronts; if it descends relative to the front it is known as a kata-front (for another use of the prefixes ana- and kata- see Figure 5.9.

Extratropical depressions

Perhaps the most important single development in the study of weather systems was the recognition of **temperate depressions**, which became the basis of weather analysis in temperate regions from 1920 onwards. These are characteristically areas of low pressure with close-spaced, generally concentric isobars and an anticlockwise wind circulation in the northern hemisphere. The early knowledge of their surface expression has been extended, so that they are now seen to be three-dimensional features extending up to, and even through, the tropopause. Surface and high-level winds are therefore related closely in a single system. The air is not merely involved in a cyclonic whirl, but passes through the depression, entering at the surface following convergence, rising and leaving at upper levels (Figure 6.7).

Several types of temperate depression are now recognised. The main variety is known as the **warm-sector depression**, in which contrasting air masses (e.g. *mP* and *mT*, or even two varieties of *mP*) are brought together. Eddies along the resultant front give rise to a bulge, or frontal wave, and warm air begins to move upwards over the cold. This wave may retain its shape without further

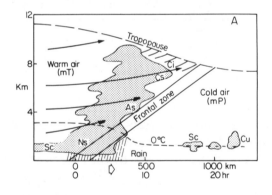

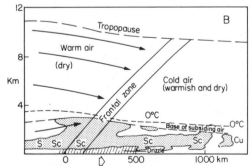

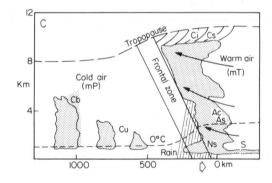

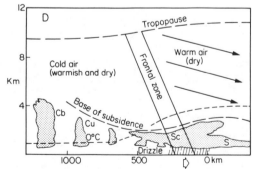

FIGURE 6:6 *Types of midlatitude fronts. Note the vertical exaggeration on the diagrams.*

(A) Ana-warm front: multilayered clouds due to variable humidity in different layers.
(B) Kata-warm front: subsidence is a major factor.
(C) Ana-cold front: note the steeper gradient of the cold front (average 1 in 75, as against 1 in 150 for warm fronts).
(D) Kata-cold front.
Compare the weather experienced (temperature, clouds, rain) for each over a period of 24 hours, assuming that the fronts travel at 50 km per hour. (After Pedgley, 1962.)

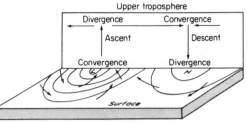

FIGURE 6:7 *Convergence and divergence (cf. Figure 5.15). Convergence at higher levels occurs with reduction of wind speeds (cf. the effect of thousands of people squeezing into a London underground station during the rush-hour); divergence is accompanied by increasing wind speeds.*

development, or may increase in amplitude until a definite warm-sector is formed (Figure 6.8). Further development leads to the warm sector being lifted off the ground altogether, when the depression is said to be occluded. The whole process, from wave formation to occlusion takes approximately three days. The occlusion brings three types of air together: the colder air in front, the warm air of the warm-sector, and the colder air behind; the relationship between the two cold air

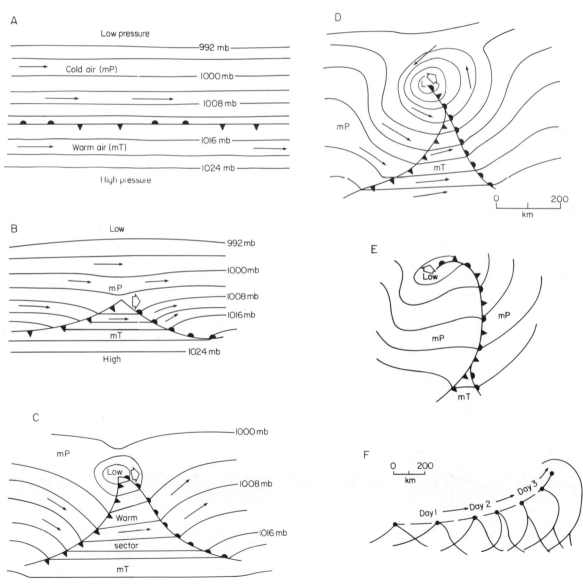

FIGURE 6:8 *The development of a warm sector depression, leading to its occlusion.*

(A) An almost stationary frontal zone.
(B) A frontal wave develops.
(C) The warm sector depression has formed.
(D) The depression begins to occlude.
(E) An occluded and decaying depression.
(F) The chronological development of the fronts.

(After Pedgley, 1962.)

masses determining the nature of the occlusion (Figure 6.9). The sequence of weather associated with the passage of a warm-sector depression in its various stages, and the relationship of the ground situation with events in the upper atmosphere are summarised in Figure 6.10. This is a model of the typical situation, but many variations occur (Figure 6.11).

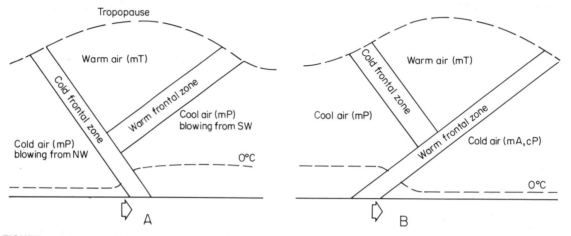

FIGURE 6:9 *Types of occlusion. They depend on the relationship of the air behind a depression to that in front of it. How will the differences shown affect the weather received?*

(A) Cold occlusion. The air behind is colder, and moves in beneath the warm sector and the warm front. This is most common when cold mP air blows from the northwest, and is colder than mP air returning polewards from warmer areas.

(B) Warm occlusion. The leading air mass is colder and denser than the following air. This often occurs when a depression reaches Europe in winter.

(After Pedgley, 1962.)

Smaller depressions may develop along the fronts of a major warm-sector depression. Thus, new waves may occur along a cold front behind an occluding depression: these may merely slow down the movement of this front, or may develop into intense, fast-moving depressions with gale-force winds. Waves may also develop along a warm front, though this is less common.

Other types of depressions develop without frontal associations and contain the weather of a single air mass type: they are smaller and less frequent in occurrence. Thus, extensive heating by day

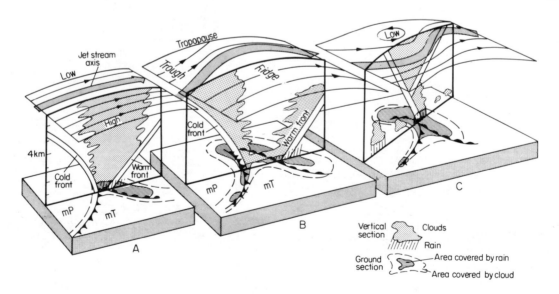

FIGURE 6:10 *A three-dimensional view of the development of a warm sector depression. Compare these diagrams with Figure 5.15 and Figure 6.8. (Data from Pedgley, 1962.)*

over a land area gives rise to centres of low pressure which may be most intense at about 4 pm. These **thermal depressions** are most common in summer, and are associated with thundery weather. **Polar depressions** may be formed as cold air moves over a warm ocean, giving rise to heating from below, and leading to heavy showers; they may drift with the mass of air, but soon fill up over land.

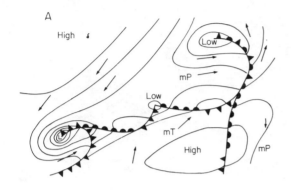
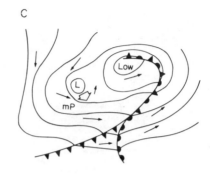
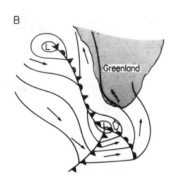
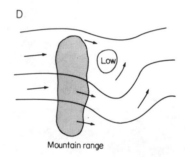

FIGURE 6:11 *Depressions may have a variety of starting points: those shown here are northern hemisphere situations.*

(A) A family of depressions may form along the trailing cold front of an occluding depression.
(B) A breakaway depression may form at the point of occlusion, especially as it is passing the southern tip of Greenland, or near a mountain range.
(C) A polar depression forming in mP air due to local summer heating within that air mass.
(D) An orographic depression formed by distortion of the pressure patterns as air flows over a mountain range.

(After Pedgley, 1962.)

The causes of depressions are not fully understood, but it seems that convergence of air at the ground and divergence aloft is a fundamental process which has to be explained (Figure 6.7 and Figure 6.10). Most of the surface depressions form in association with divergence aloft, so that the outflow of air at these levels is compensated by inflow at the ground level. Growth and deepening of a depression (Figure 6.8) occur when divergence aloft exceeds surface convergence, leading to a continuing fall in surface pressure and distortion of the surface isobaric pattern. The cyclonic circulation spreads upwards, eventually disturbing the upper air flow. At first this enhances the degree of divergence, but later it decreases it and the surface pressure falls less rapidly and finally ceases to fall. The occluding depression may then fill up as the upper air flow passes across it, isolating the centre in a zone beneath upper air convergence. Fronts are most important in the early stages of cyclone development and become less significant later.

Further association with the upper air flow is shown by the fact that the surface depressions move along with the upper winds; their movement can be predicted by studying the winds (direction and speed) between 500 and 1000 millibars (Figure 6.10).

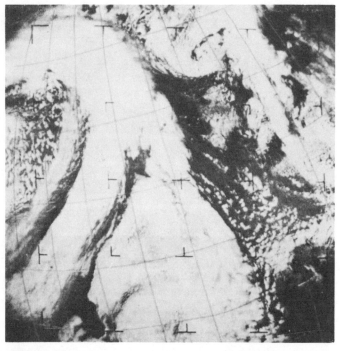

33

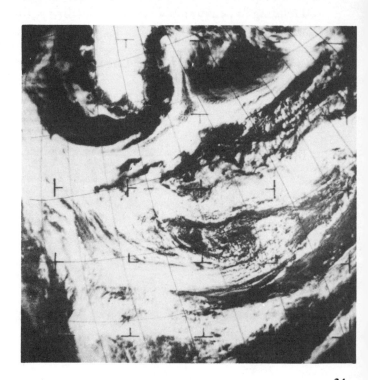

34

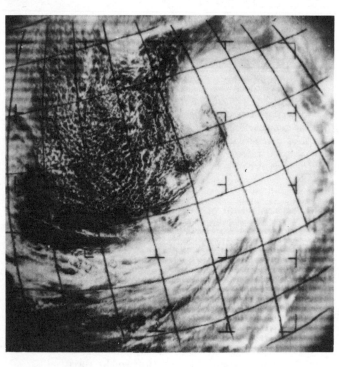

35

Depressions, as viewed from weather satellites.

Plate 33 ESSA 2 satellite, 9 May 1966: cold and warm fronts in an Atlantic depression together with a minor centre off north-west Scotland giving an occluded front across the United Kingdom.

Plate 34 NIMBUS satellite, 12 February 1970: an intense secondary depression moving up the English Channel.

Plate 35 ESSA 4 satellite, 4 February 1967: a young depression with warm and cold fronts plus extensive cumulus cloud behind. (All Crown Copyright)

36

Anticyclone over Scandinavia, as seen by ESSA 7 satellite on 14 March 1969. Note the extent of the Baltic Sea ice.

Plate 36 (Crown Copyright)

Anticyclones

Anticyclones exhibit many features which contrast sharply with depressions.
1) They have a centre of high pressure.
2) Winds circulate in a clockwise direction (northern hemisphere).
3) They are normally larger, slower-moving and more persistent than depressions.
4) Pressure gradients are weaker (i.e. isobars are farther apart, and wind speeds are lower), especially near the centre.
5) They are areas of air subsidence: the widespread slow descent of air determines the weather characteristics since it is warmed at the dry adiabatic lapse rate (10°C/km) and this reduces air humidity.

Anticyclones are thus three-dimensional features, like the depressions, and are affected by the situation in the upper troposphere. Convergence aloft causes air to descend and pressure rises at the surface if the air does not spread out as fast as it is supplied from above. Some anticyclones are also associated with the intense polar cooling of air near the surface, often extending upwards to a maximum of 3000 metres.

An important feature of the weather sequence in middle latitudes is the development of slow-moving, extensive anticyclonic conditions within the westerly flow of depressions, which are diverted to north or south. Such anticyclones are known as blocking anticyclones, and their frequency determines much of the weather in areas like Britain: the length of extremely cold winter spells or dry, sunny summer spells (Figure 6.12).

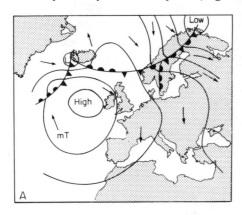
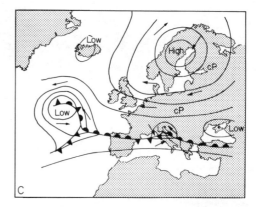
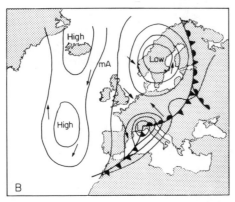
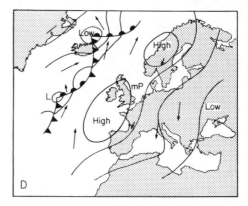

FIGURE 6:12 Blocking anticyclones, diverting depressions to the north or south of Europe, and changing the predominant west-east air flow (zonal) to north-south (meridional). Such anticyclones may persist for days or weeks.
(A) Over the North Atlantic. Cool air brings clear conditions or layer cloud to northwest Europe.
(B) An elongated feature stretching from the Azores to Greenland. Cold northerly winds dominate the weather of northwest Europe and the Mediterranean is stormy.
(C) Over Scandinavia. This is common in winter, resulting in cold easterly winds. In summer, hot sunny weather would be associated in northwest Europe.
(D) A feature stretching from the Azores to Scandinavia. This is associated with thunder in summer and snowstorms in winter. (After Pedgley, 1962.)

Anticyclones are particularly important and extensive features of the atmospheric circulation in subtropical areas. High pressure zones are normally permanent features of the oceans in these latitudes and continental areas may also be dominated seasonally by them.

Tropical weather systems

There are fewer contrasts in the regions between the subtropical high pressure zones and the equator, compared with the mid-latitude zones where the polar and tropical air masses meet. In the tropics, masses of tropical air meet, and although there is a zone of convergence, known as the Inter-Tropical Convergence Zone (ITCZ), it is much more variable in occurrence than the fronts of temperate regions. It is difficult to recognise the surface synoptic-scale features which are so commonly observed in middle latitudes, but increasing upper air investigations and satellite pictures have revealed the presence of some at higher levels. It is now thought that tropical weather systems hold the key to a fuller understanding of the worldwide atmospheric circulation, since so much energy is involved.

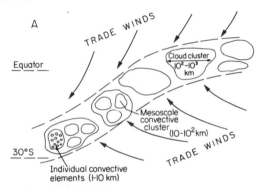
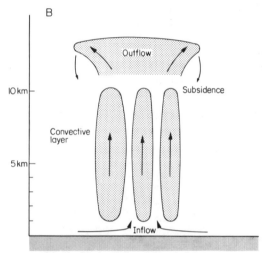

FIGURE 6:13 *Tropical cloud clusters, associated with the Inter-Tropical Convergence Zone (ITCZ). A deep, moist convective layer is established and latent heat is released during condensation. This has the effect of supplying energy to maintain the trade wind circulation. This situation may be interrupted by small, depression-type disturbances. The position of the ITCZ varies according to movements in the upper troposphere.*

(A) The different elements of the cloud clusters.
(B) Section through a typical cluster.

(After Mason, in Corby, 1970.)

Tropical cloud clusters

The trade winds blow towards the Inter-Tropical Convergence Zone where clusters of convective clouds build up, releasing vast quantities of latent heat to the outflow layer above (Figure 6.13) — energy which is vital for maintaining the surface trade wind circulation. The ITCZ does not stay long in this form, and often breaks down into easterly wave disturbances, or even hurricanes under the right conditions.

Wave disturbances

Wave disturbances occur in the trade wind belt of the North Atlantic and in the Pacific easterlies. They are relatively weak troughs of low pressure, about twice the size of mid-latitude depressions, and move at about half the speed (Figure 6.14).

The inversion associated with the subtropical high pressure area is due to subsidence of air above the surface. It seems that more active subsidence over the eastern sectors of oceans prohibits the formation of deep disturbances there, by keeping the inversion at lower levels (450-600 m). Towards the western margins of the oceans, however, the inversion is higher, and may be broken in summer to give rise to surface waves and deep convection. The rain associated makes up a large part of precipitation received in these latitudes.

Hurricanes

The hurricane ('big wind') of the Atlantic, typhoon of the Pacific, or tropical cyclone is one of the most dramatic and devastating of natural phenomena. Such storms occur over the western parts of

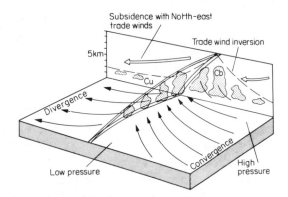

FIGURE 6:14 *An easterly wave in the northern hemisphere. A comparatively weak surface low pressure trough may be associated with veering winds and updraughts of air piercing the trade wind inversion. Cumulonimbus clouds form and rain follows.*

Area	Number of tropical storms in 10 years
North Pacific, west of 170°E	211
North Atlantic ocean	73
South Indian ocean west of 90°E	61
North Indian ocean: Bay of Bengal	60
North Pacific ocean, west of Mexico	57
North Indian ocean: Arabian Sea	15
South Indian ocean: north west of Australia	9

FIGURE 6:15 *The frequency and distribution of tropical storms. Not all of these develop to full hurricane strength.*
(After Riehl, 1965.)

oceans in tropical latitudes in late summer: Figure 6.15 shows the frequency over a period of 10 years. It is worth noting, however, that whereas there is a maximum of fifty hurricanes generated in the northern hemisphere each year (and very few in the southern hemisphere), over twenty middle latitude depressions are formed every winter's day. Hurricanes are unpredictable. In 1950, for instance, every small tropical storm seemed to develop to a hurricane, but in 1962 there were very few and none in September, normally the most important month. On average, less than 10 per cent of tropical disturbances intensify to become hurricanes.

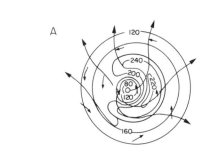

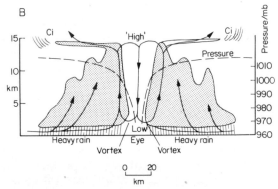

FIGURE 6:16 *The main features of a hurricane.*
(A) Windflow: the surface features are shown by the small arrows and the lines, which represent equal wind speeds (km/hr). The longer arrows show wind directions at 15 km above the surface.
(B) Section through a hurricane, depicting cloud, air currents, rain areas and pressure distribution (pecked line).
(After Riehl, 1965.)

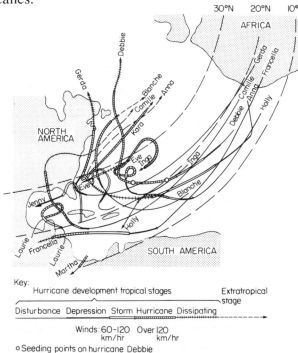

FIGURE 6:17 *The hurricanes affecting the USA in 1969. Notice the following features:*
(a) The stages in the development of each storm (see key), its source and the point where it reaches hurricane status. Is there a general pattern, and what major variations occur in this?
(b) The differences between the early season events (Anna to Gerda) and the later ones.
(c) The numbers of hurricanes which had some part of their course over land. (Data, Weatherwise Magazine, 1970.)

The main features of hurricanes (Figure 6.16) include the eye which is a zone of subsidence with reducing wind speeds, clear skies (at times), high temperatures and very low pressures. Surrounding this is the wall of convective cloud in which there is uprushing moist, warm air heated at the ocean surface and supplemented by the release of the latent heat of condensation. At the top of the storm, some 12-15 km above the surface, the cloud spreads out. The whole storm revolves in a great whirl and wind speeds in the eye wall vortex have been estimated to reach over 300 km/hour (although measuring equipment disintegrates with most other human constructions at these speeds) and heavy rain also falls. As the storm passes overhead at speeds of 16-24 km/hr, 200-250 mm of rain may fall in 24 hours, but if it lingers totals may double this giving rise to extensive flooding. Hurricanes have been known to double back on their courses, or to slow down.

What causes a hurricane? Rather special conditions seem to be necessary, combining the right situation aloft with that at the surface. The upper air currents must form an 'anticyclonic', divergent circulation to allow the development of very low pressure at the surface: an intense jet stream in the vicinity, for instance, would not permit this. At the surface there needs to be an extensive area of ocean with a temperature of over 27°C. The zone within 5° of the equator has too weak a Coriolis Force to allow the vortex development, and most hurricanes develop well to the north of this. Thus, whilst most tropical disturbances have a cold showery core, the hurricane is distinguished by its hot eye. This develops because of the release of latent heat from the towering *cumulonimbus* clouds, although the upward currents occupy less than 10 per cent of the storm area. The upper air anticyclone is intensified and upper outflow encourages further uplift of air. Downdraughts of air occur in the eye, raising the temperature at the surface. So the furnace is kept going. High wind speeds are maintained by a steep pressure gradient between the eye and the eye wall. Hurricanes will be maintained as long as this situation holds, and some have even crossed the Atlantic from America to the shores of Europe. On the other hand reduction of the surface supply of heat, or the drawing in of cool air, will lead to very rapid decay and a passage into extratropical depression.

It is instructive to examine the record of the USA hurricane season in 1969, as summarised in the map (Figure 6.17) and chart (Figure 6.18).

FIGURE 6:18 *The USA hurricane season, 1969. Notice how few caused damage. (Data from Weatherwise Magazine, 1970.)*

Name	Dates	Intensity (T = tropical storm, H = hurricane)	Lowest pressure (mb)	Highest winds recorded (knots)	Deaths and missing persons	Damage ($1000's)
Anna	Jul 23-Aug 5	T	1002	60		
Blanche	Aug 6-13	H	992	80		
Camille	Aug 5-22	H	905	175	336	1 420 750
Debbie	Aug 13-25	H	950	110		
Eve	Aug 24-27	T	995	50		
Francella	Aug 19-Sep 4	H	973	100	100	4 700
Gerda	Aug 21-Sep 10	H	979	125		
Holly	Sep 9-21	H	984	75		
Inga	Sep 20-Oct 14	H	964	90		
Jenny	Oct 1-6	T	1001	35		
Kara	Oct 7-19	H	978	80		
Laurie	Oct 16-27	H	973	95		
Martha	Nov 21-25	H	980	80	5	30 000

7

Predicting and altering the weather

Man has for long regarded it as desirable to be able to predict the future weather. Some of the sayings of weather lore built up over the centuries have more validity than others. Thus, short term sayings like 'red sky in the morning, shepherds warning' can be associated with the easterly rays of the morning Sun being reflected on clouds moving in from the west. On the other hand it is doubtful whether the saying that a wet or dry Saint Swithin's Day (July 15th) will be followed by 40 days of the same weather (Figure 7.1) has any statistical validity.

It is only now, however, that man's knowledge of weather processes and systems is enabling him to consider seriously the prospects of predicting the weather accurately enough to enable him to save money. The smaller successes of the past have been achieved without the machinery which is now available for observation (satellites and radar methods) and for the rapid processing of the measurements made (computers): national meteorological services are investing large sums in computers. In order to understand the complex, vast-scale processes of the atmosphere, it still remains for man to advance to the level where numerical models suitable for computer analysis can be devised.

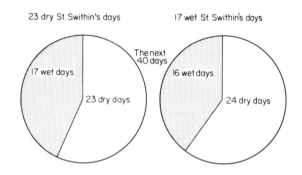

FIGURE 7:1 *One item of weather lore which is not supported by the facts. The average of dry and wet days following St Swithin's Day (15 July) over 40 years shows that the weather on this particular day does not influence what follows! There is little difference – and what there is refutes the lore. Why is this type of weather lore unlikely to be right?* *(Data from Forsdyke, 1969.)*

At present the mean error in forecasts of the positions of fronts 24 hours ahead is of the order of 100 km (i.e. average 2-3 hours). Mesoscale weather systems (Figure 6.1: e.g. small depressions, thunderstorms, squall lines) are important in meeting this gap, but they exist only for a few hours and are difficult to predict and track. National weather services are working on new schemes which should improve the position soon. The British Meteorological Office is preparing a model taking measurements from 10 levels in the atmosphere and covering an area 5000/3000 km around the British Isles with a 40 km grid. This needs a very large computer, and one of the largest computers in the world is engaged in a similar attempt for the much larger area of the USA: four different measurements are to be taken at 10 levels from 8000 meteorological stations round the world — a total of 300 000 items. It is hoped that this may give forecasts of reasonable accuracy up to a week ahead.

The forecasts on the radio or television each day must be assessed for their degree of usefulness. A particular forecast is based on observations, which have to be collected, processed and interpreted. If there is a degree of error of the order of 2-3 hours in 24 hours, it cannot be expected that a forecast at, say 6 o'clock one evening, will be adequate for the following afternoon: the 8 o'clock morning forecast will be much better for weather the same day.

Industry	Period of time covered by forecast				
	Under 48 hours	1 week	1 month	Season	Climatic changes
Agriculture	Day-to-day operations. Frost protection.	Planning of ploughing, sowing, harvesting, hay-making, spraying.	Timing of sowing, harvesting. Estimates of demand for food crops.	Forward planning of crop schedules. Choice of varieties. Forecasting yields.	Long-term land use planning. Breeding of new varieties. Investment decisions.
Aviation	Terminal, enroute weather. Flight planning. Avoiding hazards.			Traffic patterns, loads.	New routes and aircraft.
Building and Construction	Day-to-day operations. Avoiding worst effects of rain, frost, wind.	Alternate plans to suit weather. Hiring of plant, machinery.	Work schedules. Weather-proofing during construction. Hiring.	Completion dates, delays. Weather-proofing.	New towns, industries, motorways. Design of buildings, structures.
Electricity and Gas	Hourly values 48 hours ahead would assist utilisation.	Demand planned as far ahead as possible. Planning of maintenance schedules.		Seasonal demands. Abnormal spells and stand-by equipment.	Design, location of new power stations, storage and distribution systems.
Oil		Estimation of demand. Long-term planning. Production, delivery schedules.			Long-term planning.
Manufacturing, Marketing, Distribution	Assessment of weather effect on demand and delivery of raw materials. Planning of advertising, marketing.			Demands for seasonal products.	Assessment of consumer demand for existing and new products.
Road and Rail Transport	Knowledge of rail temperature in summer; of snow, heavy rain, fog, ice.	Snow and ice.		Local authority: grit and salt Traffic patterns and density. Snow-clearing equipment, flood water drainage, road heating.	
Shipping	Coastal shipping, pleasure craft warned.	Renting of ships, fishing fleets.		Ice cover and river freezing.	
Water Resources	Precipitation, evaporation, run-off, river flow. Regulation of dams and reservoir levels. Irrigation requirements. Flood forecasting.			Precipitation and drought. Water balance. Irrigation.	Location and design of dams, reservoirs. Management.
General Public	Day-to-day activities. Dress, leisure, sport. Central heating, air conditioning.	Social and sporting events. Outdoor activities, gardening.		Holidays.	

FIGURE 7:2 *The utility of forecasts to weather-sensitive industries. Consider what it would mean to know what the weather would be like a week, a month, or six months, ahead: this is in the realm of science fiction at present!*
(After Mason, in Corby, 1970.)

Forecasts are divided into three types:
1) short-range forecasts, covering the next 24 hours, or up to 3 days at the most;
2) medium-range forecasts, covering from 3 days to 3 weeks;
3) long-range forecasts, covering periods of a month or longer.

Each of these has its own methods and degree of accuracy. Perhaps the most important developments in forecasting will be to enable the gas and electricity supply industries to forecast hourly temperatures to within an error of 1°C for up to three days ahead; to give quantitative estimates of rainfall which will assist decisions concerned with levels in irrigation and water supply reservoirs; and to give warnings related to flood control (Figure 7.2).

Short-range forecasts

The most useful and valid forecasts at the moment are those given for up to 24 hours ahead, with a general intimation for up to 3 days. These are based on the fact that many weather systems have a life of this length (e.g. midlatitude depression: 3-4 days) and on the production of the synoptic weather map (i.e. maps on which are plotted the distribution of meteorological conditions at a particular

A II iii Nddff VVwwW PPPTT $N_hC_Lh\ C_MC_H$ $T_d\ T_d$ app
        (to 'arrow')
B 03534 23030 59158 07507 22401 02108
(Birmingham airport)

```
        C_H              30k 300°↑
TT      C_M      PPP     07           075           7  → 075
VVww    (N)      ppa  → 5915    (2)   081     →    59 ⌣ 08/
T_d T_d C_L      W       02            2    8       2 △
C               D                     E
```

F The sky at Birmingham Airport was 2/8 covered by cloud. A wind of 30 knots was blowing from WNW (bearing 300°). Visibility is 9 km and precipitation can be seen falling over 5 km away. Pressure is 1007.5 mb, and the temperature is 7°C. Lower clouds are *cumulus* with increasing vertical extent, and cloud base is at 1600 feet (500m); there are no middle clouds, but high *cirrus* in strands — not progressively invading the sky. Dew point is at 2°C. The pressure tendency (i.e. change in the past 3 hours) is 0.8 mb: it has risen and is keeping steady.

G

Station	Times (hrs)	\multicolumn{7}{c}{1 OCTOBER, 1970}	\multicolumn{7}{c}{2 OCTOBER, 1970}
		Nddff	VVwwW	PPPTT	$N_hC_Lh C_MC_H$	T_dT_d app	Night (06) RRT_nT_nE T_gT_g / Day (12/18) RRT_xT_xE sss		Nddff	VVwwW	PPPTT	$N_hC_Lh C_MC_H$	T_dT_d app	Night (06) RRT_nT_nE T_gT_g / Day (12/18) RRT_xT_xE sss	
Manston 797	0	52514	74012	09315	55400	12734	00150	13	33122	82020	09511	35500	05208	00100	08
	6	82817	77032	05516	4557	10719			53122	82031	11410	52500	05315		
	12	32914	69030	15816	21641	07110	00170		63121	82012	06515	626	03105	91160	
	18	72611	74021	15015	4577	08708			42915	80038	07212	48500	04308		
Kew 775	0						97140	13						97090	07
	6	82709	78022	06615	2557	10614			12909	74028	13509	18530	05209		
	12	72806	80031	16516	3253	08104	97170	060	42920	80251	07915	38638	05000	95150	052
	18	72408	62032	15415	755	09708			32707	74158	09611	34500	06218		
Mount Batten 827	0	72928	58055	16715	753	13824	97141	13	73117	74031	19812	757	05209	00120	10
	6	62926	62206	14215	65200	12714			73216	74022	20812	757	05208		
	12	73018	80025	21016	785	11210	97171	032	43130	74258	15915	22401	07012	97150	070
	18	82510	56502	20515	872	14810			33126	66011	16613	38430	07210		
Scilly 804	0	82823	59025	20115	863	14717	97130	12	73223	74011	23013	755	08004	00130	09
	6	52821	59016	18314	45420	14603			83418	74032	24613	855	08304		
	12	53011	69012	23415	58400	14211	97181	027	33023	61011	29615	38400	12206	00160	056
	18	82722	32505	22515	872	15707			23025	61020	20813	24400	09210		
Tynemouth 262	0	22832	74021	98414	20940	09606	00091	05	22922	81020	06008	00908	03220	00071	03
	6	12728	81020	98609	15500	08206			62820	81021	10007	656	02021		
	12	72819	81021	09114	31534	07400	93171	033	22830	81020	98411	21500	03602	00131	082
	18	72014	58026	03313	756	10746			23220	74020	01910	21500	05224		
Carlisle 222	0	72714	80215	00613	784	10605	01081	06	02811	80011	09108	00900	04319	00061	03
	6	52717	64808	00808	59400	06703			13011	80030	13207	18501	04224		
	12	72520	69026	10413	6837	09802	04151	009	42815	69258	01411	48500	06314	01121	049
	18	82319	62618	02815	873	13655			42812	70018	05210	48440	05222		
Wick 075	0	72924	62806	87309	784	07229	03051	04	23317	82038	04408	23400	06230	01071	05
	6	32815	58158	88805	39400	04104			43417	75028	11008	49400	06233		
	12	72616	86028	02511	38432	07803	05121	019	33122	80158	93109	33400	05227	91111	072
	18	81816	58636	90010	853	09782			23217	65808	98408	29400	06228		
Valentia 953	0	82817	40615	18414	864	13605	96121	10	33114	65018	23612	32500	08004	97111	09
	6	82920	70806	18513	825	09303			73112	65038	25712	725	09102		
	12	82518	15516	21015	861	14804	9716	012	72920	65028	20813	725	07214	9714	014
	18	72521	58035	19415	784	13614			73116	65028	22413	725	07212		

FIGURE 7:3 Weather information, as it is transmitted in code from a weather station (A, B), and the process of transforming this information into the group symbol plotted on a British map (C, D, E). The actual information contained in the code can be judged from the written account (F): 30 items replace 400!

(A) The basic code:

(B) An example of a particular report, related to the code.

(C) The code related to the basic station group symbol (e.g. Nddff group is summarised in the central arrow-style symbol).

(D) Referring the code items of the particular case to the group symbol.

(E) The final plotting of the group symbol: some items are coded whilst others are still numerical.

(F) A word description of the information contained in the group symbol (E).

(G) Code readings from 8 British meteorological stations for 1-2 October, 1970. Plot these as map group symbols (Figure 7.3) and describe the changes in weather experienced in the different parts of the country. Manston is in east Kent; Mount Batten is on the shores of Plymouth Sound; Wick is in the extreme north of Scotland; Valentia is in west Ireland.

moment in time). The Daily Weather Report, issued by the Meteorological Office, includes several synoptic maps based on information recorded at 6-hourly intervals (midnight, 0600, 1200 and 1800 hours), together with a prediction for the day ahead. It also includes the coded information from which the maps are compiled.

The synoptic weather map is thus a basic document: it is necessary first to read the symbols used, secondly to be able to analyse the situation, and thirdly to predict what may develop.

a) **Reading the information.** The weather map plots isobars and fronts, but the fullest information comes from the weather station records. These are plotted on the map, and are listed on the reverse of the Daily Weather Report in the coded form which is transmitted by telephone or teleprinter to the Meteorological Office from the individual weather stations (Figure 7.3).

b) **Analysing the situation** which has been plotted involves the relationships of many factors. Certain patterns are associated with particular weather systems, and can be related to the model of, say, a warm front or anticyclone. Thus the information supplied must be mapped, an explanation made, and an account presented in useful form so that it can be communicated to others: precise terms have to be used. The significance of the various elements, and their relationship to weather systems, have been themes of this part of the book, but a more detailed treatment, for those who are interested in going more deeply into the principles of analysis, is found in *The practice of weather forecasting* by P.G. Wickham (1970).

c) **Making the forecast** is basically a straightforward process: knowing the state of the atmosphere at a certain time, and the physical laws which govern changes in the state, should lead to a prediction of the future. Difficulties arise, however, because knowledge of the atmosphere as a whole is still incomplete, and because the complex interactions of physical factors are still imperfectly understood. Two approaches are adopted. For a long time the individual weather forecaster, basing his interpretations on a mixture of theoretical knowledge and past experience, has been the most successful, even though his judgments are qualitative rather than quantitative. It has long been hoped that more rapid methods of processing will make objective and numerical approaches increasingly possible, achieving eventually a higher degree of accuracy. Work began on mathematical models in the 1920's, but the lengthy and laborious calculations necessary in the pre-computer age were unsatisfactory: the mathematical formulae, to which natural processes were reduced, were too simple but the calculations were still too prolonged for any effective forecasts to be made. Advances have taken place in the theory behind numerical prediction, isolating certain vital factors, such as the pressure patterns at various levels in the atmosphere. Forecasters now make increasing use of future circulation patterns suggested by these methods, but still have to predict the weather which will result from computed systems. Weather maps may reflect different situations (Figure 7.4): these include the synoptic maps based on observations one day; the forecast chart for the following day; and the actual synoptic map for the second day.

Medium-range forecasts

Forecasts for periods of 3 days to 3 weeks are amongst the most important for human use, but will probably be the last to be made satisfactorily. Knowledge of the development of atmospheric processes and weather systems over such periods of time is still inadequate, although it appears that a greater knowledge of the habits of upper troposphere movements will help. Some success has been obtained in Britain and the USA with 5-7 day forecasts based on a combination of numerical predictions for days ahead and weather experienced in the past with similar upper troposphere conditions. Fundamental research is, however, needed into the ways in which weather systems originate before progress is made in locating accurately the position of a depression 2-3 weeks ahead.

Long-range forecasts

As the range of forecast increases it becomes more and more general in its nature, but it is helpful to a farmer if he can know that a certain month will be more or less rainy or frosty than normal. At present, forecasts of the weather expected a month ahead are made using quite different methods

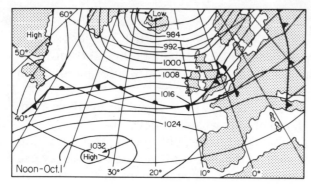

FIGURE 7:4 *Daily forecast maps printed in a national newspaper (i.e. from forecasts compiled up to 24 hours earlier), and daily charts compiled from readings of the actual conditions. Compare the differences between the two maps for each day. (From 'The Guardian' and Meteorological Office Daily Weather Reports – the latter Crown Copyright.)*

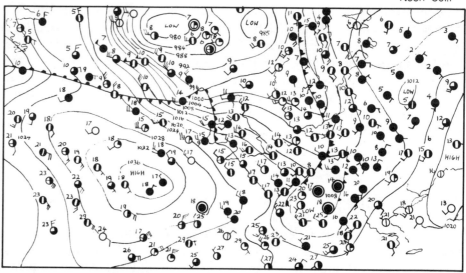

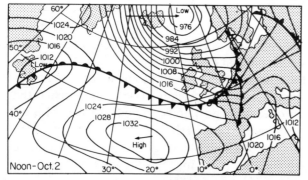

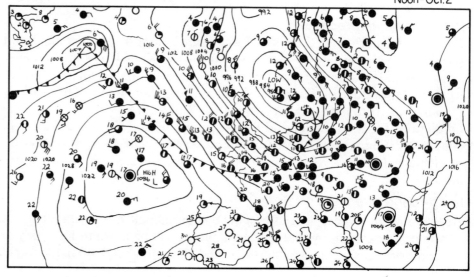

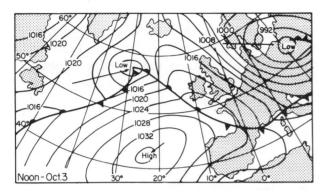

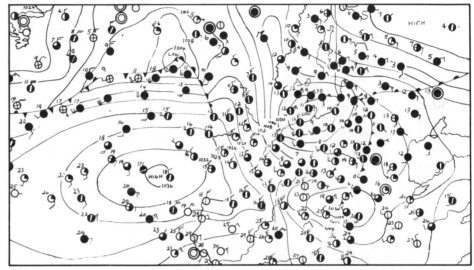

from those for shorter periods. The essential features of the present situation are determined, and then past records are examined to find situations which were essentially similar. It is then presumed that the future development will bear some resemblance to what happened in the past and a forecast is made. Three main methods are used for selecting similar past situations. Charts of the mean monthly temperature anomaly (i.e. differences from the average for each month) are available back to 1881 for most of the northern hemisphere, and to 1761 for the British area (Figure 7.5). The

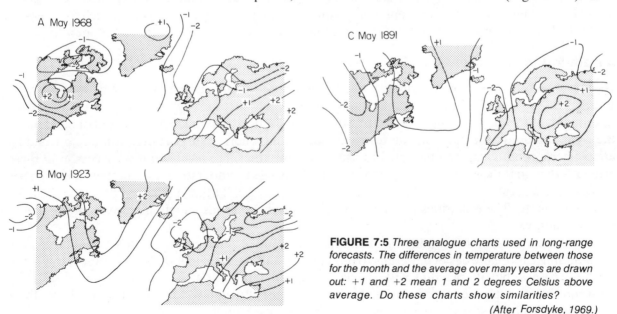

FIGURE 7:5 Three analogue charts used in long-range forecasts. The differences in temperature between those for the month and the average over many years are drawn out: +1 and +2 mean 1 and 2 degrees Celsius above average. Do these charts show similarities?

(After Forsdyke, 1969.)

second method is the examination of monthly sea level pressure charts, which have a similar availability in time-range. A link has been established between surface temperatures over the North Atlantic and atmospheric pressure patterns in Europe. Thirdly, a study of daily weather types and sequences is carried out.

When these processes have been completed, groups of more likely years are selected and a group of meteorologists discuss and decide on the possible course of events. Such forecasts have a moderate degree of success on approximately three-quarters of the occasions for which they are made.

This sort of method is adopted by insurance companies when a person wishes to insure against a rainy holiday, or when the organisers of a local fete wish to do the same. The company will calculate a premium based on the frequency of rain in the area for the season over a period of years and the financial loss which might be sustained. This premium will be payable a month or so ahead, and if it does rain within a specified period the agreed compensation will be paid.

Statistical methods are also used in predicting large-scale phenomena like the Indian monsoon: forecasters use a variety of factors, some of which seem to be quite unconnected with the situation (e.g. pressure conditions in South America).

Problems which remain

Weather forecasting is obviously at an early stage in its development, although it can point to real achievements. The problems which remain can be solved — at least partially — by research into the critical scientific and technical fields which may be most useful. Thus, work is proceeding on the numerical simulation of atmospheric processes, particularly the processes affecting weather systems which are clearly very much involved in the atmospheric circulation, since they receive the bulk of solar energy. It is true, however, that relatively little is understood about the processes. Amongst the technological problems, there must be an increasing number of observations made at higher levels in the atmosphere so that the records from such regions approach the density of ground observations; a worldwide weather station density of 250 km is necessary; increasing speed of communicating resultant observations around the world is required; and so are improvements in data processing. The USA computer already mentioned (ILLIAC IV) should be ready by 1980. In addition there is a need for more rapid dissemination of the forecasts once made. The Global Atmosphere Research Project (GARP) scheduled for the mid-1970's, and the increasing use of satellite information, should lead to further progress in all these realms. The early years of weather forecasting have been linked closely to the requirements of aviation. These are becoming less dependent on the weather, but it is expected that industrial and governmental planning may come to depend more on such forecasts, so that it will be worth investing in the costly machinery and increased international co-operation.

Modifying the weather

As well as his attempts to predict future weather, man has also felt it is in his interest to try to alter the weather he receives. From primitive societies, where witch doctors attempt to produce rain in time of drought, to the most advanced sector of western civilisation, where farmers are still dependent on the right amount of sunshine or rain for their yields, and where transport may still be impeded by atmospheric conditions, man has tried to change the weather. He has been less successful in these attempts than in forecasting. This is partly because he needs to understand the complex atmospheric processes to a greater extent, and partly because the vast scale of atmospheric events is too great for man to emulate. The quantities of energy involved are beyond his capabilities, even in the age of nuclear energy.

At the same time, man is modifying the atmospheric and oceanic environments without thinking. He is adding vast quantities of materials to both realms and affecting not only the composition of the air he breathes, but its reaction with the incoming solar radiation. It is necessary to distinguish between the atmospheric modifications caused consciously by man — and rather puny efforts these are — and the larger-scale changes he is accomplishing accidentally.

Intended modifications

The greatest efforts have been expended in the realm of **rain-making**. Charlatans made a lot of money from gullible individuals in the United States of America before the true nature of rain formation was understood. By 1940 it was known that ice crystals and certain types of dust nuclei are important in the formation of large rain drops in a cloud, and aeroplanes carried out experiments by dropping dry powdered ice, or silver iodide smoke into supercooled clouds with temperatures of $-5°C$ to $-15°C$. These experiments were on both commercial and non-commercial scales, but results were disappointing and largely discontinued after 1956. Small clouds grew into large clouds, but rain did not always fall unless the cloud temperature conditions were right, and when it did fall it was often in the wrong place. Experiments near mountains have been the most successful, but these are often away from the areas where extra rain is needed. The production of rain needs the right atmospheric conditions before seeding is worthwhile; such clouds would probably have produced rain in any case!

Hail-storms are notorious for ruining crops, especially of soft and delicate fruit. For some time it was thought that a loud noise would reduce the size of hailstones. Guns were fired and church bells were rung, but this only led to the deaths of the bellringers by lightning. Hail is a particular danger to the Po Valley vineyards of northern Italy, and over 50 000 rockets containing silver iodide are fired into thunder clouds each summer before they develop to their fullest extent: the moisture is precipitated as rain rather than hail, or the hail comes down in smaller stones. The process is not fully understood, but the continued investment seems to be justified.

Lightning flashes are another hazard associated with the development of thunderstorms, and have caused extensive forest fires in the western U.S.A. and eastern Australia. One attempt to suppress the discharge of lightning involves emptying electrically charged aluminium foil strips into the thundercloud, but, as with many similar experiments, the processes are understood imperfectly, and the results difficult to assess.

Fogs are becoming an increasing hazard, particularly in the car-ridden, urbanised parts of the world, where smog cuts visibility alarmingly. Airports have experimented with devices like oil burners to raise the temperature and evaporate the atmospheric moisture or prevent fog forming, and some freezing fogs have been seeded with dry ice to encourage precipitation from the fog just before a plane lands. Both types of operation are expensive, and airlines have tended to invest in 'blind landing' systems rather than the attempt to overcome the fog hazard.

Only one experiment has tried to tamper with the **larger scale weather phenomena**. The hurricane storms experienced around the Caribbean cause extensive damage every year. Project Stormfury was instigated in 1962 by the U.S. Weather Bureau and U.S. Navy. The clouds and the storm eye are seeded with silver iodide in an attempt to alter the dynamic structure of the storm: latent heat is released as the water cloud changes to ice and it is hoped that wind speeds will be reduced. Such projects demonstrate the scale of problem facing man in his attempts to make alterations in the weather. All that he can hope to do is to tip the balance in a system already on the verge of instability.

In short there is little evidence that precipitation can be produced to order, or that man can accomplish very much at all in changing the type of weather. Whilst the world is a place where the science fiction of one decade becomes the accomplished fact of the next, it is salutary to recognise that man is a long way from controlling all the forces at work on this planet.

Unintended modifications

It is now realised that it is more important to overcome the dangers of polluting the ocean-atmosphere system than to try to alter the weather it produces. When seeming problems arise the majority of people react by confessing ignorance and inability to act. The 'doomwatch brigade' look at the worst possible outcome, but most people take too complacent a view and forget the threat to the environment almost as soon as it is raised.

37

39

40

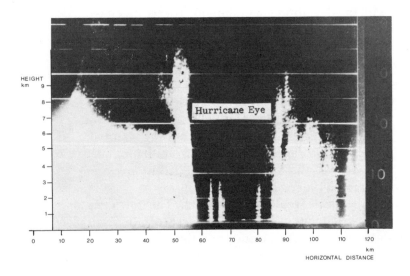

Hurricanes

Plate 37 The effect of high winds in a hurricane.

Plate 38 A radar profile of hurricane Donna (1960). The clear eye is bordered by clouds reaching over 10km high.

Plate 39 Satellite picture showing a tropical storm developing off West Africa as Hurricane Celia approaches the USA Gulf Coast.

Plate 40 Three satellite photographs following the progress of hurricane Agnes in June 1972. (All NOAA)

PREDICTING AND ALTERING THE WEATHER 83

Predicting and modifying weather

Plate 41 Skywarn is the system for forecasting severe storms in the USA, and it is particularly important in the mid-west, where tornadoes occur regularly. Over the last few years there has been increasing success in forecasting them and in warning the local inhabitants. The photographs show local observers at Amarillo, Texas, following the storm's progress on radar, and a policeman broadcasting a warning of approach. Over 1100 tornadoes were experienced in central USA in 1973, including one of the most powerful ever recorded near Salina, Kansas. The total death toll was under 100, largely due to the forecasts and warnings, and despite widespread destruction.

41

Plate 42 The modern weather forecaster in the USA relies upon data from the National Meteorological Center: this includes upper air radiosonde information, radar, satellite photographs, and automatic ground recording instruments. In addition, ships at sea are used and forecasts are relayed back to them.

Plate 43 Aeroplanes seed sections of clouds with silver iodide smoke (canister at wing tip), and this leads to the 'raining out' of a cloud (middle distance). (All NOAA)

42

43

The Study of Critical Environmental Problems (SCEP) gathered eminent scientists at the Massachusetts Institute of Technology to discuss the magnitude and menace related to such problems, and to plan for action (Figure 7.6). In most cases it seems that the greatest need is for an international body to monitor the levels, and to promote further research into the effects of increasing quantities of certain materials in the ocean or atmosphere. The tone of the report is of watchfulness, rather than panic.

Man has a unique trust. Never before has a creature been so much in command of his environment, and this brings responsibility, as well as freedom. He can use all the resources of the Earth, but must take care not to poison the environment for himself or for those who follow. It is to be hoped that the salutary mistakes — like the death of Lake Erie — will open the eyes of all to the need for a greater understanding and consideration of a provision often taken for granted.

Problem	Comment	Recommended action
Atmosphere		
1 Carbon dioxide increase	Fossil fuel burning will lead to increase from 320 to 379 ppm by 2000 AD, raising surface temperature by 0.5°C. Direct climatic changes unlikely in this period, but estimates inaccurate. Long term effects could be serious.	Improve estimates of change. Study changes in mass of living matter. Continuous measurement in remote areas. Systematic study of CO_2 in atmosphere, ocean.
2 Fine particles	Affect heat balance (absorption, reflection) and cloud-formation. Man releases few in total, but especially sulphur dioxide, nitrates. Magnitude of effect unknown.	Determine optical properties and effect on radiative transfer; distribution; sources. Improve methods of measuring insolation.
3 Jet contrails	Jets in upper troposphere may give rise to cirrus clouds, upsetting radiation balance and influencing precipitation.	Determine increase in cirrus and properties of contrails.
4 SST's in stratosphere (Super Sonic Transports)	SST's will fly at 20km: rarified zone with little mixing. 500 SST's may be operating by 1985-90 at Mach 2.7. Seems there will be little effect from adding CO_2, or decline in ozone. But little known about particles from exhausts.	Uncertainties about SST contaminants to be examined. Lower stratosphere programme needed.
Ocean		
1 Pesticide concentration	DDT most used and studied, though concentrations and effects in open oceans unknown. Collects in marine animals — some now unfit for human consumption.	Reduce use of DDT, subsidising developing countries. Programme of measurement. Greater support for integrated pest control.
2 Mercury, heavy metals	Many toxic to life stages of organisms. Mercury comes from industrial processes.	Curtail pesticidal uses of mercury and control industrial wastes. Examine effects on selected organisms. Monitor totals.
3 Oil in oceans	2 million tons enter every year. Little known about effects, but could be severe.	More research to examine what happens. Make transport safer. Possibilities of re-cycling.
4 Nutrients	Overfertilisation of coastal waters with phosphorus and nitrogen due to town wastes, fertilizers. Leads to organic matter decaying, reducing oxygen, killing fish.	Improve technology for recycling wastes. Avoid use of nutrients in materials dumped in large quantities. Improve monitoring of water quality.
5 Nuclear energy wastes	Monitored carefully, but increasing.	Independent surveys needed to maintain safe management.

FIGURE 7:6 *Some of the problems arising from man's activities which may affect the future well-being of the ocean-atmosphere system. What is most striking about the types of action regarded as necessary? (Adapted from SCEP, 1970.)*

8
Climate

The climate of a place on the Earth's surface is not the 'average weather'. It can be defined more accurately as the sum of weather changes experienced at that place. This will include the general progression of seasonal changes in terms of measurements of temperature, rainfall, wind strength and direction, air pressure, humidity, cloud cover, hours of sunshine, duration of snow cover and general weather systems; it will also take into account the extreme conditions and the range of variability over the years. It is desirable therefore to obtain weather records for a period of several years: at one time it was calculated that 35 years was an adequate length of time and this is still the general standard, but it is now realised that changes of emphasis can occur within such a period (chapter 14), and shorter periods (e.g. 10 years) are regarded as significant.

Climatic factors play an important part in determining the character of the soils within an area, and also the type of natural vegetation and land use possibilities. Hence they help decide whether a particular area can support a large rural population. Areas of climatic difficulty, where it is very cold or very dry, demand the use of special and expensive techniques to overcome the initial disadvantage, and often provide an uncomfortable environment for human endeavour. Thus, 5 per cent of the national income of Sweden is spent making life more comfortable in the harsh winter by such means as the heating and insulation of buildings, snow clearance and ice breaking. Irrigation is also extremely costly for the dry lands and can be justified only where the sun's warmth and soil quality lead to high crop yields and more than one harvest per year. On the other hand the tropical areas of high humidity and temperature encourage plant growth, but the soils have to be used very carefully: human beings find it difficult to live in such conditions where disease-carrying organisms multiply rapidly (Figure 8.1).

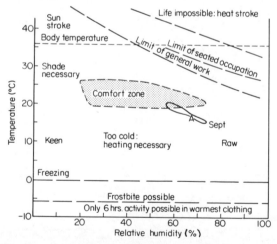

FIGURE 8:1 *Climate and comfort. Notice when the climate becomes dangerous for man without some form of protection. How has man's development of clothing, shelter, domestic heating, etc. affected the range of climates he can tolerate? Make a chart of the range of humidities and temperatures experienced in your own area: the range plotted at A is based on the September conditions for New York.*
(After Olgyay, in Barry and Chorley, 1968.)

The study of world climates is basic to an understanding of the factors affecting the distribution of plant and animal life on the Earth, and has an important bearing on human occupations. The world-scale patterns, however, tend to obscure important local effects, and the study of climatology means much more if observations of local conditions and interpretations of their causes are made.

The study of climates on a small scale is sometimes called **microclimatology** to distinguish it from the world-scale study, but this title has been given to at least two different types of investigation. One group of microclimatological studies is based on the climate of a relatively small area, such as a woodland or city. Individual studies concentrate on distributions of temperature or wind speed or fog occurence within that local environment, and contrast it with the surrounding area. In this way it is found that each distinctive local environment — field, woodland, coast, built-up area, mountain — has its own climate, and these make for variety within the major world divisions. The second group of studies investigates in detail the zone of the atmosphere closest to the Earth's surface: many microclimatologists confine their attention to the layer 2 m above the ground surface or to 30 cm into the soil; others would include the lowest 50-100 m of the atmosphere. Marked daily changes, which are important for the understanding of plant and animal life, take place in this zone.

The study of climatology thus begins properly with the study of local areas, and also provides the basis for biogeographical investigations. Studies of two particular environments — the city and woodland — will demonstrate the type of work which is possible.

City climates

Most people in industrial nations (e.g. over 80 per cent of the UK population) live in large towns with their tall office blocks, large factories and high percentages of paved road which have been developed over the last two centuries. These towns have climates of their own, contrasting with the surrounding countryside. One aspect of such climates is known as the 'heat island' effect. The towns depicted in Figure 8.2 have a variety of sites — hilly, lowlying, coastal, inland — so that this effect must be due to the influence of the city itself rather than the natural features. The town also affects other aspects of the climate as well as the temperatures (Figure 8.3). The best-documented variations are those concerned with temperature, winds, visibility and precipitation.

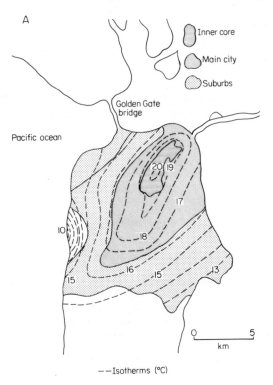
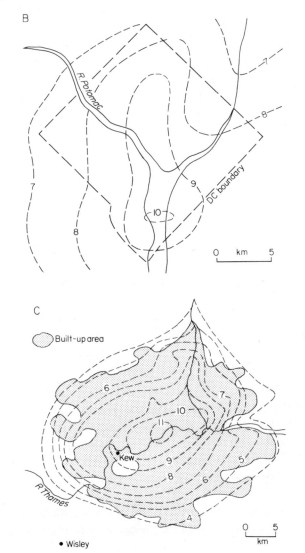

FIGURE 8:2 The 'heat island' concept, illustrated in three ways.
(A) Temperature distribution in San Francisco on a spring evening.
(B) Washington DC, mean annual temperatures (1946-60).
(C) London, minimum temperatures on a May night.
In each case the highest temperatures coincide with the most densely built-up areas. ((A) and (B) after Lowry, in Ehrlich, 1971; (C) after Chandler, 1965.)

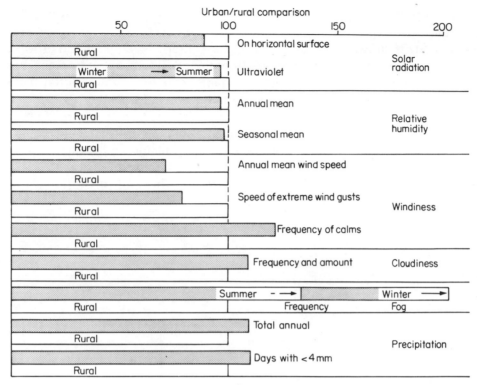

FIGURE 8:3 *Urban and rural climates compared. In what ways do the two differ? By how much?*
(Data from Landsberg, after Lowry, in Ehrlich, 1971.)

The temperature differences between town centre and country are particularly marked (Figure 8.2). A number of factors are responsible for this.

1) The town is composed of a great variety of structures: houses, blocks of flats and offices, flat and sloping roofs, streets and gardens. This presents a maze of reflecting and absorbent surfaces to the incoming solar energy, so that much less is reflected than in the countryside (Figure 8.4). There are more surfaces which absorb the heat, and the turbulence caused by the winds, being slowed down by the buildings or speeded up through the canyon-like streets, distributes this heat as it is re-radiated from the warm bricks or concrete.

2) The surface materials of the buildings and streets include bricks, stone, concrete, asphalt. All

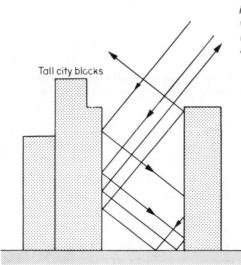
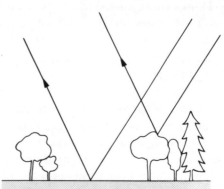

FIGURE 8:4 *Contrasts between urban and rural surfaces in reflectivity and absorption of radiation from the Sun. At what time of the day would there be (a) the greatest, and (b) the smallest differences? N.B. bricks and stone take up and store heat more readily than plants.*
(After Lowry, in Ehrlich, 1971.)

conduct heat more efficiently than soil or plant cover, so that more heat is stored during the day for release at night.

3) Towns generate their own heat, especially in the winter when homes and offices are heated, but also in summer, since vehicles and factories still continue to liberate heat from mechanical processes. It has been calculated that the burning of coal in Hamburg before 1956 resulted in a heating factor of 17 joules/cm² day — more than that received from the Sun in December.

4) Towns dispose of surface water quickly by way of drains and sewers, and soon clear away snow from roads and pavements. This means that less solar energy is involved in processes like evaporation or melting of snow in the towns and is thus available for heating the air.

5) In addition, the air above a town is heavily charged with fine particles, which scatter some of the short wavelength sunlight back to space, but also absorb more of the longer wavelength heat radiated back from the city surfaces.

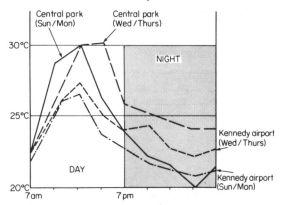

FIGURE 8:5 *New York temperatures: differences between the city centre and the outskirts at weekends and in midweek during August.*
When and where is the temperature range greatest, and why? *(After Lowry, in Ehrlich, 1971.)*

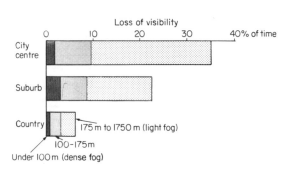

FIGURE 8:6 *Fog in Paris. What are the differences in the various parts of the city?* *(After Lowry, in Ehrlich, 1971.)*

Temperature differences between town and country thus vary from one season to another, with maximum differences on week days and in winter when most heat energy is supplied by man. Figure 8.5 illustrates this effect. The combined results for people living in towns are that they have somewhat lower heating bills than those in country districts, fewer snowy days, earlier flowering and up to three or four weeks longer without frosts. On the other hand the build-up of heat on summer evenings can become insufferable in overcrowded housing conditions, a factor which has been put forward as influencing the eruption of violence in recent years in the cities of eastern U.S.A.

Towns have become notorious as places where **fog** and dangerous smoke fog ('smog') develop (Figure 8.6). A period of winter cold weather is marked by increasing consumption of fuel for heating purposes, pumping out smoke and dirt into the slow-moving air. The particles of smoke act as nuclei for the condensation of water vapour and a persistent fog develops. The London smogs of the early 1950's were caused in this way and thousands died after respiratory illnesses. The Clean Air Act of 1956 restricted the amount of dirt and combustion gases passing into the atmosphere. Already, the London airports and central London experience fewer and less dense fogs (compare the two charts of Figure 8.7); Sheffield, once notorious for its smoky atmosphere, recorded a fourfold drop in smoke pollution and a twofold drop in foggy days between 1955 and 1970; and perhaps the greatest difference has been experienced in the 'Potteries' of Stoke-on-Trent, where the change to electrically-fired kilns has cleared the atmosphere for the first time in 200 years.

The problem continues in cities like Los Angeles, where a million cars and large oil refineries add over 12 000 tons of pollutant to the atmosphere each day and give rise to almost continuous smog. This is intensified by the bowl-shaped local relief and the dominance of clear skies and light winds. The photochemical reactions between the ultraviolet radiation from the sun and the hydrocarbons and oxides of nitrogen released by car exhaust fumes results in a hazy, yellowish-brown glare, which irritates the eyes and throat, damages crops and causes rubber to deteriorate. It is small wonder that the city authorities are trying to encourage the invention of a battery-operated car, although there would then be a problem due to the increased smoke emitted from electrical power stations.

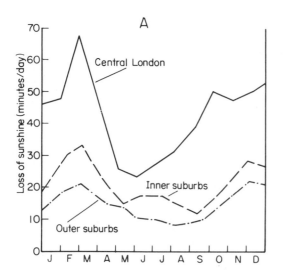
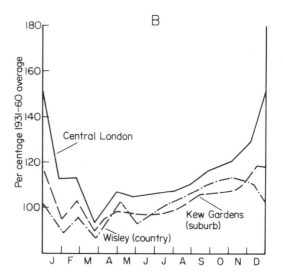

FIGURE 8:7 Sunshine in and around London.
(A) Mean loss of sunshine, 1921-50.
(B) Average monthly sunshine records for 1958-67 as a percentage of the 1931-60 average.
Study these graphs carefully. What changes are recorded? (After Lowry, in Ehrlich, 1971.)

Town pollution adds to the condensation nuclei: in Britain the concentration may average 9500/cm³ in the country, but from 150 000 to 4 000 000/cm³ in the town. This reduces visibility and the entry of sunlight, as well as impeding the nightly loss of heat; fogs increase in regularity and intensity, cloudiness is greater, and even rainfall is probably heavier. At this point, however, the study goes outside the factors which can be measured in terms of the strict confines of the city limits, although it has been shown that London is more susceptible to thunderstorm rain than many of the surrounding areas (Figure 8.8).

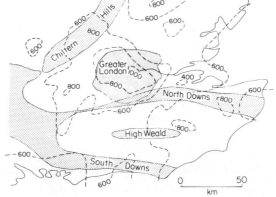

FIGURE 8:8 Total of thundery rain (1951-60 in mm). Of these amounts, 90 per cent fell in the summer months (May-Sept). The city environment encourages steep observed lapse rates at night, when many urban thunderstorms occur. Which other areas experience high rates of thunderstorm rainfall? (After Atkinson, Transactions of the Institute of British Geographers, 1968.)

The town also provides a peculiar environment in connection with the moisture content of the atmosphere. Surface moisture disappears down drains and there is relatively little vegetation to transpire, whilst the warm house surfaces do not encourage condensation. There may be a 30 per cent difference in relative humidity between the warm air trapped in a town street 'gorge' and the cooler air of the surrounding countryside.

Towns have their own local climate, and although certain atmospheric phenomena like cyclones and frontal systems are too large in scale to be greatly modified by passage over a town, the local conditions may give rise to more intense thundery rain along a front. Perhaps the rougher surface and greater heating induce upward currents and more rainfall, but it is difficult to assess such an effect.

Certainly the local climate has its effect on the population, and the physiological consequences of life in a city climate must be considerable.

Woodland climates

The town can be regarded as one type of surface which imposes a degree of modification on the features of the climate. There is a great variety of local surfaces, all of which have some effect: thus, ploughed fields, lakes, the various facets of a mountain or hill, and wooded areas will all experience particular conditions.

The woodland area has the densest and tallest cover of vegetation, and so represents an extreme case in terms of these small-scale climatic environments. Distinctive climatic features can be related to the size of the woodland, the height of the trees involved, their density and their nature (evergreen or deciduous, small or large leaves) (Figure 8.9).

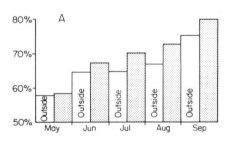
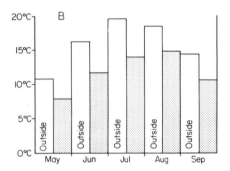
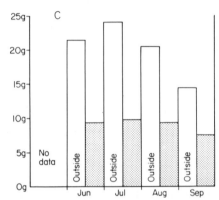

FIGURE 8:9 *The microclimates of woodland (stipple) and nearby open field environments: differences between the two during summer months. The woodland is temperate deciduous (birch-beech-maple); the records are averages over a 10 year period.*
(A) Average relative humidity (N.B. trees are fully in leaf by late May).
(B) Average daily soil temperatures to a depth of 15 cm.
(C) Average daily evaporation.   (After Riehl, 1965.)

Woods have a wide range of albedos (i.e. the amount of solar radiation reflected directly back to space), ranging from 10-20 per cent in temperate woodlands to 30-38 per cent over desert shrubs, but most of the solar energy is trapped at treetop level. In the densest woodlands, little insolation, especially in the ultra-violet wavelengths, reaches the floor: only 0.01 per cent of the total radiation reaches the floor of German elm woods, 0.1 per cent in the Congo, 20-40 per cent in pine woods and up to 75 per cent in an open birch-beech woodland. This is why it is so cool beneath the leafy canopy.

Woodland presents a barrier to airflow, and the air in the deep interiors of large stands of trees is often quite still. This factor has a practical significance in the planting of windbreaks: denser trees give greater immediate shelter, but give rise to greatly increased turbulence downwind. It is often better to use a tree which filters the wind rather than excludes it altogether.

Humidity characteristics are also modified: there is little evaporation from the forest floor. Transpiration from the millions of leaves, however, is often as great as evaporation over a water surface: it is controlled by such factors as the length of daylight, the leaf surface area and temperature, the species of tree, its age and the general atmospheric conditions. The supply of moisture from the trees of a forest may be sufficient to give rise to local rainfall. When rain does fall on woodland, the leaves at the top may intercept the drops, which are then either evaporated or run slowly down to the ground: leaf debris accumulating there is not disturbed by any surface run off.

Temperatures in woods are subject to lower extremes than the surrounding areas: it is cooler at midday and warmer at night. The woodland climate is one of fewer extremes in all directions.

9
World climates

The studies in microclimatology emphasised that there must be a multitude of climatic types throughout the world. No one thing in nature exactly repeats any other, and there are gradual changes from the climate in one area to that in the next. It is thus difficult to make a classification of climates, and especially to draw the boundaries when the classification of climates on a world scale will group a number of only generally similar regions together under a particular heading. Any line drawn on a map as a natural boundary is really drawn across a zone of transition. In addition, many parts of the world do not have records which are sufficiently extensive for boundaries to be drawn with any sense of statistical accuracy.

In spite of these difficulties many attempts have been made to classify the climates of the world, and to map them in terms of distinct regions. This is because men like to place all kinds of phenomena in convenient pigeon-holes, and also because the making of regional divisions has been a favourite pastime with geographers. Earlier attempts were based on rainfall and temperature records, or even on the distribution of vegetation and soil types — which are closely related to climatic conditions, and use commonly measured criteria. These were relatively easy to map, but were inadequate in a number of ways as a basis for climate classification. As more was understood about the meteorological processes, such classifications showed fewer relationships with the dynamic atmospheric activity. Although it was realised that the oceans are important contributors to climatic conditions, few climatic maps extended their divisions across these areas because of the lack of long-term records. More recent attempts at climatic classification have tried to relate the divisions to the atmospheric conditions and weather system types (chapter 6), rather than to arbitrary figures of isolated measurements such as temperature and rainfall. The difficulty which such schemes meet is that the criteria are less quantitative, and the authors have often hesitated to draw lines on maps to delineate the distinctive areas they have described.

The scheme adopted here is the result of an attempt to present the simplest meaningful division of climates related to the global atmospheric phenomena described in the previous section of this book. Such features are governed basically by the heat and water budgets of the atmosphere, and the consequent movements of air and transfers of energy. Thus the high latitude climates are more affected by the horizontal transfer of energy, whilst the tropical areas experience a preponderance of vertical movements. There is also the major difference between the oceanic areas together with the lands bordering them, and the continental interiors which do not have direct access to the oceanic heat and water store.

Before the climatic types which have been delineated on this basis are studied in detail, some of the factors on which climatic classification has been based in the past will be examined.

1 Temperature distribution patterns may form the basic means of defining the climatic regions. Thus the Greeks divided the world into torrid, temperate and frigid zones, and it is still common to use such terms as polar, temperate, tropical and equatorial in the same way. Another approach is to divide the world into climates defined as winterless (i.e. tropical areas with no month of mean temperature below 18°C), midlatitude summer-winter, and summerless (i.e. polar areas with no month of mean temperature above 10°C) climates.

2 Precipitation is important particularly in terms of its effect on vegetation and soils, and is also a good indicator of the atmospheric conditions. Rain is associated with surface convergence and ascent of air; lack of rain with descent and surface divergence. Divisions can be made on the basis of the total annual rainfall, or according to whether the distribution throughout the year is uniform or seasonal. In such seasonal groupings, however, temperate and tropical wet regions with rain throughout the year are put together, as are the tundra and hot deserts owing to their deficiency of moisture: this emphasises the point that more than one characteristic should be involved in a climatic classification.

3 Vegetation and soils have been used as a basis for some climatic classifications on a world basis. Vegetation distribution is related closely to conditions of rainfall or drought, high or low temperature (chapters 20-23). Soil groups are also seen to be more closely associated with climatic elements than any other factor, including the parent rock material (chapter 17). Here, then, is one method of defining climatic zones: the isotherm of the 10°C maximum mean monthly temperature follows closely the boundary between the tundra and boreal forest in the northern hemisphere. It must be seen, however, that soils and vegetation are the result of climatic factors and not the cause of them, that there are many departures from a simple climate-soils-vegetation association, and, moreover, that such an approach limits the consideration of climates to the continental surfaces.

The **Köppen system** of climatic types is one such classification which has been more widely used than most, and was extended to cover the ocean areas. It is based on temperature and rainfall characteristics related to vegetation and soil boundaries. Thus it is rooted in recorded and observed facts rather than in the causes behind these figures. It restricts the basis to the two most easily and commonly measured factors, and, whilst this has the virtue of simplicity, it masks the importance of other factors. In short, this sort of classification was adequate in the early days of climatology, when less was understood about the working of the atmosphere, but it should be possible now to devise a classification based on genetic factors. Köppen used five major climate groups, some of which were divided into subgroups and even closer definitions within these (Figure 9.1).

Major climate groups	Subgroups Second letter	Third letter		
A Tropical climates Every month ave. temp. > 18°C No winter. Heavy annual rainfall (> annual evaporation)	m f w		Code: f adequate rain all months F ice cap: perpetual frost m rain forest despite short dry season s summer dry season S steppe: 360-760mm rain in low latitudes T tundra w winter dry season W desert: <250mm rain per year	a hot summer, warmest month > 22°C b warm summer < 22°C c cool, short summer: 4 months only >10°C d very cold winter < −38°C h dry-hot: mean annual temperature >18°C k dry-cold <18°C
B Dry climates Potential evaporation > annual precipitation No permanent streams	S W	h k		
C Warm temperate (mesothermal) climates Coldest month 18°C to −3°C Summer and winter seasons	s w f	a b c		
D Snow (microthermal) climates Coldest month < −3°C Warmest month >10°C	w f	a b c d		
E Ice climates Warmest month <10°C No summer	T F			

FIGURE 9:1 *The Köppen system of climate classification. Discuss the basis of this classification, its advantages and drawbacks. (Data from Strahler, 1969.)*

A **genetic classification** of climatic regions attempts to relate the divisions to explanations of their origins. It is less possible to be quantitative because the processes involved are difficult to measure, and ideas about them are still largely theoretical. Theories are likely to change, and so the classification itself becomes liable to modification. Although the genetic and empirical approaches are so different it is possible to reconcile them in broad terms: the classification used in this book will attempt to give an explanation for the observations collated by Köppen and others.

The bases of the genetic classification are the positions of the air mass sources in major high pressure zones (which are relatively stable and from which winds diverge), and also the positions of the frontal zones where air converges in cyclonic circulations. In the sub-tropical regions, air descends in high pressure cells and blows by way of the surface trade winds towards the equatorial low pressure convergence zone (the Inter-Tropical Convergence Zone, or ITCZ). The middle latitudes experience seasonal weather caused by the positional changes of the overhead sun, and the west-east movement of the belt of cyclonic storms. The high latitudes are dominated by air from the polar regions and from the cold winter continents. Within this pattern there are some special climates which are related to other factors. Thus the tropical Asian monsoons may resemble the transitional zones between the equatorial and subtropical desert climates in that both have distinct summer maxima of rainfall, but the monsoon regime is much more intense and is motivated differently. This is a particular point at which the genetic classification is able to contribute more than the mere consideration of rainfall and temperature figures might suggest, since it points to the dynamic factors at work.

The major mountain ranges and the oceans are also rather special climatic zones. The mountains often cut across a series of climatic zones and create their own zones with ascending altitude. The oceans are really part of the general pattern of climatic zones, but have not normally been included in classifications. This is not only because the lack of data makes it an uncertain process, but also because the nature of the water surface forms a marked contrast with the continents. The flatness of the sea surface, and the absorption of a large amount of solar radiation in heating the waters to depth and in evaporation, means that there is much less daily and seasonal temperature variability, and that there is less rainfall over the oceans: the water cycle diagram (Figure 4.1) and the map showing the distribution of the climatic regions (Figure 9.2) demonstrate this. Arid climates are far more extensive over the oceans, and so are summer maxima of rainfall (due to greater rates of evaporation and vertical instability at that season).

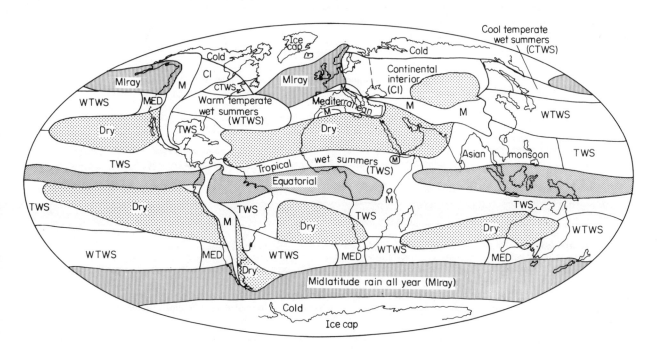

FIGURE 9:2 *World climate regions. Refer to Figure 9.3 for the key to the abbreviations used.*

Figure 9.3, giving the general details of the climatic zones used in this book, shows that the rainfall-temperature regimes are significant in association with the movement of the climatic belts in seasonal accordance with the overhead Sun. Thus the two major uniform divisions, 'all wet' and 'all

Regime of precipitation	Climatic type (abbreviations used on Fig.9.2)	General atmospheric conditions	Features of climate	Chapter
Wet all year	Midlatitude Rain all year (Ml ray)	In belt of 'polar front' cyclones all year.	Changeable, moist weather all year. Extremely equable on coasts. Cool winters, though temperatures high for latitude.	10
	Equatorial	Continually affected by Inter Tropical Convergence Zone (ITCZ)	Cloudy, very wet, very hot all year. Often a few months slightly drier.	10
Dry all year	Arid (Dry)	Varied conditions: (a) Subtropical high pressure zone: surface divergence. (b) Coastal area with offshore winds. (c) Rain shadow, continental interior: isolated from rain-bearing winds.	Dry all year; few days with rain.	11
Seasonal rain	Winter rain: Mediterranean (Med)	Summer: Subtropical high pressure zone Winter: 'Polar front' cyclones	Dry, hot summers Wet, mild and changeable winters.	12
	Summer rain: tropical wet summers (TWS)	Summer: ITCZ rain belt. Winter: Subtropical high pressure zone.	Hot all year. Summer rains, winter drought.	12
	Summer rain: Asian monsoon	Strong summer indraught of moist air; reversal of flow in winter.	Heavy summer rains; dry and hot winters.	12
	Warm temperate wet summers (WTWS) Cool temperate wet summers (CTWS) Continental interior (CI)	Summer: moist air drawn in off oceans. Winter: cold air flows out from land.	A mild 'monsoon' effect. Warm, wet summers; cool-to-cold and dry winters.	12
Cold climates and ice cap		Dominated by polar air.	Cold all year	13
Mountain (M)		Varied conditions, extending through other zones.	Increasing altitude gives lower temperature and pressure; more rain and wind.	13

FIGURE 9:3 *The major world climatic zones as described in chapters 10-13.*

WORLD CLIMATES 95

dry' are centrally based in terms of these seasonal movements, whilst the 'seasonally wet' areas lie between and are essentially transitional. This pattern is confined largely to the west coasts of continents, and contrasts with the east coasts. These experience a monsoonal effect, reversing the circulation with the seasons, almost throughout their length unless the continent is sufficiently narrow for the westerly air movements to penetrate across.

There are distinct differences between the climatic distributions in the northern and southern hemispheres. This is a result of the greater areas of land in the northern hemisphere, breaking up the zonal arrangement of the pressure systems; in the southern hemisphere these are arranged more consistently in east-west belts, and effectively prevent so much north-south exchange of cold and warm air.

Climatic statistics

It is most usual to find the climatic situation at a particular station expressed in terms of the mean monthly figures of temperature and rainfall, and a great deal can be gathered from an intelligent study of such statistics. These mean monthly figures are calculated from observations over a

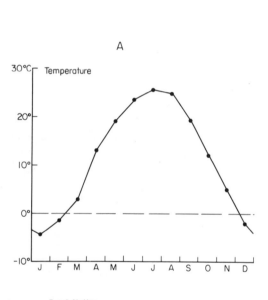
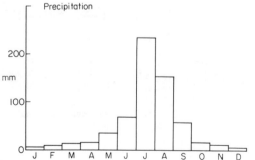
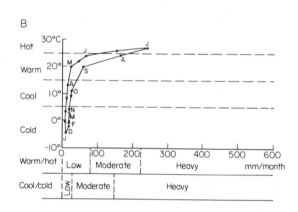
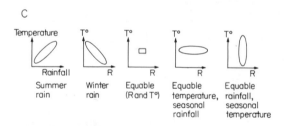

FIGURE 9:4 *Plotting climate statistics. The most common sources quote mean monthly figures for rainfall and temperature.*

(A) The traditional method of visual representation: a line graph for temperature and a bar graph for rainfall. The station plotted is Peiping, China (40°N).
(B) A combined plot of rainfall and temperature – a hythergraph. A key is provided for the general interpretation of temperature and rainfall figures: notice the different level of effectiveness of rainfall in warm/hot lands compared with cool/cold lands.
(C) Some typical hythergraph patterns.

standard period of 35 years, although this is not possible for some parts of the world. Doubt may be cast on the validity of such statistics if the climatic conditions can be shown to fluctuate (Figure 14.1). It is an advantage to represent the situation diagrammatically, and the line and bar graphs of temperature and rainfall are familiar in junior geography texts. The system used in this book

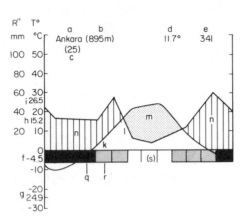

FIGURE 9:5 *A climatic diagram. Not all the data will be available or necessary for all stations. (After Jena, in Walter, 1973.)*

a station
b height above sea level (m)
c duration of observations (yrs): if 2 figures, first = temperature, second = precipitation
d mean annual temperature (°C)
e mean annual precipitation (mm)
f mean daily minimum, coldest month
g lowest temperature recorded
h mean daily maximum, warmest month
i highest temperature recorded
j mean daily temperature variations
k curve mean monthly temperature
l curve mean monthly precipitation
m relative period of drought (dotted area)
n relative humid season (vertical shading)
o mean monthly rain > 100m (black scale, reduced to $^1/_{10}$)
p reduced supplementary precipitation curve (10°C = 30mm) + dry period (dashes)
q months with mean daily minimum below 0°C (black) = cold season
r months with absolute min < 0°C (diagonal shading) = late/early frosts
s mean duration frost-free period in days
(N.B. Not all data will be available or necessary for all stations.)

illustrates the seasonal and annual range features of certain stations within each climatic region (Figure 9.4). Distinct patterns can be seen at once by this method. More detailed climatic diagrams are also used (Figures 9.5 and 9.6).

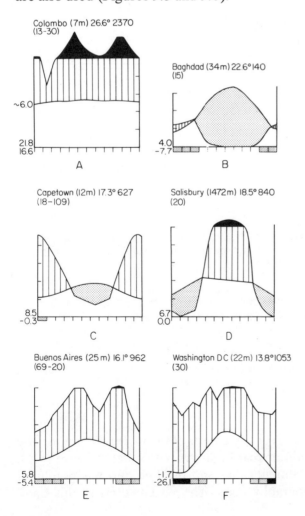

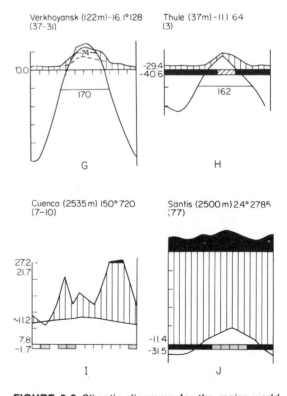

FIGURE 9:6 *Climatic diagrams for the major world regions. That for a British station is left out: add details from your own locality. (Walter, 1973.)*
(A) Colombo, Sri Lanka (B) Baghdad, Iraq (C) Capetown, South Africa (D) Salisbury, Rhodesia (E) Buenos Aires, Argentina (F) Washington, DC, USA (G) Verhoyansk, Siberia, USSR (H) Thule, Greenland (I) Cvenia, Ecuader (J) Säntis, Switzerland.

10

Climates which are wet all year

A fall of rain may be caused by local convection currents, or by moist air being forced to rise over mountains or along frontal zones (chapter 4). When the atmosphere is considered on a larger scale, however, it is the conditions of atmospheric circulation which are associated with the mean distribution of precipitation. The annual amount of rainfall, together with its average seasonal distribution, becomes a good indicator of the atmospheric situation in a particular area.

Climates which are wet throughout the year are situated in areas where even the north-south movements of weather phenomena with the overhead Sun do not upset the almost continuous reception of rain. There may be periods of drier conditions, or even a regular season of lower rainfall, and such areas are flanked by the transitional seasonal climates where a dry season without rain becomes a major feature. Convergence of winds at ground level is a common feature at every season in wet regions, although other atmospheric conditions may break in for periods of a few weeks, or, in exceptional years, for 2-3 months.

Two main zones experience such conditions (Figure 10.1).

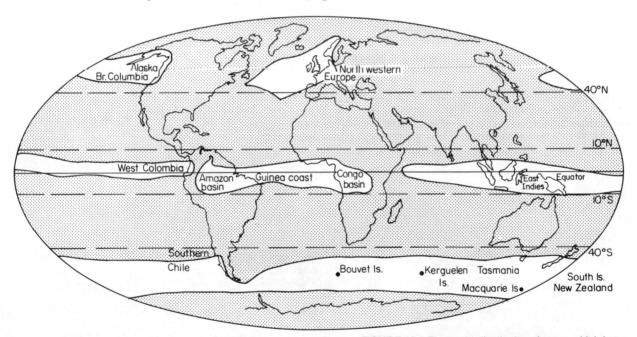

FIGURE 10:1 *The world distribution of areas which have climates receiving rain at all seasons of the year.*

1 In the temperate middle latitudes (approximately 45-60° North and 40-55° South) a position beneath the jet stream causes air to be drawn upwards, leading to surface convergence of cold polar air and warm tropical air. Associated cyclonic disturbances (depressions) and frontal zones give rise to heavy rains, moving from west to east and affecting the west-facing coasts to the greatest extent: as the depressions move across a continental area the air in them often suffers a decrease in moisture which is not replaced by evapotranspiration, but they can still be detected in the pressure patterns.

2 The equatorial zone (approximately 5° North to 5° South, but sometimes 0-10° North) is another area of surface convergence, where the tropical easterly winds (i.e. the north-east trades and south-east trades) meet, and air is forced to rise. Clearly defined fronts, like those of middle latitudes, are rare, since both types of air meeting along this zone have been warmed in tropical latitudes. As in

44

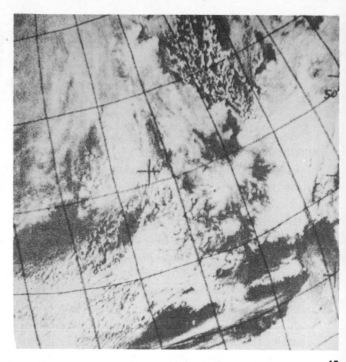

45

Humid temperate areas

Plate 44 Cloud patterns in the southern hemisphere, taken by Apollo 17. Describe the weather systems between the southern tip of Africa and Antarctica. Which other part of Africa was cloudy at this time (7 December 1972)? (NASA)

Plate 45 10 February 1969. ESSA VIII photograph of western Europe. Scotland and north-eastern England are covered by snow resulting from a cold northerly airstream. The North Sea has a convective cloud cover. Can you explain this? Wave clouds can be discerned on a northerly jet stream south of Wales, and a belt of frontal cloud is moving in from the west.

Plate 46 A similar view, seven days later. The snow still persists in the north. In southern Britain a depression moves north-eastwards, bringing strong easterly winds to England on 19 February.

Plate 47 Maritime polar air reaches the western coast of France in an unstable condition, giving rise to cumulus clouds over the land. This photograph was taken from 8000m.

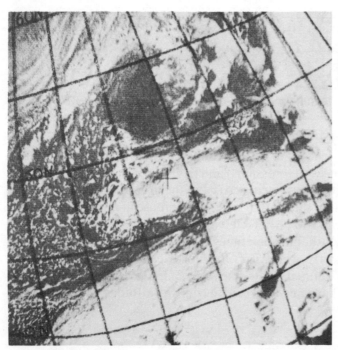

46

47

CLIMATES WHICH ARE WET ALL YEAR

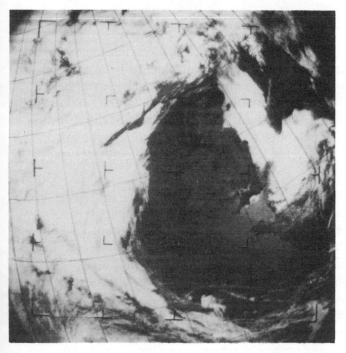

48

49

Plate 48 An ESSA II view of early summer over Britain (late May, 1966). An anticyclone is centred off the north-west coast of Ireland, and an easterly airstream brings stratus cloud and fog to the east coast. (45-48 Crown Copyright)

Tropical humid areas

Plate 49 The cloud patterns in the equatorial zone, as seen by Apollo 15 on its return to Earth in early August 1971. (NASA)

Plate 50 Tropical stratiform clouds over Penang, May 1969. (B Wales-Smith)

Plate 51 Cumuliform clouds over the Ceram Sea, April 1971. (J W Burton)

50

51

the temperate latitudes, it is often the coastal areas which are most affected, and the extension inland of very rainy climates may be limited. The higher rates of evapotranspiration and the greater quantities of moisture possible in the warmer air, mean that precipitation can be much heavier near the equator than in temperate latitudes.

Each of these rainy climatic types is distributed widely around the world, but each locality has its own special set of circumstances. These may be influenced by the shape of the land area, or by the nature of the local relief. The interrelations between the relief and climate are extremely important on this scale.

The middle latitude west coast rainy climate

The 'Polar Front' is one of the most conspicuous features of any world climatological map. It is not a permanent or continuous line around the world, as it is often depicted, but is made up of a whole series of frontal and cyclonic disturbances moving continuously from west to east, in which tropical and polar air masses are brought face to face. This contrast of surface air mass conditions can be observed and plotted on weather maps. The associated weather is extremely variable. Changes occur every few hours, and rain is common from the steady uplift of air along the warm front or from the more sudden rise induced along the cold front which is associated with clearing showers (chapter 6). Temperatures fall in winter, but the annual ranges of temperature are the lowest in the whole belt lying within these latitudes, and the cyclonic disturbances can occur throughout the year. Although these middle latitude climates are in a zone with great contrasts of seasonal insolation, the atmospheric circulation reduces this effect to such an extent that a warm winter's day may be very similar to a cool day in summer. This type of climate is thus extremely equable (i.e. the temperature and/or the rainfall vary little throughout the year).

The outstanding example of such middle latitude equability is experienced in the **southern hemisphere**, where the westerly circulation is almost uninterrupted by land and the difference between the warmest and coldest month is as little as 4°C. Only the narrow South Island of New Zealand, Tasmania and the tip of South America extend into this zone between 40° and 50° South (Figure 10.1). There is a constant succession of cyclones and fronts associated with strong winds and heavy rainfall. Cold winds descend from the Antarctic ice-cap throughout the year, and the heavy sea ice of the southernmost South Atlantic, South Pacific and South Indian Oceans provides an extension of low temperature surface conditions and intensifies the circulation. Thus Bouvet Island (54°S — cf. York 54°N) has glaciers reaching almost to the coast. Islands, like this one, standing in the way of the passage of disturbed weather, have rain on 300 days of the year.

Southern Chile, the South Island of New Zealand and Tasmania all record even temperature distributions through the year, and high totals of precipitation as shown in Figure 10.2. The even distribution of temperature is more marked and precipitation totals are higher farther to the south. Totals of 3000 to 5000 mm annual precipitation are experienced in these stormy regions, which are all mountainous. The moist air from the ocean is forced to rise, and there is a marked 'rain shadow' effect on the lee of each upland region: the plains of eastern South Island, New Zealand, receive only 20 per cent of the total annual rain falling on the western slopes of the island. Thus, Hokitika at sea level on the west coast receives an average of 2900 mm rainfall per year, whilst Dunedin on the east coast receives only 630 mm.

The **northern hemisphere** situation is different: the greatest widths of Europe and North America stand athwart the belt of converging air along the Polar Front. The disturbances within this belt move from west to east, as in the southern hemisphere, but have a much more restricted length of passage over the oceans. As the disturbances cross the continents, the winds decrease in force due to friction with the rougher surface, and they lose much of the moisture picked up over the ocean. Some seasons are marked by the development of high pressure areas over the land, blocking the passage of the cyclones (chapter 6). Like the southern hemisphere, the west coasts of the continents face into the oncoming cyclonic storms, and are thus most affected by the weather they produce. A special feature of the northern hemisphere is the poleward diversion of warm tropical surface waters, which

A Station	Latitude	Height above sea level (m)		J	F	M	A	M	J	J	A	S	O	N	D
Hasselbough Bay, Macquarie Island	54° 30'S	7	T (°C)	6.7	6.1	5.6	4.4	4.4	2.8	2.8	2.8	3.3	3.3	4.4	5.6
			R (mm)	101	89	104	96	83	74	81	81	96	84	71	99
Heard Island (Indian Ocean)	53° 01'S	5	T	3.3	3.3	2.6	2.2	1.1	0	−0.6	−0.6	−1.1	0	0.6	2.2
			R	146	146	144	154	146	99	91	55	63	94	101	129
Valdivia, Chile	39° 48'S	5	T	16.7	16.1	14.4	11.7	9.4	7.8	7.8	8.3	9.4	11.7	13.3	15.6
			R	65	68	114	213	376	414	373	303	213	119	121	106
Puerto Montt, Chile	41° 28'S	1	T	15.0	15.0	15.0	11.1	9.4	7.8	7.8	7.8	8.9	10.6	12.2	13.9
			R	88	103	139	180	236	256	210	197	157	119	129	124
Dunedin, New Zealand	45° 42'S	80	T	15.0	14.4	13.3	11.7	8.9	6.7	6.1	7.2	9.4	11.1	12.2	13.9
			R	86	83	83	81	89	83	78	76	74	81	86	96
Bergen, Norway	60° 24'N	45	T	1.7	1.7	3.3	6.7	11.1	14.4	16.1	15.6	12.8	8.9	5.0	2.8
			R	200	152	137	111	99	106	131	185	233	233	203	205
Brest, France	48° 19'N	20	T	7.2	7.2	8.3	10.6	12.8	15.6	17.2	17.8	15.6	12.8	9.4	7.8
			R	89	76	63	63	48	51	51	55	57	91	106	111
Vancouver, Canada	49°N	45	T	2.2	3.3	5.6	8.3	12.2	15.0	17.2	16.7	13.3	9.4	6.1	3.3
			R	218	154	134	83	75	68	33	42	103	149	280	197
Portland, USA	46°N	50	T	3.9	5.6	7.8	10.6	13.9	16.1	19.4	18.9	16.1	12.2	7.8	5.0
			R	170	139	121	78	57	40	15	15	48	83	165	175

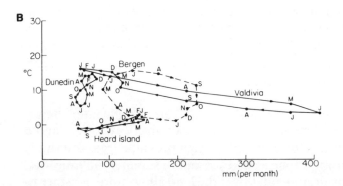

FIGURE 10:2 *Statistics for humid temperate climatic stations.*

(A) Lists of rainfall and temperature records: mean monthly totals. Compare conditions in the northern and southern hemispheres; contrast these conditions with those in other climatic types on approximately the same latitude.

(B) Hythergraphs of humid temperate climatic stations. What do these demonstrate about the nature of such climates and the variation within one world climatic type?

transfer vast quantities of energy towards the higher latitudes and cause marked positive temperature anomalies (Figure 3.6).

Western Europe, because of its peninsular nature, is the largest land area affected by this equable temperate climate and there is a very gradual transition eastwards into the contrasting interior continental conditions of European USSR. Few mountain ranges lie across the path of the Atlantic depressions, and the continent is penetrated deeply by arms of the sea, including the North Sea/Baltic Sea area in the centre, and the Mediterranean-Black Sea group in the south.

The climate of this region is marked by the passage of depressions to the north of the subtropical high pressure zone centred on the Azores (Figure 10.3). Continuity in terms of a succession of

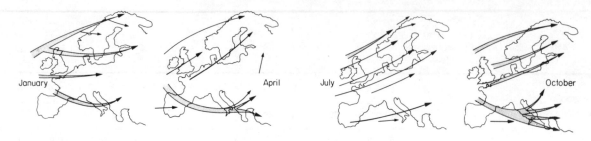

FIGURE 10:3 *The main tracks followed by depressions passing over western Europe. (After Rumney, 1968.)*

depressions is most marked in summer and autumn when the continental interior of Eurasia heats up and a surface low pressure area is induced by the heating — i.e. a mild 'monsoon' effect emerges, encouraging the penetration of Atlantic air. In winter and spring, however, the Siberian high pressure centre may extend westwards over Europe, blocking the passage of the cyclones and changeable, mild weather by sending them north of Norway, or south through the Mediterranean. Such 'anticyclonic blocking' occurs on 30-40 per cent of the days between December and June and affects profoundly the weather of northern and central Europe. Britain may be subjected to lengthy periods of severe cold winter weather with easterly or northerly winds, or may enjoy calmer sunny weather in early summer (Figure 6.12).

The air brought across the Atlantic and drawn into a depression is much affected by its journey over the ocean waters. The warm Gulf Stream current flowing northwards along the east coast of the USA transfers enormous quantities of heat to the cold air flowing off North America in winter. Thus warmed, the air continues to Europe taking a few days above the North Atlantic Drift, whose waters take 9-12 months to cross the ocean. Cold, unstable *mP* air from the north, and stable, warmer *mT* from the south converge over this zone to the north-west of Europe. The former type of air is warmed by surface friction and heat transfer from the ocean, whilst the latter is cooled. Uplift of the warmer air over the cooler along fronts takes place over the ocean, and, on arrival in western Europe further uplift over coastal mountains will add to periods of heavy rainfall. The west-facing mountains of north-west Spain, Wales, Scotland and Norway receive annual totals of 3000-5000 mm. However, the lowland totals of rainfall received across north-central Europe are relatively low (500-750 mm/yr) and the climate is almost subhumid in places: farmers in the lower Thames basin, for instance, would profit by the application of irrigation waters in nine years out of every ten. This area is very low lying and much of the moisture brought by the Atlantic air is 'rained out' on the western hills of Britain. The low totals in areas such as this show that a much more vigorous 'trigger' effect is necessary to induce rain in a midlatitude position compared with the tropics, where heavy rains fall from much less intense situations.

The mild, maritime influence can extend far to the east in Europe. Even in midwinter the icy temperatures and snow of Moscow (the mean for January is −10.6°C) can give way to rain and a rapid thaw when an occasional depression penetrates the dominantly high pressure area. But this is rare in virtually the eastern extremity of the transition zone at that season. In summer the transition zone extends farther east, since higher evapotranspiration rates over the land allow more moisture to be brought to the Asian sector of the USSR.

Even more notable is the extension of these equable conditions to the north along the coast of Norway. Temperatures above freezing, and ice-free seas, extend well into the Arctic Circle. Tromso, at 70° North, has a February (coldest month) mean temperature of −4°C. This area lies at the northernmost extent of the North Atlantic Drift and the coastal temperatures are as much as 25°C above the average for the latitude at the winter season. The Scandinavian peninsula also presents the most obvious and clearcut climatic divide in Europe, for the Kiolen Mountain range effectively separates the continental and maritime influences. The steep west-facing slopes may have milder, wet winters, whilst the eastern slopes are in the grip of the Siberian winter (Figure 10.4).

CLIMATES WHICH ARE WET ALL YEAR   103

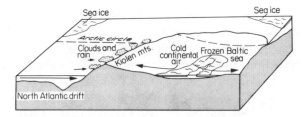

FIGURE 10:4 *Scandinavia, showing a common winter situation. Note how the Kiolen Mountains form a marked climatic divide.*

Western Europe, in particular, is a region of low annual temperature ranges and of no dry season. Only the extreme western coasts have the majority of their rain in winter, and most of the region has an autumn maximum when the anticyclonic blocking is at its lowest intensity. Attempts have been made to discern definite patterns of weather through the year, but in almost every case the exception is more common than the rule. The British situation has been characterised in this way (Figure 10.5 and Figure 10.6), but most people would point out that this pattern is not experienced in any single year, and it is thus unreliable as a basis for prediction.

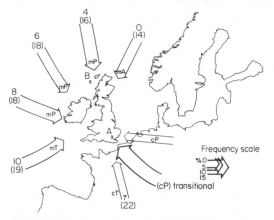

FIGURE 10:5 *Air mass frequencies at Kew (A). Under 30 per cent of the air masses arriving there are mP in type, compared with 38 per cent in the Hebrides (B). The figures by each arrow are the average temperatures associated with each air mass type in winter (summer). The chart shows the main weather characteristics of each type. (After Belascon, in Barry and Chorley, 1968.)*

Code	Airmass characteristics
mP	Cool and showery at all seasons, especially on windward coasts. Drier in lee.
mA	More extreme than mP. Cold winter weather bringing snow to east and south-east.
mT	In warm sectors of depressions. Mild with stratus cloud: drizzle, light rain.
cP	Only December to February, bringing very low temperatures and snow in north.
cT	Rare and only in summer. Hot, settled weather, but often unstable aloft bringing thunder.

Season	Months	Weather conditions: likelihood of longer spells
Spring to early summer	April May	Variable weather: least likely period for long spells of one type
High summer	June July August	Long spells of air from west, north-west with succession of cyclones. Anticyclonic conditions less common
Autumn	September October November	Long spells often anticyclonic in early part; later cyclonic
Early winter	December	Fewer long settled spells. Often westerly mild flow with storms
Late winter	January February March	Long spells may be either cold anticyclonic with easterly flow, or mild and cyclonic from the west bringing early spring

FIGURE 10:6 *The general pattern of weather experienced in Britain: long spells of one type at various seasons. (After Lamb.)*

The **Pacific coast of North America** between 45 and 60 degrees north of the equator experiences a similar type of climate to north-west Europe. The layout of the land-sea margin and the arrangement of the relief features are, however, very different from Europe. An almost unbroken coastline runs from north to south and is backed by mountain ranges which are much higher than those along the coasts of north-west Europe, providing a partial barrier to the eastward movement of weather systems. In addition, the warm water drift across the North Pacific is not so extensive or warm as that in the North Atlantic and winter temperatures along the coast are only 10°C warmer than the latitudinal average. The flow of this drift is diverted southwards by the Aleutian Islands, and the west-facing coast of Alaska, covering the latitudes of Norway, has an ice-blocked coast in winter.

The high cordilleras of western North America slow down the eastward progression of depressions, and force the moist air to rise. This combination leads to long periods of precipitation along the coast, but also restricts the width of the zone affected by heavy falls to a narrow coastal strip. The mountain ranges also seem to affect the movement of the subtropical high pressure area, which maintains an unusual poleward bulge along the north Californian coast in summer. Northerly coastal winds result, with an associated northern extension of the cool coastal current, leading to drier-than-usual summers for this latitude.

There is thus a considerable degree of variety within the broad heading of 'midlatitude west coast rainy climates'. Much of this is attributable to variations in reactions of the atmospheric circulation with the ocean-continent distribution, the ocean currents and the relief features. This is a pattern which recurs throughout world climatic zones. Whilst the overall yearly regime is similar, the atmosphere-surface interaction gives rise to many distinctive local conditions.

The Equatorial rainy climates

Although the areas with rainy climates are grouped together, great contrasts exist between the two major divisions, both in the conditions experienced at the surface, and in the causes lying behind these conditions.

In the middle latitudes, the intense, vortex-like wind pattern within a depression, and the fronts associated with the meeting of contrasting air masses, are the typical situations which cause masses of air to be lifted off the ground and precipitation to fall. The high quantity of kinetic energy involved in the jet wind movements in the upper troposphere is necessary to produce vertical lifting in these cooler regions and this results in strong surface winds. By contrast the equatorial latitudes commonly experience light, variable surface winds and atmospheric disturbances which seem to be of less significance when plotted on a weather map. The air, however, is warmer than that of middle latitudes and so can hold much greater volumes of water vapour in which latent heat is trapped: it is also generally unstable so that the slightest degree of convergence will lead to vertical uprushes of air and the formation of convective clouds leading to heavy rain storms.

For a long time, meteorologists, accustomed to the middle latitude weather systems, were deceived by the quiet surface conditions. Observers measured the light winds and noticed the lack of temperature contrasts between the air masses of the region. Rainfall was attributed solely to the convergence associated with local daily heating and uplift of the air. Trade winds were seen to be drawn towards the equatorial 'front' (or ITCZ) which was almost impossible to map, but which was assumed to be present along the line of their convergence. Increasing knowledge of the upper air conditions and of the periodicity of rainfall, however, suggests that these explanations were far too simple: 90 per cent of the rain can now be linked to clearly defined disturbances, although they may be apparent only from observations above ground level. Local conditions may have a great, but not dominant, influence, and the idea, repeated in so many junior geography texts, that there is a daily regime of clouds building up to afternoon storms in all these equatorial areas, is a myth: it is a local feature when it occurs at all.

The belt of equatorial rainy climates is found between 2° and 8° North over the oceans, but widens to as much as 7° North to 8° South over the continents and the East Indies, where the trade wind system is not so strong (Figure 10.1). One major break in the worldwide belt is in East Africa, where parts of the high plateau surface in these latitudes reach nearly 2000 m above sea level and rise into a different climatic zone. In South America the high Andes present a barrier to the atmospheric circulation and are zoned vertically in terms of climates (chapter 13).

The location of these areas so close to the equator means that the Sun is almost always overhead, and that average daily temperatures are uniformly high throughout the year (c. 27°C). Outstandingly high temperatures are, however, rare, because the cloud cover reduces the insolation below the potential limit. Humidity is high because of high evaporation and transpiration rates: at times these may be greater over the land than over the sea. The diurnal range of temperature, which may be as much as 10°C, gives rise to night dews and early morning ground fog: a slight drop in temperature brings the humid air to saturation point. Any small disturbance of the atmosphere with such high temperatures and humidities leads to convection, enlarged *cumulus* clouds and heavy showers. Some of the weather systems which give rise to disturbances are clearly associated with the ITCZ, and when this is at least 5 degrees from the equator small cyclones may develop: the Coriolis Force is too weak nearer to the equator to give rise to a circular arrangement of winds. Easterly waves and hurricanes also contribute to the rainfall in these regions.

The main belt of rains migrates with the overhead sun, but only in the most general sense, and seldom moves far south of the equator (Figure 10.7). Places near the equator may have two periods of maximum rainfall in the course of the year, separated by drier seasons. There is thus more variation in the timing of seasonal rainfall than in the temperate latitudes, but there is still no season which could be called dry (Figure 10.8).

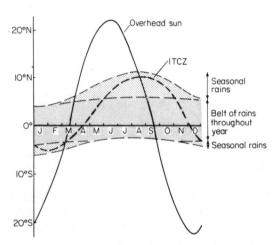

FIGURE 10:7 *The equatorial rain belt and its slight movement through the year compared to the ITCZ and the position of the overhead Sun.*

The daily regime of fine mornings followed by cloudy afternoons and rainy evenings is true only of some of the island localities in the East Indies because of the strength of the land-sea breezes which impose a regular series of meteorological events: it is not a basic feature of the climates of equatorial regions. During the day, air is drawn into an island from the sea, and is forced to rise and converge with air entering from the opposite side. Heavy clouds build up in the afternoon and lead to thunderstorms: one place in Java records such weather on 300 days of the year. At night the air drains back to the sea and clouds develop there. Similar diurnal regimes occur at places like Lake Victoria in central Africa: nocturnal katabatic movements (chapter 5) of air cause convergence and upward movement over the centre of the lake, which has an annual rainfall of 2200 mm, compared with only 800-1500 mm on the shores. In general, however, it cannot be stressed too greatly that the rain of equatorial regions is caused by disturbances of the types described in chapter 6, and may come at any time of the day or night. Periods of a few rainy days may be followed by several days of clear, fine weather.

The **Amazon Basin** in South America is the largest land area subject to a climate of this type, and most of the region drained by the world's longest river receives over 2000 mm of rain each year. The coastal regions and the upper basin at the foot of the Andes receive up to 3000 mm, but the narrower

Station	Latitude	Height above sea level (m)		J	F	M	A	M	J	J	A	S	O	N	D
Andagova, Colombia	5° 05'N	60	T (°C)	27.2	27.8	27.8	27.8	27.8	27.2	27.2	27.8	27.2	27.2	27.2	27.2
			R (mm)	530	577	538	526	600	628	582	605	543	582	495	536
Manaos, Brazil	3°S	50	T	26.7	26.7	26.7	26.7	26.7	26.7	27.2	27.8	28.3	28.3	27.8	27.2
			R	233	228	244	216	177	92	55	35	51	104	139	195
Para, Brazil	1°S	12	T	25.6	25.0	25.6	25.6	26.1	26.1	25.6	26.1	26.1	26.1	26.7	26.1
			R	317	357	357	320	259	170	149	114	88	86	66	154
Lagos, Nigeria	6° 27'N	3	T	27.2	28.3	28.9	28.3	27.8	26.7	26.1	25.6	26.1	26.7	27.8	27.8
			R	28	45	101	149	269	454	279	63	139	206	68	25
Libreville, Gabon	0° 23'N	40	T	26.7	27.2	27.2	27.2	27.2	26.1	24.4	25.0	25.6	26.1	26.7	26.7
			R	221	292	246	282	249	45	3	10	96	363	431	280
Mombasa, Kenya	4°S	16	T	26.7	26.7	27.8	27.2	25.6	25.0	23.9	24.4	25.0	25.6	26.1	26.7
			R	20	22	57	197	347	91	88	55	48	86	126	55
Singapore	1° 19'N	3	T	26.1	26.7	27.2	27.2	27.8	27.2	27.2	27.2	27.2	27.2	27.2	26.7
			R	251	173	192	188	173	173	170	195	178	207	254	256
Tarakan, Borneo	3° 19'N	13	T	26.1	26.7	26.7	27.2	27.2	26.7	27.2	27.2	27.2	27.2	26.7	26.7
			R	277	259	356	353	342	320	261	315	294	363	386	340

FIGURE 10:8 *Statistics for climatic stations in the equatorial zone.*

basin of the east has less than 2000 mm. This heavy rainfall over such a large area drained by one major river system means that the Amazon carries nearly twice as much water to the sea each year as any other river in the world. The North-east Trades blow into this region after a long passage over the North Atlantic, and the convergence zone (the ITCZ) is marked by wave and eddy disturbances which have most effect in the air movements several thousand metres up. Other disturbances are caused by surges of polar air entering the basin from the south along the Parana corridor between the Andes and Brazilian Highlands. These surges occur only from 1 to 3 times each southern hemisphere winter and take place below shallow inversions, lowering the temperature by 4-5°C. The Amazon Basin is, however, a region of little human settlement and few weather records, and so little is really known or understood about the detailed weather patterns. Rainfall over the basin is markedly seasonal, unlike many African stations, with less in the 'winter' season as the South Atlantic subtropical high pressure belt penetrates into the south-eastern sector, but there is no marked dry season.

On the west coast of South America **the Pacific slopes of the Colombian Andes** also have a very rainy climate, with totals of 4000-8000 mm per year. The position of the ITCZ varies little from a line between 3° and 4° North, and this is one of the few parts of the world with a westerly wind circulation in this latitude. The winds are therefore onshore and the steep rise forced by the mountains leads to the high rainfall totals. The local westerly circulation may be due to a land-sea breeze effect, and it results in afternoon rain on the higher cordillera slopes, evening rains on the lower slopes, and continuous night rains on the coast. It is thought that daytime upslope winds give rise to the afternoon rains, followed by downslope winds during the night accompanied by offshore convergence, which cause the rains at that period.

The **Congo basin of Africa** is another large river basin astride the equator, but it has a different position with respect to the main airflow compared with the Amazon. The trade winds leave the land at this point instead of arriving from the ocean, and one consequence of this is decidedly lower rainfall totals when compared with the Amazon area. Most of the Congo Basin has over 1500 mm but large areas around the margins have much less.

At the surface there is a south-westerly flow of humid maritime air blowing from the South Atlantic, but this is overlain by upper easterly winds, mostly from the south, but including some from the north during the southern hemisphere summer. The south-westerly and easterly wind circulations meet at the surface along the ITCZ, which moves north-and-south with the seasons. It is not certain, however, which of these air flows is responsible for the rain-causing disturbances, although it does seem to be established that rainfall is lower in the eastern areas when the south-westerly flow is less vigorous. Some of the disturbances causing the rainfall are along the convergence zone, but others are along indefinite lines within the south-westerly flow. This south-westerly air enters the basin in a stable condition, but the cover of *stratus* cloud it brings soon breaks up to *cumulus* with heating, and these individual heap clouds develop into thunderstorms. The lower rainfall totals at certain seasons and in certain areas are due to the effect of the very dry air entering from the Sahara and Kalahari desert areas in the easterly flows: there is no comparable adjacent source in South America. In addition, the East African plateau tends to keep out the moist air of the Indian Ocean.

East Africa itself (i.e. Kenya and Tanzania) is an area of very low rainfall, and is almost arid in parts. This is unusual for the equatorial zone, as is the markedly seasonal distribution of rainfall through the year. The area is discussed in chapter 12.

The **East Indies** are the most complex and unusual of the equatorial regions. There is no continuous land area, but a series of mountainous islands and peninsulas. The patterns of weather associated with the convergence zone of the tropical easterlies is interrupted seasonally by the invasion of the Indian monsoonal winds from the north-west (Figure 10.9). Seasonal changes are most marked in terms of reversing wind directions, which turn around the windward-leeward relationships of the island coasts: a lee position can result in aridity. As in other parts of the

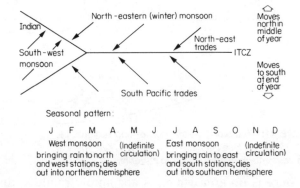

FIGURE 10:9 *Indonesia: seasonal changes are due largely to variations in the predominant wind directions. This is a meeting point for conflicting air types and wind systems.* (After Trewartha, 1961.)

equatorial zone, the Coriolis Force is too weak to give rise to revolving storms, or even to wave depressions. Rainfall occurs following convergence and uplift along the air stream boundaries, especially when the ITCZ is moving southwards during the latter part of the year. Local convergence and uplift in the lee of islands, as bifurcating air streams come together, also leads to rain. Thundery rain is especially important: Bogor on Java had an average of 322 thundery days a year from 1916-1919. The East Indies record the greatest regularity of local diurnal regimes, and the greatest development of small-scale climatic variations within the equatorial climatic zone. It is therefore almost impossible to pick a representative climatic station.

11
Climates which are dry all year

The atmospheric conditions which give rise to arid climates are very different from those which determine the rainy climates studied in the last chapter. Whereas rain is associated, particularly in the middle latitudes, with areas of surface wind convergence and upward vertical movements, aridity is found in areas where conditions of subsidence and surface divergence occur. Whilst the rainy climates are often near coasts which have onshore winds bringing a plentiful supply of moisture, many dry areas are far from the sea, or in the 'rain shadow' zone on the leeward side of a mountain range. The 'rain shadow' may be due to the release of moisture as an airstream passes over mountains, followed by adiabatic warming and decreasing relative humidity as it descends, or to the existence of a mountain barrier which prevents the moist airstream crossing. There are, however, some important exceptions to the general case. Westward facing coasts in subtropical regions, which are bordered by a cool current, and experience offshore winds, are amongst the driest areas of the world. Figure 11.1 shows the distribution of the dry regions. The general lack of water means that these areas, covering 20 per cent of the Earth's land surface, are all areas of great difficulty for human habitation.

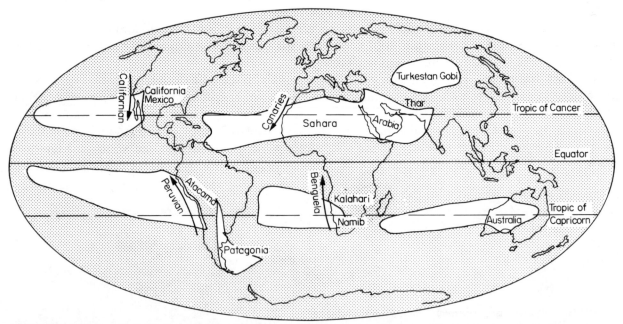

FIGURE 11:1 *The world distribution of arid regions: i.e. those areas with less than 250 mm rain per year. The four ocean currents marked are all cold currents.*

There is no area which is absolutely rainless, and there is a gradation from the extremely dry regions through the semiarid into subhumid and humid climates. It is often difficult to draw the boundaries between arid and humid, or arid and semiarid, climates along simply computed lines: in addition to the problems of drawing lines across transitional zones, criteria of aridity are disputed and records from these regions are sparse.

The definition of 'aridity' is made more complex by the fact that evaporation rates are greater in the warmer parts of the world: thus the 250 mm annual isohyet would delimit arid regions and 500 mm semiarid in temperate regions, but the quantities could be doubled in the tropics (Figure 11.2). Perhaps the main characteristic of the rainfall in these dry areas is that it often seems to be variable

CLIMATES WHICH ARE DRY ALL YEAR   109

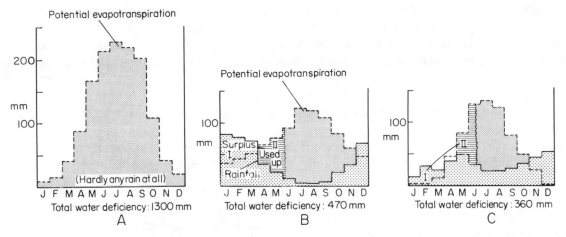

FIGURE 11:2 *Three stations showing a major excess of evapotranspiration over precipitation. Not all are included in the 'arid' areas as defined here.*
(A) Beni Abbes, Algeria (30°N): a truly desert area with hardly any precipitation.
(B) Los Angeles, USA (34°N): strictly within the seasonal 'Mediterranean' climatic zone, but with a marked water deficiency in summer. The surplus at I is used up during period II, leaving the rest of the summer arid.
(C) Ankara, Turkey (40°N): an upland site with small quantities of rain throughout the year. These are insufficient to compensate for the summer rates of evapotranspiration.

(After Strahler, 1969.)

and unpredictable in occurrence (Figure 11.3). This affects the extent, as well as the timing of storms: several storms 100 km across may occur every year in an area 1000 km across. Particular places in arid areas may receive heavy rainstorms — once every few years! As in the equatorial

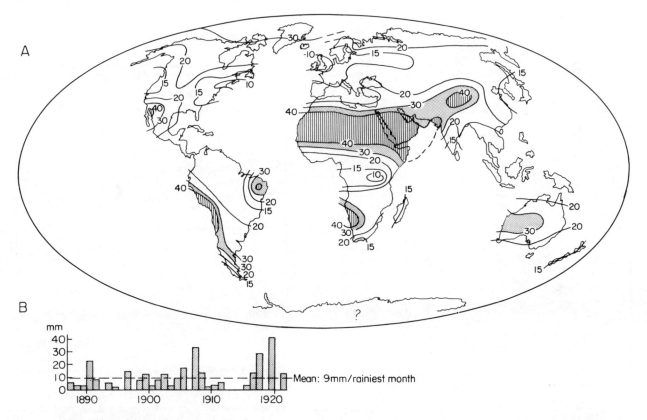

FIGURE 11:3 *Rainfall variability.*

(A) A map of world rainfall variability: the numbers refer to percentage deviation from the mean annual precipitation. Compare this map with Figure 11.1.

(B) Variations in the rainiest month in each year at Cairo, Egypt, from 1877 to 1922. How many times in this period of 35 years was the mean for the month exceeded? In how many years was there no rain at all?

(After H.H. Clayton and E. Biel, in Strahler, 1969.)

regions, however, it is becoming clear that the rain which falls comes in definite spells occurring over wide areas, and this suggests that there are disturbances which cause these periods of heavier rain.

An investigation of a spell of heavy rains over the western part of Saudi Arabia in mid-April, 1968, illustrates several important points. This area receives on average 60-70 mm rain per year on the Red Sea coast, rising to 500 mm in the mountains behind. Records for the whole area were sparse until a series of rain gauges was installed as part of a survey of the water resources of the region. During 15-20 April, 1968, amounts of rainfall exceeded the annual mean totals for most areas (Figure 11.4(A)).

A study of the surface weather map showed that there had been a south-easterly air stream over Arabia and the Red Sea, bringing warm, moist air from the Indian Ocean (Figure 11.4(B)), and

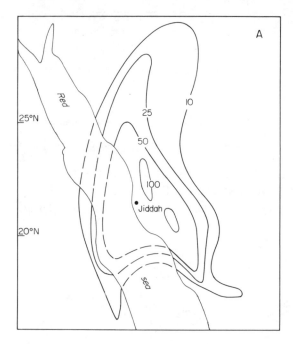
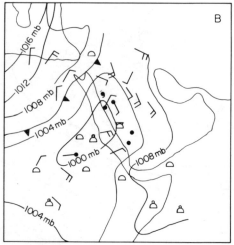
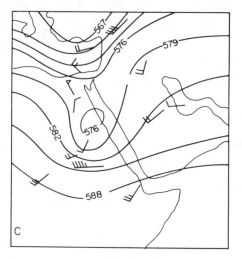
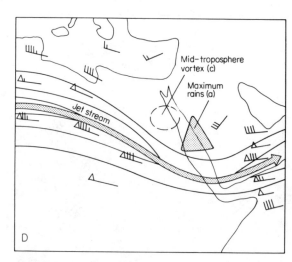

FIGURE 11:4 *An atmospheric disturbance leading to rainfall in a desert area.*
(A) *The total rainfall over 5 days: 9 a.m. on 15 April 1968 to 9 a.m. on 20 April 1968. Totals in mm.*
(B) *Surface pressure, winds and clouds at 12 noon on 17 April 1968.*
(C) *Contours (metres) and winds at the 500 mb level. Compare the wind speeds and directions here in the middle troposphere with those at the surface (B).*
(D) *The upper troposphere conditions (250 mb level with winds at 11 km). Notice the relationship of the jet stream to other atmospheric features at lower heights.*

(After D.E. Pedgley.)

leading to the development of *cumulus* and *cumulonimbus* clouds. The rather confused pattern in the rainy areas may have been due to downdraughts of air. A trough extended northwards from Sudan, bringing north-easterly winds to Egypt. The middle troposphere map (500 mb level) records the development of a weak vortex above the surface trough. The upper clouds at this level formed a sheet extending along the line of the Red Sea (Figure 11.4(C)), and were associated with much of the rain. At the top of the troposphere (250 mb level — i.e. 11 000-12 000 m) it can be seen (Figure 11.4(D)) that the subtropical jet stream was deflected southwards over the Red Sea: this movement reached a maximum by 15 April, retreating northwards to a west-east flow by 19 April.

The rain had thus been experienced in a situation where there was a surface northward flow of air, a mid-tropospheric cold core vortex to the west, and a displacement of the subtropical jet to the south. This situation is associated commonly with torrential rains and thunderstorms in India, the Middle East and North Africa. It is due to a combination of increased instability on the east of a mid-tropospheric 'low' and the induced upward flow caused by the jet stream flowing in this position. Deep convection is released, and amounts of rainfall can be correlated with the instability induced: thus, local showers (10-20 mm) took place on 15 April when instability was encouraged, but heavy rains fell on 16-17 April when it became extreme.

This study illustrates a number of important points which have become clear as more has been discovered concerning arid climates.
1) Too little is known at present to enable reasonable generalisations to be made about arid climates.
2) Events in the middle and upper troposphere are most important in the understanding of arid weather patterns particularly in the subtropical zones.
3) The general aridity of these regions is broken only by exceptional events which give rise to sufficiently deep convection to induce rainfall.

The lack of rainfall is accompanied by a general lack of cloud in the inland deserts, although a layer of lowlying *stratus* cloud may persist at certain seasons along coasts where there is an upwelling of cold water. Insolation by day is extremely high through the cloudless skies, and the world's highest temperatures have been recorded in the low latitude deserts: 58°C was recorded at Al'Aziziyah (Libya) in the northern Sahara on 13 September 1922, and 56°C in Death Valley, California on 10 September 1913. At night, radiation is rapid, and diurnal fluctuations of 20°C are common: one of 37.8°C (i.e. from -0.6°C to 37.2°C) occurred immediately south of Tripoli in Libya. The annual range is also much greater than for other areas in similar latitudes. Maximum insolation is received when the Sun is overhead, and there is maximum radiation from the surface through the lower atmosphere in the cold season.

The largest area of desert in the world is the **Sahara** of North Africa, together with its eastward extension across the Arabian Peninsula and into Pakistan. It has recorded large daily and seasonal temperature contrasts, and the highest surface temperatures in the world. Such an extensive area with drought conditions is controlled by the large-scale subsidence of air associated with the subtropical high pressure belt which moves north and south with the seasons, but covers the Sahara-Arabia area throughout the year.

The low pressure recorded at the surface (see the world map of sea-level isobars, Figure 5.17), due to the excessive local heating with its attendant convection, is shallow (i.e. a maximum of 500 m) even in summer. Above this temperature inversion prevents deep convection. Any air which rises contains a very small amount of moisture and in general cannot rise far enough for condensation to take place. Even when rain is formed it may evaporate before reaching the ground. A little more rain is received along the northern and southern margins of this desert, where rain-producing disturbances are more frequent. Even in the marginal zones, however, the rains are unreliable and several years with hardly any rain may occur. The mountain ranges of the Sahara (e.g. the Ahaggar and Tibesti massifs) are associated with more frequent convergence and uplift since heating of the air above them results in a zone of greater warmth than the surrounding air and this facilitates deep convection. *Cumulonimbus* cloud forms in contrast to the clear skies or shallow *cumulus* cloud around. Rain is also produced by disturbances in the higher troposphere, which may not be reflected in surface pressure patterns (Figure 11.4).

The driest desert of the world is the **Atacama** desert of northern Chile and the Peruvian coast. The northern parts of Chile have received no rain for at least 400 years. Rainfalls of under 50 mm per year are recorded from 5° to 30° South, but most areas have much less than this figure. Many stations have not recorded rain in living memory, and beds of very soluble nitrate minerals are preserved on the floors of ancient, dried-up lakes. This extreme aridity is surprising at first sight in view of the proximity of tropical ocean waters and the high humidity of the air. There seem to be four major factors which not only limit the dry climate to the coast, but intensify the aridity (Figure 11.5).

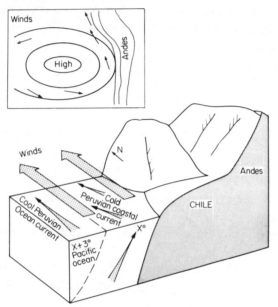

FIGURE 11:5 *Winds and currents along the north Chilean coast. There is little change throughout the year from the situation depicted on the inset map: the Andes prevent the anticyclone ('High') from moving to the east, and the winds blow virtually parallel to the coast. How do they affect the ocean and coastal currents? (Hint: relate this to the Ekman spiral.) In the water temperatures X° is often 12-14°C, compared with surface temperatures of over 20° at the same latitude in the central Pacific Ocean.*

1) The Andean mountain ranges undoubtedly have a strong influence. In other parts of the world the subtropical high pressure zone represented on mean isobar maps reflects the west-to-east passage of a series of anticyclonic pressure systems, which are interrupted occasionally by short-lived, low pressure disturbances. The Andes rise to heights over 7 km, and restrict this movement: the eastern Pacific is the site of an almost permanent and unmoving anticyclone. Winds around the eastern end of the anticyclone blow from the south (the anticyclonic wind pattern in the southern hemisphere is anticlockwise), and drag the cooler waters of high latitudes with them.

2) The southerly winds blow along most of this coast. Thus, the cool surface waters of the Peruvian ocean current flow towards the north-west (Figure 5.5), and even colder water chills the surface air,

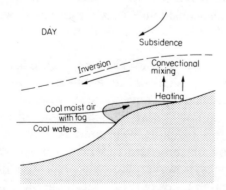
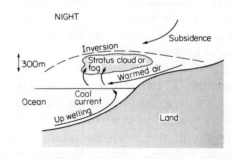

FIGURE 11:6 *Fog and cloud formation on the Chilean coast. A shallow surface circulation occurs during the day due to the heating of the land and the colling of the sea. A contrast in temperature is intensified by the heating of the bare desert landscape and the upwelling of cold waters offshore. Describe what happens, and how the coastal, inland and offshore climates will be affected.*

reduces the observed lapse rate and makes the air more stable. Rainfall totals on the coast are the lowest, whether there are hills behind or not: one might have expected coastal hills to facilitate local convection. On the other hand, the coastal areas are often subject to fogs or, more commonly, a layer of *stratus* cloud some 300 m thick with its base at 500 m above the surface (Figure 11.6). Fine drizzle may come from this in Peru during winter, when a cold front pushes its way northwards.

3) The trend of the Andean ranges is responsible for the way in which the desert climate comes so close to the equator. The mountains bend westwards, almost parallel to the circulation as far as northern Peru, and maintain the conditions which give rise to the aridity. Farther north, the coast and mountain ranges bend away and the cool current and aridity are continued out into the Pacific Ocean. The dry climate thus extends to within 5° of the equator and results in the northern margin being subject to occasional periods when the drought is broken by torrential rains. These occur when the subtropical high pressure zone is weaker and allows the equatorial westerly and northerly winds to extend southwards: this happens once every few years and most commonly in the local summer. Warmer waters flow southward into the area, giving rise to intense evaporation and heavy rainstorms on the land as the prevailing inversion is suddenly broken. Not only do the local inhabitants suffer from these storms, but the warmer waters kill vast numbers of fish in what is one of the world's major fisheries (chapter 24), releasing hydrogen sulphide and killing the sea birds.

4) Intense and regular sea breezes are also a feature of these coasts, but instead of reducing the aridity they also intensify the situation (Figure 11.6).

This combination of factors also causes severe aridity in other coastal areas of the world. The **Moroccan and Rio de Oro coast** of the Sahara has the Canaries Current offshore; the Benguela Current off the **Namib-Kalahari desert** of South-west Africa and the Californian Current off **southern California and north-west Mexico** also have the same effect. But the areas are not so extensive because of the local changes in trends of coastline.

The **Australian desert** illustrates another set of conditions which give rise to aridity. Here, however, the aridity is much less severe. The continent has few relief barriers which affect the climate, and the main mountain range is along the east coast. Thus the eastward passage of subtropical anticyclones moving across the land — for Australia lies in this belt — is not interrupted. Subsidence and surface divergence of the air is dominant, and much of the continent is arid. The anticyclones move across Australia at an average rate of 800 km per day. Still conditions are experienced in the central regions, but greater variations of weather occur around the margins of the anticyclones especially on the northern and southern edges of the desert. Troughs of lower pressure between the anticyclones are often identified on weather maps as frontal lines (Figure 11.7). Something like 40 of these 'fronts' pass across the coast of Western Australia each year, and occasionally a wave depression or even a tropical cyclone reaches the area. Such disturbances bring more rain to the Australian desert than to most, but the high pressure subsidence dominates the circulation. The coasts have the highest totals of rain (averaging 225 mm as compared with 150 mm at inland stations), and there is no offshore cool current. In fact, the water near the coast of Western

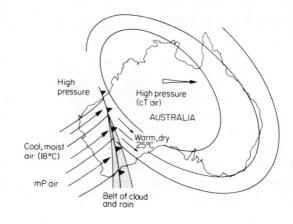

FIGURE 11:7 *Anticyclones moving across Australia in winter, with a front between. The dominance of subsiding air gives Australia a desert climate, except on the north and east coasts, and on the southwestern tip.*

(After Trewartha, 1961.)

Australia is warmer than that out to sea, which is drifting northwards from the Antarctic.

Patagonia is an east coast, temperate region which has a dry climate. Situated in the south of Argentina, it is in the midlatitude belt of cyclonic storms, but the Andean range makes it into a 'rain shadow' zone: Santa Cruz, on the coast at 50° South, has a mean annual rainfall of under 150 mm. The Canterbury Plains on the eastern side of South Island, New Zealand, experience a similar effect, but the rainfall there is four times that in Patagonia and the climate becomes subhumid. The rainfall in Patagonia decreases from the Andean foothills in the west to the coast in the east. The air crossing the Andes drops most of its moisture on the western flanks of the range, and very little is carried across to the eastern foothills. Then the descending air is warmed adiabatically and the strength of the circulation in the southern hemisphere maintains the drying downdraughts throughout the year. New cyclones form over Uruguay and move south-eastwards into the South Atlantic as they develop, continuing as part of the 'Roaring Forties' disturbances.

A similar effect is felt to the **east of the Rockies** in North America, where the warm wind flowing down the eastern slopes is known as 'the chinook' (Figure 11.8). In this case, however, the continent is much wider, and cyclonic disturbances form again in the westerly circulation over the prairies. The **interior areas of Eurasia** do not suffer such a clearcut 'rain shadow' effect in the lee of a prominent mountain range, but disturbances lose some of their vapour in the west, and this is seldom replaced by evaporation from the land, especially in the very cold winter season. The southern margins of the USSR are arid because the belt of mountains running from Turkey, through Iran and eastwards restricts the inflow of moist air from the south, and also gives rise to lee subsidence when winds are from that direction.

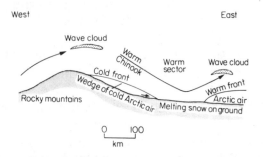
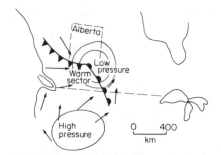

FIGURE 11:8 *The chinook wind of southern Alberta. A winter depression moves from the west, and the warm sector air melts snow: temperatures may rise from −20°C to above freezing point within a few hours as the warm front passes. The warm descending winds may, however, cause damage as they are channelled down valleys, since they not only thaw the snow but dry out the soil and then blow it away. In summer the associated drought brings a fire hazard to the area. What will happen as the cold front crosses the area in winter?*

There are thus several factors which lead to aridity on the continents, and they may be combined in several ways in both temperate and tropical regions. The **dry oceanic areas** cover more than double the area of arid climate on the land: they are associated mostly with extensions of the cool ocean currents (above which the air is extremely stable) or with the atmospheric subsidence around the eastern margins of the subtropical high pressure belts.

The semiarid regions

The margins of the humid and arid regions become semiarid when there are seasons of the year which can be termed as definitely arid or humid. These areas are tempting to settlers: whereas nobody will try to live in the desert unless irrigation water is freely available, many have attempted to farm the desert margins. Long periods of drought and the unpredictability of the rainfall from year to year have led to the tragedies of the 'Dust Bowl' in south-western USA in the 1930's and of the Virgin Lands of Siberian USSR in the 1960's. In both areas the land was extensively ploughed in years of higher rainfall, but then had to face years of drought and wind erosion. The western areas of Queensland, Australia, also suffered 8 years of continous drought between 1962 and 1970 and once-fertile fields are covered with blown sand.

CLIMATES WHICH ARE DRY ALL YEAR

Station	Latitude	Height above sea level (m)		J	F	M	A	M	J	J	A	S	O	N	D
Deaver, Wyoming, USA	44° 53'N	1300	T (°C)	−8.3	−5.0	0	7.2	12.8	17.2	22.2	20.6	14.4	7.6	−0.6	−5.6
			R (mm)	5	3	5	12	23	30	15	8	12	10	5	5
Yuma, Arizona, USA	32° 40'N	65	T	12.8	15.6	18.9	22.8	26.7	31.1	35.0	34.4	31.1	24.4	17.8	13.9
			R	8	8	8	3	0	0	5	12	15	8	3	15
Sanluis Potosi, Mexico	22° 09'N	2000	T	12.8	15.0	17.2	20.6	21.7	21.1	19.4	19.4	18.3	17.2	15.0	13.9
			R	12	5	8	5	30	72	55	42	86	17	10	12
Lobitos, Peru	4° 27'S	22	T	25.6	26.7	26.7	26.1	23.9	22.2	20.6	20.6	20	21.1	21.7	22.8
			R	10	37	3	0	0	0	0	0	0	0	0	0
Lima, Peru	12° 02'S	115	T	21.7	22.2	22.2	20	17.8	15.6	15	15	15.6	16.1	17.2	18.9
			R	0	0	0	0	3	3	5	8	5	3	0	0
Antofagasta, Chile	23° 29'S	100	T	20.6	20.6	18.9	17.2	15.6	13.9	13.3	13.9	14.4	15.6	17.2	18.9
			R	0	0	0	0	0	3	3	3	0	0	0	0
Commodoro Rivadavia, Argentina	45° 47'S	65	T	18.9	18.3	15.6	13.3	9.4	7.8	6.7	7.2	10	13.3	15.6	17.8
			R	10	15	17	17	30	25	22	17	15	10	15	12
Walvis Bay, SW Africa	22° 50'S	8	T	18.9	19.4	18.9	18.3	17.2	16.1	14.4	13.9	13.9	15.0	16.7	17.8
			R	3	5	8	3	3	0	0	3	0	0	0	0
Cape Juby, Spanish Sahara	27° 56'N	6	T	16.7	16.7	17.2	18.3	18.9	20.0	20.6	21.1	21.1	20.6	19.4	17.2
			R	8	5	5	0	0	0	0	0	8	0	15	8
Timbuktu, Mali	16° 46'N	260	T	22.2	25	28.3	31.7	34.4	33.9	31.1	29.4	31.1	31.7	28.3	23.3
			R	0	0	0	0	3	20	58	81	35	3	0	0
Aswan, Egypt	24° 02'N	130	T	15.6	17.2	21.1	26.1	30.6	32.8	33.3	32.8	31.1	28.3	22.8	17.2
			R	0	0	0	0	3	0	0	0	0	0	0	0
Jiddah, Saudi Arabia	21° 3°'N	6	T	22.8	22.2	25	26.7	28.3	29.4	30.6	31.1	30	28.9	26.7	25.0
			R	22	0	0	0	0	0	0	0	0	0	40	15
Karachi, Pakistan	24° 48'N	4	T	18.9	20	24.4	27.2	30	30.6	30	28.3	27.8	27.2	23.9	20
			R	12	10	8	3	3	17	81	40	12	0	3	5
Astrakhan, USSR	46° 15'N	−15	T	−7.2	−5	0.6	8.9	17.8	22.8	25	23.3	17.2	9.4	2.2	−2.8
			R	12	12	10	15	15	20	12	10	15	10	15	15
Alice Springs, Australia	23° 38'S	600	T	28.9	27.8	25	20	15.6	12.2	11.7	14.4	18.3	22.8	26.1	27.8
			R	42	33	28	10	15	12	8	8	8	17	30	38

FIGURE 11:9 Climatic mean statistics for arid regions. T = mean monthly temperature (°C), R = mean monthly precipitation (mm). Where possible, climatic statistics are taken from observations over periods of 35 years, but at times the most reliable figures have to be given for shorter periods: this is often necessary in regions of difficulty and sparse settlement, like deserts.

Dry regions

Plate 52 The Atacama desert, southern Peru (Gemini XI, 12 September 1966). The coast is obscured by stratus cloud, and a few rivers drain across the desert from the Andean ranges behind. (NASA)

Plate 53 The Dust Bowl of the USA. A dust storm in the south-west of Boca County, Colorado, 14 April 1935.

Plate 54 A clump of bluestem grass in Roosevelt County, New Mexico: the grass held the soil in place beneath it whilst 1.5m was removed.

Plate 55 An abandoned farm south of Brownsfield, Texas. The topsoil has been blown away and the farm is now useless. (53-55 USDA Soil Conservation Service)

CLIMATES WHICH ARE DRY ALL YEAR 117

56

57

58

Seasonally wet areas

Plate 56 Coastal fog at San Francisco: a Skylab 2 view of the central Californian area, May 1973. The coastal fog contrasts with the clear skies of the inland San Francisco Bay area.

Plate 57 The Indian peninsula. A Gemini XI (mid-September 1966) view at the end of the summer monsoon season. The clear zone along the west coast results from a convectional cell with cumuliform clouds over the land. This forms brighter (larger) masses of cloud over the western mountains and smaller 'streets' of clouds elsewhere over the land.

Plate 58 The Himalayas and the Tibetan massif, taken from Apollo 7 in mid-October 1968. Mount Everest is in the central part of the snow-covered Himalayas, and Kanchenjunga in the foreground. The snowline was at 5500m. This massif forms a major climatic divide, and has a major effect on the surrounding climates. Recent ideas suggest that the Tibetan mountain region plays a part in causing the jet stream shifts which have been related to the monsoonal flows. The plateau has a low proportion of cloud cover and is heated intensely through the summer, giving rise to steep temperature gradients in the upper troposphere. Satellites can help in flood forecasts, by monitoring the snow levels in winter. (56-58 NASA)

Plate 59 Cumulonimbus cloud over New Delhi in July. (A C Wall)

59

12

Climates wet at one season

The seasonally wet climates are partly transitional in nature due to the transfer of the rainy climatic belts associated with the seasonal movements of the overhead sun. At one season of the year the dry belt moves in to dominate the weather of the region, and in the other half of the year it is replaced by a belt of rains. In other places the seasonal contrast results from a major reversal of the wind circulation: moist winds from the ocean give way to dry winds blowing outwards from the continental interior. The transitional types of seasonal climates occur mostly on the west coasts of continents, and the circulation-reversal climates on the east coasts (Figure 12.1).

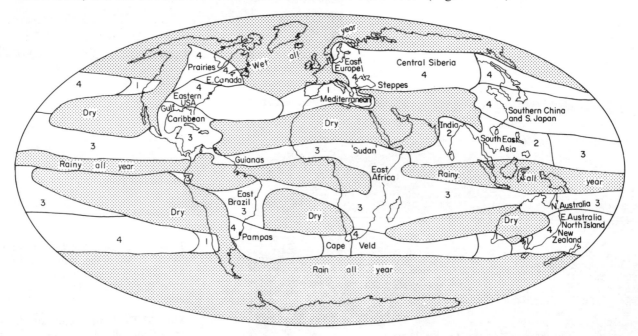

FIGURE 12:1 *The world distribution of regions with seasonal rains.*

1 Winter rains: 'Mediterranean' type regions.
2 Summer rains: Asian tropical monsoon lands.
3 Summer rains of other tropical areas.
4 Summer rains of temperate areas: east coasts (warm and cool temperate) and interiors.

Most seasonal climates have the maximum rainfall during the summer season. These include all the circulation-reversal areas: moist winds tend to be drawn into the land areas during the summer because of the heating and uplift of air over the continental areas; in the colder season the land is dominated by subsidence and outward-blowing winds. On the west coasts the transition between desert and equatorial climates is such that the rains move into the transition area with the overhead sun, giving another summer rainfall maximum. The poleward margins of west coast deserts, therefore, are the only areas having seasonal climates with rainy winters and dry summers. These occupy less than 2% of the world's land surface, and are often known as the areas of 'Mediterranean' climate because that sea is the largest area exposed to the winter rain type of regime.

Wet winters: the Mediterranean climate

The **coasts of the Mediterranean Sea** between 30° and 45° North are noted for their long periods of uninterrupted dry, sunny summer weather. In the past the bordering lands have been the home base for the development of many important civilisations, and today they have become the main tourist area in Europe, drawing people to the sun and ancient ruins. The coasts of eastern Spain, southern France, Italy, Yugoslavia, Greece, Morocco, Algeria and Tunisia rely on sun-seeking tourists for their main source of income. Similar climatic regimes occur elsewhere in the world, but in each case the area affected is relatively small in terms of latitudinal extent and inland penetration on the scale of world climatic regions. The general atmospheric conditions resulting in the typical 'Mediterranean' regime are similar in all the regions, although the local relief and land-sea distributions impose significant variations.

The Mediterranean Sea area is distinctive because of the unusual extent around what is virtually an inland sea, and also because of the hilly and mountainous relief of the peninsulas extending out into that sea. These conditions lead to considerable variations of climate: maximum periods of rainfall may be in autumn, winter or spring; the western sides of the peninsulas receive more rain and experience milder winters than the eastern; and the eastern basin of the Mediterranean Sea is much drier throughout the year than the western sector. The Sea is a deep reservoir of warm water, preserved from the entry of colder Atlantic water by the threshold at the Gibralter entrance. Evaporation rates are high and the surface waters are warmer than the overlying air for nine months of the year. Thus the air blowing across the Mediterranean Sea is warmed in winter, and receives supplies of evaporated moisture throughout the year along a zone extending well into the continental mass.

The summer drought is due to the northward movement of the Azores-Sahara high pressure cell. This begins to assert itself over northernmost Africa in mid-April and covers the whole region as far north as southern France by late June. Subsiding, dry air is dominant during the summer, although some of the wider peninsulas (Iberian, Balkan and northern Italy) have shallow low pressure centres at the surface because of local heating. Local thunderstorms arise frequently where this is the situation, and late summer is very rainy in northern Italy with crops suffering from hail: the Po valley does not have a Mediterranean climatic regime. Even at this season, small cyclonic disturbances may interrupt the long, fine spells. Low pressure cells may become detached from the Sahara surface low pressure system, and intensify into small depressions over the Mediterranean Sea, leading to local thunder and heavy rainfall. Others may enter from the west, but are much weaker than the winter disturbances coming from that direction, and are most common in the northern areas.

Winter sees a succession of Atlantic depressions passing through the Sea area (Figure 10.3). The winds become strongly convergent in areas like the Gulf of Genoa, where the northerly winds blowing via the Rhône-Saône corridor bring cold air into the backs of depressions which have formed in the lee of the Pyrenees: 70 per cent of all the Mediterranean depressions are intensified in this way, and may receive renewed vigour as they pass the Aegean Sea, where further draughts of cold air blow in from the north. The Mediterranean Sea track has one of the major concentrations of cyclonic flow in Europe during winter, rivalling the northern path via Iceland for numbers of depressions. The upper tropospheric jet stream commonly splits at this season and one branch follows the line of the Mediterranean Sea. Activity is particularly marked when there is a surface flow of air from the north giving rise to surface temperature contrasts. The cold fronts of the depressions bring especially heavy rainfall and strong winds.

Whilst winter is the most disturbed season of the year there are also long periods of settled and fine anticyclonic weather at this time. In the western part of the Mediterranean basin they may be experienced for up to half the winter season, and occur when the jet stream takes a single path aloft leading to a sucession of depressions crossing Britain. Thus the winter rainy season brings, on average, only 6 days of rain per month to Libya, and 12 to the west coast of Italy. This fact can also be expressed in terms of hours of sunshine.

	Hours of sunshine	
	Year	December
Oxford	1400	43
Rome	2300	108

Rainy days are fewer than in northern Europe, but totals may be similar: individual falls are therefore heavier, and an important feature of the rainfall in the Mediterranean Sea area is its irregularity, especially in the drier — and virtually semi-arid — regions of the eastern and southern margins. Athens received 850 mm of rain in 1883, but only 110 mm in 1898, and has an annual mean of 390 mm.

Throughout the year, local winds may dominate the climate of a particular area: the mistral (Rhône valley) and bora (northern Adriatic) are strongest in winter and spring, bringing cold air from the north, whilst in summer the scirocco (Algeria), leveche (south-eastern Spain) and thamsin (Egypt) are hot, dry winds drawn from the Sahara into the warm sectors of depressions moving through the Sea. So, whilst the general regime is recognisable throughout the Sea basin, there are also very strong local influences at work leading to differences in temperatures and rainfall totals.

Central California is the only other region experiencing this regime in the northern hemisphere, being on the only other west-facing coast in the hemisphere. The winter maximum of rain is carried much farther north than in Europe for the reasons described in chapter 10, but the summer drought is confined to central and southern California. The mountain ranges of the Western Cordilleras of North America restrict the inland extent of this climatic type, and halt the passage of cyclones in winter, when winds are mostly northerly or north-westerly, bringing rain only with the occasional front. Rainfall amounts decrease towards the desert conditions on the southern margins: San Francisco (38°N) has 550 mm mean annual rainfall; Los Angeles (34°N) has 375 mm; and San Diego (33°N) has only 250 mm and can be considered as being in a desert.

The Coastal Range, rising to heights of 1000 m, keeps many of the marine influences on the climate out of the central valley, which is drained by the Sacramento and San Joaquim rivers. Such protection results in the valley becoming extremely hot and dry in summer (Figure 12.2), since it is not affected by the fogs which often obscure the coastal parts. These fogs are associated with the offshore cold current extending northwards from the desert region in the summer season. The fogs and the strong sea breezes funnelled through the Golden Gate gap in the coast range together lower the July-August mean temperatures at San Francisco to 15°C, a marked contrast with the 30°C a few miles inland. Hot winds from the south may be drawn into depressions and these *Santa Annas* may damage spring blossoms.

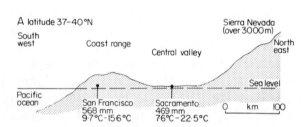
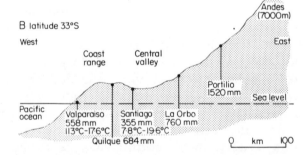

FIGURE 12:2 *Temperature and rainfall conditions on the coast and inland for (A) central California, and (B) central Chile. Rainfall totals in mm; temperatures (°C) of mean lowest and highest months. What are the differences between the two areas? Can you explain these differences?*

Central Chile is similar in many ways to central California, having high mountains parallel to the coast; but the coast range is more broken, allowing marine influences to penetrate more easily, and the interior valley is higher (Figure 12.2). This area has the most marked winter rainy regime of any 'Mediterranean' region: 90 per cent of Valparaiso's rainfall comes in the four wettest winter months (cf. 50 per cent at Lisbon, Marseilles, Rome; 75 per cent at San Francisco). The climate is not so foggy as that in California, and has a very marked summer drought with several rainless months. Rainfall once again increases towards the pole, but more sharply than in North America: Valparaiso (33°S) has 550 mm; Concepcion (37°S) has 1325 mm; and Valdivia (40°S) has 2625 mm. The contrasts here are largely due to the position of central Chile between the world's most arid desert (chapter 11), and the most intensive zone of mid-latitude cyclonic activity (chapter 10).

Only the **southernmost tips of Africa and Australia** extend into the latitudes of the Mediterranean regime, and neither is able to show the poleward transition to cool temperate rainy climates. No part of New Zealand has this climate type. In winter, the cyclonic disturbances affect these areas to a greater extent than in summer, but the totals of rain are extremely variable from year to year. The Cape Town coast of South Africa has cooler summer temperatures than normal because of the offshore Benguela Current. South-western Australia, however, has a more reliable rainfall and much higher temperatures due to warmer offshore waters. Adelaide is another area which is sometimes included in the Mediterranean group because it receives a winter maximum of rainfall, but it has no summer drought. The region of south-west Australia is particularly interesting because it has no backing mountains, and there is a gradation inland towards the area of lower rainfall, and, in addition, no interruption to the passage of low pressure troughs. This situation gives rise to a more varied passage of moister weather types.

Wet summers: the tropical Asian monsoons

A maximum of summer rainfall coupled with a winter drought is much more common. Some such climatic regimes are due to a transitional position between desert and equatorial rains, but most continental east coasts are typified by climates involving a seasonal reversal of atmospheric circulation: this effect has been termed **'monsoonal'**.

A monsoonal climate is essentially one where the seasonal circulation has been considered to be controlled by differences in heating. It is suggested that in summer intense heating of the dry continental air causes it to rise and that moist air is thus drawn in from the ocean (Figure 12.3). The process is reversed in winter as the air over the land cools rapidly: in short, the effect has been seen as a gigantic land-and-sea breeze situation. The Indian subcontinent and South-east Asia have for long been regarded as the outstanding examples of this type of climatic regime, and many parts receive 80 per cent of their total rainfall within 3-4 summer months. Cherrapunji, Assam, still holds the world record for the wettest month (9155 mm) and year (26 300 mm) in 1860-61.

FIGURE 12:3 *The Indian Monsoon: a traditional view. This suggests that the wind-reversal is due merely to differences in surface heating. In summer this would lead to rising air and an indraught from the sea; in winter air it would subside over the cooling surface and blow outwards. This is typical of the 'simple' explanations given before extensive investigations were made of conditions aloft.*

It can now be shown, however, that the cause of the tropical Asian monsoon is not a simple surface reversal of winds due to summer heating and winter cooling. Studies of the middle and upper troposphere conditions above India have indicated that there is a strong link between seasonal conditions there and those at the surface. Any purely thermal effect is extremely shallow, and the main features of the climate are dependent on the local manifestation of larger-scale atmospheric processes. Much less is known about the climates of south-east Asia (i.e. Burma, Thailand, Indochina) than about that of the Indian subcontinent, but it seems that these areas may be affected by similar factors.

The Indian conditions have been studied intensively because the onset of the seasonal rains is such a matter of life or death. The summer monsoons break suddenly in mid-June, bringing heavy rains in the hottest part of the year. These may last until September, or October, and gradually give way to the winter monsoon and drier conditions, although the temperatures in this part of the world, sheltered from the Siberian cold by the Himalayas, never fall below tropical warmth. The early part of the following year experiences a gradual build-up of heat, culminating in a renewed burst of monsoonal rains.

The seasonal atmospheric circulation and weather are closely related.

a) The **winter circulation** at the surface in India is dominated by anticyclonic conditions, both to the north and to the south of the Himalayan-Tibetan massif which forms an effective barrier to communication between sectors of the surface circulation. The Indian subcontinent is thus isolated from the rest of Asia. Winds blow outwards from the land, over which air is subsiding; the air dries out and rainfall is largely prevented. Over India the subtropical high pressure cell gives rise to dry winds blowing down the Ganges valley and then changing to a north-easterly (i.e. trade wind) direction across the Bay of Bengal.

Above 3000 m the circulation is quite different (Figure 12.4). A strong westerly flow, including the jet stream, flows across the continent and bifurcates around the Tibetan mountain mass at this season. The southern branch is much stronger and is a permanent feature from November to April because of the strong thermal gradient above northern India. This upper air flow has the effect of intensifying the surface anticyclone, and also brings a string of surface cyclonic disturbances into the north-west from the Middle East and Mediterranean areas. The general pattern of clear skies and warm, dry weather is interrupted by some cloud and rain, which becomes snow on the Himalayan slopes, and even by cold spells as *cP* air spills into the Indus lowlands. In the extreme north-west these winter rains are generally greater than those received in summer. As the airflow passes over the Bay of Bengal it picks up moisture and becomes involved in weak tropical disturbances which bring rain to south-eastern India and the east coasts of Ceylon.

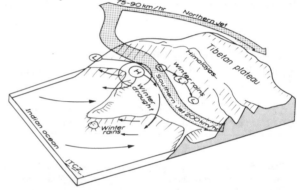

FIGURE 12:4 *The Indian Monsoon: winter. The southern jet is very strong and maintains the surface high pressure as well as bringing in a series of cyclone storms along its northern margin. Relate this pattern to the discussion of jet stream and surface effects in chapter 5. The dominant surface circulation consists of winds blowing outwards from the land: this is a dry season for most of India, but note the exceptions to this rule.*

Thus, although most of India is dominated by a winter drought, it may be interrupted by welcome light rains in the extreme north and south, and even elsewhere if the northern cyclones draw in *mT* air from the Indian Ocean. Temperatures remain high because of the protecting effect of the Himalayas and Tibetan Plateau: Calcutta is 7°C warmer in February than Hong Kong (both 22½°N).

b) **Spring** becomes extremely hot and dry. The subtropical jet circulation begins to weaken relative to the northern branch, and eventually moves to a position entirely north of the Tibetan mountain mass as the summer monsoon breaks. India is still dominated in spring by a subsiding and outward-blowing anticyclonic circulation, and the clear skies allow maximum and increasing insolation. High temperatures, a heavy heat haze, and drought are the characteristic features of the weather, especially in the central areas. The equatorial low pressure trough begins to move

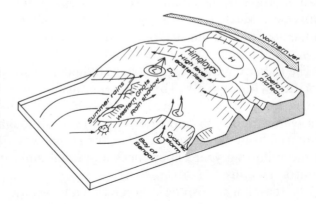

FIGURE 12:5 *The Indian Monsoon: summer. The jet moves north of the Himalaya-Tibet massif, allowing massive updraughts of air to take place to the top of the troposphere over the peninsula. Heavy rains occur along the Western Ghats, and when cyclonic disturbances enter the Ganges Basin. Other regions receive less rain, and its arrival is much less certain. Where are these regions?*

northwards, bringing rain to the south, and the increased heat intensifies some of the squally disturbances in the north. The weakening of the jet allows convection and thunderstorms due to explosive *cumulonimbus* growth in the afternoon. In the north-west, such conditions give rise to the 'andhis' — dust storms forming in the absence of atmospheric moisture.

c) The **summer monsoon** breaks suddenly, but progressively from the south. Burma receives the heavy rains from May onwards, but they are delayed until late June over India. This change to wet conditions is thought to be associated with the upper air changes: the westerly jet moves north of the Himalayan-Tibetan mountains due to the formation of a thermal high pressure zone over the Tibetan plateau, which shifts the zone of maximum contrast to this northerly position. Easterly winds are established around the southern margin of this feature due to subsidence on the southern margin of the jet (Figure 12.5). It is thought that the onset of rains is related to the northward movement of the jet, which has impeded convection through the winter and spring.

A surface convergence zone is established in summer over India beneath the easterly winds aloft, and moist winds are drawn in at the surface from the expanses of the Indian Ocean. Again the position is unique, for no other continental area has an open tropical ocean to its south: a comparison can be made with the reverse situation in the equatorial Amazon basin, another region of high rainfall, which is almost as open to the North Atlantic Ocean. Thus, air reaches India which has had a long passage over the water and is extremely moist (although some observations have suggested that it is relatively dry until it reaches the Arabian Sea). The south-westerly flow of air brings a favourable set of circumstances for the production of rain, but atmospheric disturbances are necessary to trigger the major storms, since the basic temperature stratification of the air masses does not favour upward movement. These disturbances include the south-westerly surges in which speeds rise due to convergence near the equator and lead to instability and heavy rain as the flow reaches the mountainous Western Ghats facing the Arabian Sea. The Bay of Bengal area is associated with the formation of monsoon depressions, assisted by the easterly flow aloft and the easterly waves moving into the area from the Pacific. These move westwards and north-westwards over the land approximately once every 6 days, and the intensified surface heating there leads to widespread instability and heavy rain. In addition, some of the cyclonic troughs still move into the **northern areas, and sudden, uprushes of air are responsible for heavy showers on the west coast.**

The summer, from mid-June to September, is a season of heavy rains over much of the subcontinent: most places receive 80 per cent of their annual total of rain. Even at this season there are considerable spells of fine, sunny weather, and the amounts vary from place to place, and from year to year. Large areas are semi-arid. Some of these are in the rain shadow of the Western Ghats, and some in the north are influenced by a wedge-like inversion which allows the south-westerly flow to penetrate on the surface, but largely prevents vertical convective activity. In some years the westerly circulation aloft is re-established during the summer season and rainfall totals are extremely low. This is disastrous to the peasant farmers, and famine results. It is now clear that the rainfall is associated closely with a series of disturbances, and these in turn with events in the upper troposphere, rather than an inexorable seasonal reversal of airflow.

d) In the **autumn** there is a transition from the wet season to the dry, as the equatorial trough moves back southwards across India. It brings rains to the south of the peninsula (which had little in summer). Easterly winds from the Pacific replace some of the south-westerly airflow, and enable severe, hurricane-type storms to reach the Bengal coast. Two-thirds of this type of disturbance arrive between September and November. The most notable of these in recent years struck the Ganges delta in November, 1970, causing a disastrous tidal wave to sweep across the lowlying area. Finally the westerly jet stream reappears and the disturbances entering from the west increase in number. The winter drought returns.

The tropical Indian monsoon, which used to be thought of as very distinct from the other climates of the world, can be seen as a rather special case of the summer maximum rainfall type. The unusual features are related once again to the factors of relief and land-sea distribution, rather than to local atmospheric heating effects. Other tropical areas show relatively slight changes in the atmospheric situation from those encountered in India.

Other tropical wet summer climates

The situations in the Caribbean, central Brazil, west Africa and east Africa will now be compared.

a) **Central America and the West Indies** occupy an area with a great variety of relief including many islands, a high mountain chain and local areas of lowlying plains. On the other hand the atmospheric conditions are fairly constant. The easterly 'trade' winds associated with the subtropical anticyclones are dominant at all seasons and there is no invasion by the equatorial trough. Summer rains are caused largely by the seasonal removal of a strong inversion which is associated with the subsident air flow around the subtropical oceanic anticyclone: this is evident particularly over the western sector of the Atlantic Ocean. The Pacific coasts of central America are dry in winter, since they come under the influence of the eastern edge of the subtropical high pressure system in that ocean, and the desert conditions of north-western Mexico extend southwards at this season. In summer this area receives rain from easterly waves and hurricanes generated in the Pacific, and air ascent is assisted by the mountainous topography.

The Caribbean is more complex. The summers are the most rainy season, especially on the east-facing coasts, but onshore winds may bring rain in winter as well, when disturbances moving northwards across the area are invigorated by incursions of cold air from North America. Late summer is characterised by the violent hurricanes (chapter 6), which are a feature of the western margins of tropical oceans at this time of year. The trade wind inversion rises under the influence of surface heating, and gives greater scope for vertical convective activity. The storms are often accompanied by a large proportion of the annual rainfall, and their variability is important to the local economies. They not only cause damage and loss of life, but supply moisture to the crops.

b) **Central Brazil**, south of the main part of the Amazon Basin, also has a maximum rainfall in the summer, and it is thought that this is due to factors similar to those in the Caribbean. The trade winds dominate the surface air flow and the lower winter inversion brings very dry conditions to the area, especially in north-eastern Brazil, an area of great precipitation irregularity. The wedge shape of the land mass here juts into the anticyclonic cell of the South Atlantic Ocean. In summer the circulation and the inversion is weaker, convective activity is greater, and humid air from the north reaches the area to bring some rain.

The situation is complicated by the northward movement of cold front waves along the east coast of South America — one of the few places in the southern hemisphere where exchange takes place at the surface between colder higher latitude air and that from the warmer tropical latitudes. These disturbances are less frequent in summer, but may lead to heavier rains in the interior. In winter they may penetrate farther north, but their effects are limited to the coast and even offshore areas by out-blowing winds. This is another semi-arid region, where variation in rainfall from year to year means life or death to the people.

c) **West Africa** is the classic area for demonstrating a transition between the equatorial rainy climates and the subtropical deserts. A major reversal of winds at the surface is experienced from the dry northerly trades of the winter season to the wet southerly flow of the summer (Figure 12.6). This is due to the seasonal movement of the ITCZ in response to the overhead sun, although it is displaced an unusual distance to the north in this case, and is not here associated with strong vertical movements and heavy rain. The extreme southern coast does not experience the reversal, and has onshore winds throughout the year, but this may be due partly to the effectiveness of the sea breezes, extending up to 200 km inland even in winter.

The south-westerly flow reaches its maximum invasion of the land in summer. The relationship of this moist air to the overlying northerly flow is of great importance in determining the weather, but not enough is known about it. Much rain is also associated with disturbances of the easterly wave variety, where vertically ascending air breaks into the upper air and results in heavy thunderstorms.

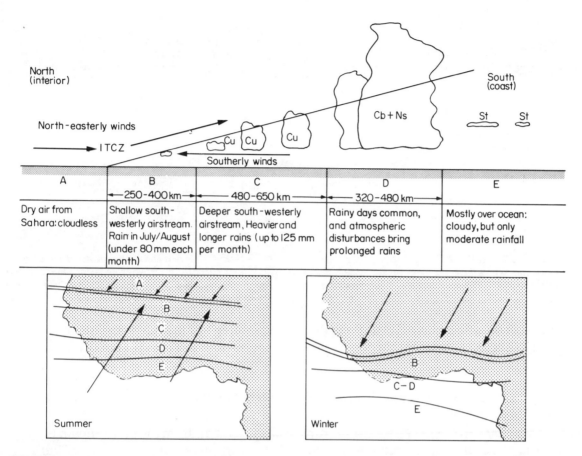

FIGURE 12:6 West Africa. The general succession of climatic conditions from the coast to the interior is shown above. These zones can be related to their seasonal distribution shown on the maps. (After Trewartha, 1961.)

Long periods of steady rainfall occur as frontal disturbances move in from the ocean.

The Harmattan is a dry wind from the north, which is most frequent in the winter months, bringing intense drought and dust storms to the southern regions, and also becoming involved in coastal disturbances because of the great contrast it provides with the moister air.

d) **East Africa** also has a summer maximum of rain, corresponding to the movement of the overhead sun and equatorial trough, but the most notable characteristic of this area is the low amount of rainfall received in a virtually equatorial situation. During the northern summer (April to October) the winds come from the south, and in the southern summer from the north. Near the equator there is a double rainfall maximum, but the maxima are separated by months of drought: many areas are semi-arid. Much of the rain which does fall occurs in hilly regions, or where local convergence takes place. The rain-generating processes of the tropics, which give rise to such heavy falls elsewhere, are weakly developed here.

A number of reasons have been suggested to account for the low rainfall of East Africa, and several of these may work together, since the situation is obviously complex.

1) Both seasonal flows are largely over the land: there is thus a small proportion of ocean-to-land moisture transport, and most rain comes in the transitional seasons when this is in force.

2) The plateau relief affects the situation by cutting down the thickness of the troposphere affected by the surface flow (Figure 12.7). This reduces the effectiveness of convective activity and of the air forced to rise up the coastal scarp.

3) The upper air flow is north-easterly with a very low moisture content and a low-level inversion. Any clouds which form remain as *stratus*. The Nairobi weather forecasters depend on measurements of humidity in the upper layers: no rain will fall unless there is sufficient moisture there.

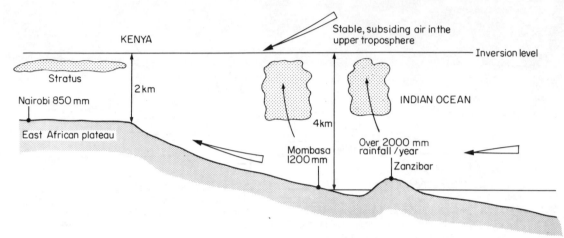

FIGURE 12:7 *East Africa. The situation on the plateau, near Nairobi, where the atmosphere beneath the dominant inversion is much more shallow, gives rise to a much lower rainfall than on the coast: totals are given for mean annual rainfall in mm. Rain falls at Nairobi only when the upper air conditions change. This is why forecasters in East Africa pay particular attention to records of conditions in the upper troposphere.*

4) Even when the atmosphere is disturbed, the rainfall is often low in total. Some of the disturbances form over the ocean as belts of rain perpendicular to the airflow, but the rain rarely penetrates more than 80 km inland.

5) Although the maximum rains are associated with the passage of the ITCZ, it is difficult to recognise the type of disturbances which are associated with the zone, since the air characteristics vary so little on either side.

6) Malagasy may be a further factor in interrupting the flow of air from the moist Indian Ocean area. It stands in the path of the south-easterlies, and the north-western part of the island is a dry, rain-shadow area in the lee of the eastern mountains.

e) **Summary**. Studies of tropical wet summer climates result in the following conclusions:

1) Knowledge of the causes of the observed weather conditions is often slight. It is becoming clear, however, that a knowledge of the vertical structure of the atmosphere is at least as important as that of the horizontal circulation in determining what happens in these regions.

2) Rainfall in tropical areas is an uncertain affair, dependent upon unpredictable and seemingly weak disturbances rather than on large-scale convection.

3) Seasonal variations are due to movements of the rainfall belts in tune with the apparent movement of the overhead sun. These belts are associated with zones of convergence, but may also be controlled by the upper level airflow.

Temperate east coasts and interiors

The climates of the east coasts of North and South America, Asia and Australia in temperate latitudes (i.e. 30-60°N and S) are characterised by wet, warm summers and cold, relatively dry winters. The coldest winters are experienced in North America and Asia, which are the widest land masses, but the winters are much milder in the southern hemisphere continents, South America and Australia, where the continent is much narrower in the appropriate latitudes. Eastern Asia and south-eastern Australia illustrate these contrasts.

a) The **winter weather of eastern Asia** is dominated by the westerly circulation common in these latitudes (Figure 12.8(A)). A succession of cold waves moves out from the Siberian anticyclonic area, bringing very low temperatures. Convergence may occur over the land as such a wave meets older and modified *cP* air, but this process is most effective along the coast, where the *cP* air is modified in contact with the sea and gives rise to cloud and drizzly rain. Western Japan receives rain from these winds when they have crossed the sea which separates it from the mainland. There is also a coincidence of the heaviest belt of winter rain with the position of the high level jet stream, and the

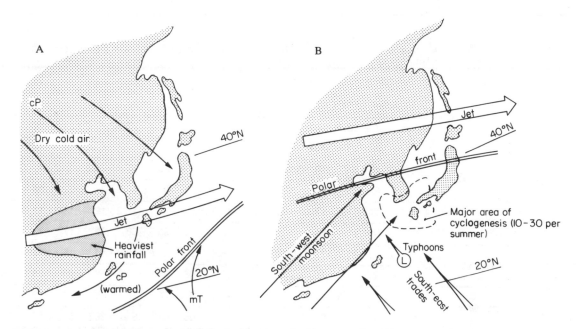

FIGURE 12:8 *Eastern Asia. Why does this area have most rain in summer whereas winter rain is restricted to a small zone? Notice the relationship between the jet, the so-called 'Polar Front' zone, and the zone of maximum cyclone formation.* (After Trewartha, 1961.)

jet fluctuates little in its position here. Small-scale disturbances entering from the south-west (i.e. from northern India) intensify as the two branches of the Asian jet join after separation around the Tibet-Himalayan massif, in an area of conflict between air types.

This cold, but occasionally disturbed and drizzly, weather continues into spring as late as April or May, when the low-level cold air stream is replaced by mT air. Westerly depressions become more frequent as the Siberian anticyclone breaks down and the northern branch of the jet stream intensifies. Nearly half of the annual total of depressions pass over China between March and June, and up to one-third of the annual rainfall total may come in these months.

In **summer** the westerly flow moves northwards, and southern and central China are dominated by a south-westerly monsoon flow extending into the area from the Indian Ocean and forming an extension of the flow across India (Figure 12.8(B)). The winds above this surface flow and up to 3000 m are weak, and much rain comes from thundery showers associated with weakly developed low pressure disturbances. These interrupt hot sunny weather. Later in the summer, storms may reach the land from the Pacific, and the most intense are of the hurricane variety, known here as typhoons. In the north there is considerable cyclonic activity along the 'Polar Front': cyclones are regenerated over the sea area to the south and west of Japan and bring heavy rains to the southern coastal areas of that country. The summer pattern eventually breaks up in October and the winter conditions are re-established.

b) **Eastern Australia** lies between a zone of moving anticyclonic cells, which reach it after travelling from the west across the continent and the eastern range, and the almost stationary east Pacific anticyclone. The contrast between the dry cT air from the continental area to the west and the moist mT from the north-east gives rise to disturbances as they meet, and these bring rain to the coast. There is not so much variation through the year as along other east coasts because there is no winter interruption to the circulation: Sydney receives 42 per cent of its total rain in the four wettest (autumn) months, and nearly 25 per cent in the four driest; Brisbane receives 50 per cent of its total in the four summer months, and reflects the greater summer concentration to the north. The rainfall depends on the frequency and nature of the disturbances, and can be extremely heavy; on the other hand there may be considerable periods of the year without any rain.

Station	Latitude	Height above sea level (m)		J	F	M	A	M	J	J	A	S	O	N	D
Algiers, Algeria	36° 46'N	60	T (°C)	12.2	12.8	14.4	16.1	18.9	22.2	25	25.6	23.9	20	16.1	12.8
			R (mm)	114	83	73	40	45	15	3	5	40	78	129	137
Beirut, Lebanon	33° 54'N	40	T	13.9	14.4	16.1	18.9	22.2	25.6	27.8	28.3	27.2	24.4	20	16.1
			R	182	162	89	57	15	3	0	0	8	48	126	188
Sacramento, California, USA	38° 31'N		T	6.7	10	12.2	14.4	17.8	21.1	23.9	22.8	21.1	17.2	11.7	7.8
			R	68	71	53	35	12	3	0	0	3	22	38	75
Valparaiso, Chile	33° 01'S	42	T	17.8	17.8	16.7	15	13.3	12.2	11.7	12.2	12.8	13.9	15.6	17.2
			R	3	3	5	17	96	129	86	68	30	17	8	3
Capetown, South Africa	33° 56'S	12	T	21.7	21.7	20.6	18.3	15.6	13.9	12.8	13.9	15	16.7	18.9	20.6
			R	17	15	22	48	93	108	93	83	58	40	25	20
Bombay, India	18° 58'N	12	T	23.9	23.9	25.6	27.8	29.4	28.3	26.7	26.7	26.7	27.2	26.7	25
			R	3	3	3	3	20	465	618	350	267	55	10	3
Hanoi, North Vietnam	21° 03'N	8	T	16.7	17.2	20	23.9	27.8	29.4	28.9	28.9	27.8	25	21.7	18.3
			R	22	35	45	91	218	259	340	340	268	111	50	28
Havana, Cuba	23° 08'N		T	22.2	22.2	23.3	25	26.1	27.2	27.8	27.8	27.2	26.1	23.9	22.2
			R	73	45	45	58	119	165	124	134	149	175	78	58
Acapulco, Mexico	16° 50'N	3	T	25.6	25.6	26.1	26.7	28.3	28.3	28.3	28.3	27.8	27.8	27.2	26.1
			R	10	0	0	0	30	428	215	246	360	170	30	12
Matogrosso, Brazil	15°S	275	T	25	25	25	25	22.2	21.1	20.6	22.2	25	25	25.6	25
			R	231	210	218	99	55	30	20	17	30	99	129	200

FIGURE 12:9 Mean climate statistics for regions with a wet season in the year.

c) **The temperate continental interiors.** These regions experience the most extreme temperature changes of anywhere in the world, and only occur in North America and Eurasia. At Verkhoyansk in Siberia an annual range between −70°C and 36.7°C has been experienced; and at Browning, Montana (USA) in the 24 hours from midday on 23 January 1916 the temperature ranged between 6.7°C and −48°C. In Eurasia the westerly flow of cyclones is not impeded by the relief, and precipitation amounts fall off as the disturbances penetrate further across the land surface: the deep interior of Siberia includes the increasingly arid steppes and Gobi desert. Without moisture, the cyclones have little effect on the weather and weaken until they reach the east coast of Asia, where they are regenerated by contact with the sea along the convergence line of contrasting air masses. For much of the winter season the surface anticyclone, which forms due to the excessive cooling, prevents the passage of such cyclones, although they occasionally break through and bring milder conditions to Moscow, and even to Siberia. During the summer the anticyclonic blockage is present less frequently, and the cyclones pass through. They are associated with the heaviest rains of the year. Once again it is the atmospheric disturbances, and not the simple local convection, which gives rise to the rain.

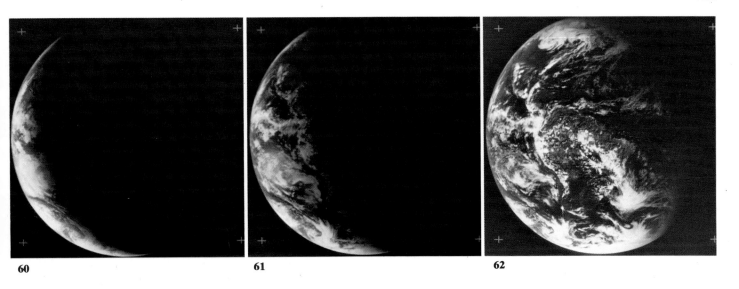

60 61 62

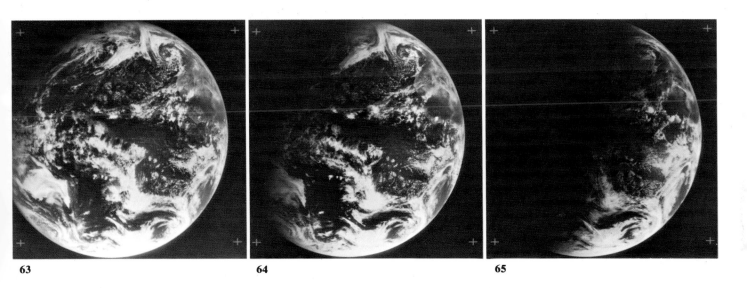

63 64 65

66

Earth from sunrise to sunset. A series of views taken from 07.37 to 23.30 hours on 18 November 1967 by ATS-III (Applications Technology Satellite), stationed 35680 km above the mouth of the river Amazon. How does the season of the year affect the duration and distribution of darkness in the northern and southern hemispheres? Compare the views of the Arctic and Antarctic areas. Notice how the cloud patterns change during the course of the day.
Plates 60-66 (All NASA)

67

68

The Earth and Moon. Contrast the features of the Earth and Moon in terms of the extent of ocean and atmosphere. Compare the landing methods used by the Apollo crews on the two bodies.

Plate 67 Airglow above the Earth, photographed by Skylab 3 in 1973. Airglow is luminescence caused by the dissociation of ions in the uppermost atmosphere due to reaction with sunlight. Clouds can be seen at lower levels.

Plate 68 The Moon surface with the Earth in the background, taken from Apollo 10.

Plate 69 The return to Earth of Apollo 14, 9 February 1971, splashing down in the Pacific Ocean 1000km south-east of Samoa.

Plate 70 Clouds cover 50 per cent of the Earth on average: the patterns are related to movements within the atmosphere set off by differences in heating by the Sun. The oceans, covering 71 per cent of the Earth's surface, show up as blue, and the land as brown. (All NASA)

Platforms of observation. Observations of weather and ocean characteristics are derived from a variety of sources.

Plate 71 The ground observer can record temperature, pressure, wind, sunshine and precipitation changes, but is limited by being confined to the base of the atmosphere.

Plate 72 The high-flying plane can take a variety of instruments to investigate characteristics (winds, temperatures, humidities) in the upper troposphere and lower stratosphere. At lower altitudes planes also explore rain clouds.

69

70

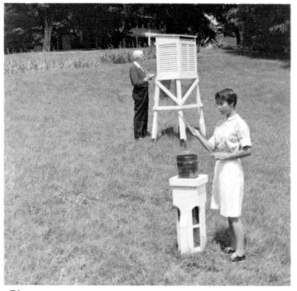
71

72

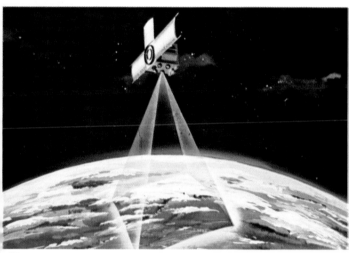
73

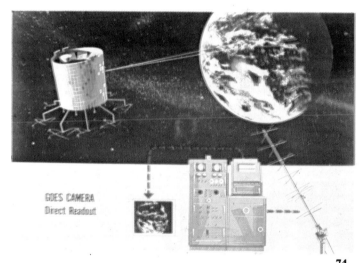
74

Plates 73/4 Weather satellites have been important since the first was launched in 1960. The first generation of those operating in the 1960's were known as TIROS (Television Infrared Observational Satellite), and were sent up in ESSA (Environmental Survey Satellite) spacecraft. These were Earth-orbiting satellites, and the second generation series for the 1970's is known as ITOS (Improved TIROS Operational System — **Plate 73**) providing world coverage every 12 hours instead of every 24. In addition the GOES (Geostationary Operational Environmental Satellite — **Plate 74**) will provide a continuous view above one point, enabling a watch to be kept for destructive natural events. (71-74 NOAA)

75

Plate 75 An Apollo 9 view of the Gulf Stream (March 1969). The clouds are above the current, demonstrating its offshore position as cold winds blow offshore and are warmed over the current. Notice also the sediment entering the sea through gaps in the offshore bars at Cape Hatteras, North Carolina. (NASA)

 76
 77
 78

 79
 80
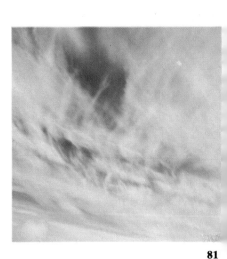 81

 82

Clouds

Plate 76 Small cumulus clouds formed by rising convectional currents over the Solent estuary, southern England. (Aerofilms)

Plate 77 Cumulus castellatus. (Aerofilms)

Plate 78 Cumulus humilis over south-eastern USA. Relate Plates 77 & 78 to patterns of atmospheric temperature structure and winds. (NOAA)

Plate 79 Apollo 6 photograph of stratocumulus over the Pacific Ocean. What are the differences on either side of a line running north-south through the centre? This may be due to the boundary of water masses. (NASA)

Plate 80 Altocumulus clouds. (Aerofilms)

Plate 81 Cirriform clouds. (Aerofilms)

Plate 82 Heavy rain falling from a cumulonimbus cloud. (NOAA)

Tornadoes and waterspouts. Three examples of tornado-type ('twister') features, which may cause considerable local damage.

Plate 83 Scotts Bluff, Nebraska, 27 June 1955.

Plate 84 Enid, Oklahoma, 5 June 1966.

Plate 85 Waterspout. (All NOAA)

Plate 86 World wind patterns. How can this satellite photograph be used in reference to the world patterns of wind and pressure? (NASA)

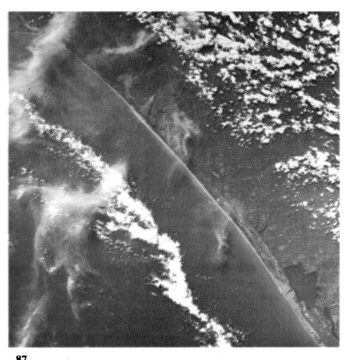
87

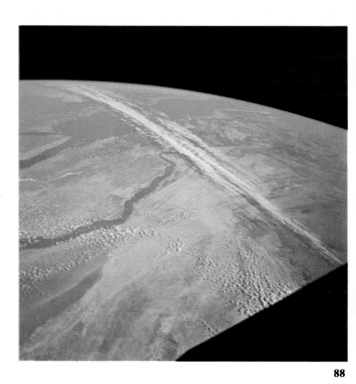
88

89

90

Movements in the atmosphere

Plate 87 Sea breezes along the West African coast, as seen from Apollo 6. The breeze from the Gulf of Guinea was strong enough to prevent cloud formation for 30-40km inland. The area shown extends just into Ghana in the north-west corner, and into Nigeria at the other end. Offshore another line of cloud is associated with the vertical pattern set off by sinking air. The wisps of cirrus cloud between are many thousands of metres above the lower clouds of the convectional circulation.

Plate 88 Clouds associated with high-level jet stream winds as they pass over the Red Sea and Nile Valley. (87 & 88 NASA)

Plate 89 The 1968 hurricane Gladys, seen from Apollo 7 about 250km southwest of Florida. Wind speeds were over 120km/hr at this stage.

Plate 90 Damage along the Gulf Coast of the USA caused by hurricane Celia.
(89 & 90 NOAA)

91

92

93

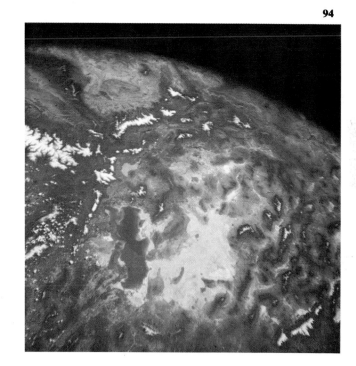
94

Satellite photos of climatic zones

Plate 91 Convective cloud in the Amazon basin (Apollo 9, March 1969). Several current thunderstorms can be seen with cloud tops rising through the general cloud cover to the tropopause. At the top of the photograph is one cloud mass with concentric rings, reflecting a series of updraughts spreading out as the rising air fails to penetrate the tropopause.

Plate 92 The Arabian desert (Gemini IV), looking south to the Gulf of Aden. The area is quite barren, but the pattern of valleys leading to the sand-clogged Wadi Hadramawi suggests that rainfall has been heavier in the past.

Plate 93 The deserts of South America (Skylab 3, 1973). In the foreground are the salt-flats of northern Argentina, whilst on the right the coast of Chile is covered by stratus cloud.

Plate 94 Salt Lake City, Utah (Skylab 1, 1973), showing the extent of the former Lake Bonneville by the white salt flats. These were formed by the evaporation of lake waters which were extensive during the glacial periods. The migration of climatic belts towards the Equator caused present desert areas to become humid. (All NASA)

95

Plants, nutrients and water

Plate 96 Various nutrient deficiency symptoms are shown by 1-year cuttings of poplar grown in sand. Lack of nitrogen (N) leads to a stunted plant with yellowing leaves, scarcely different from a plant with no nutrients at all. Without phosphorus (P) the plant is also short in stature and begins to lose its lower leaves. Without potassium (K) it is slightly taller, but its lower leaves are blotchy. Absence of magnesium (Mg) does not affect the height, but the leaves suffer an interveinal loss of colour.

Plates 97/98 Water in plants. The moisture content of an oak tree (*Quercus sp*) and a Douglas fir (*Pseudotsuga taxifolia*) in gallons. How does this affect the types of area where these trees can exist?
(All Forestry Commission)

Mountain climates

Plate 95 A view of the Alps of western Switzerland, with Lake Geneva clearly visible (Skylab 3, 1973). Notice the pattern of features on the folded Jura north of Lake Geneva, the cloud patterns over the central lowlands of Switzerland, the ice-capped peaks and glaciers, and the differences of land use in the major valleys. (NASA)

96

97

98

Clay in soils is important as a binding medium and in the mixture with humus.

Plate 99 Clayskins, rich in organic matter, show up as dark, glistening features in this Oregon soil. The white marker is 2mm long.

Plate 100 A photomicrograph of a soil from Washington, USA. The dark, egg-shaped mass is an iron oxide pellet, approximately 0.3mm long. It is surrounded by a thin layer of silicate clay, which also fills a pore to one side. The clear grains are mostly quartz and the voids are black.
(Both USDA Soil Conservation Service)

99

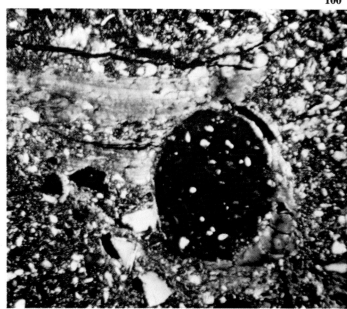

100

101

102

103

104

British soils. The first two photographs depict the most common soil types in Britain; the next three demonstrate varieties related to local conditions.

Plate 101 A humus-iron podzol in the New Forest, Hampshire, developed on Barton sand and some Plateau Gravel flints.

Plate 102 A forest brown earth profile at Weston Common, near Alton, Hampshire, developed on thick loamy material above Chalk. At the top the A-horizon has 5cm of dark brown loam with close grass roots, and below this there is 30cm of yellowish-brown fine loam with weak angular blocky structure and the roots of coppice sycamore. This rests on a B-horizon of reddish sandy clay loam. It is difficult to distinguish the different horizons, in contrast to the podzol.

Plate 103 A peaty gley soil with a mottled blue-brown appearance due to the presence of water.

Plate 104 A thin peaty soil on gravels on the Haldon Hills, Devon. In places the humus layer cements the gravels. Lower gravels have more clay between, and a gley-podzol results.

Plate 105 Up to 75cm peat over glacial drift in northern Scotland. Only poor heather grows over large areas of this type, which cannot be planted with trees.
(All Forestry Commission)

The US Soil Classification system. Relate Plates 106-126 to the outline classification in Figure 17.10. Can you suggest how these soil types can be fitted into the classification adopted in this book? Plates 106-108 illustrate three epipedons (surface horizons).

Plate 106 Mollic epipedon. A dark, thick surface horizon on calcareous silt loess soil in Nebraska; grassed; undulating relief. Scale in feet (1 foot = 30.5cm).

Plate 107 Umbric epipedon has a lower base status. This profile is developed on alluvium from a sandstone rock in Washington, USA, on a steep slope at 1500m. Black ticks on the scale are 15cm apart.

Plate 108 Ochric epipedon near Waiwera, North Island of New Zealand. Its light colour reflects a low organic content.
(All USDA Soil Conservation Service)

The US Soil Classification system: subsurface horizons.

Plate 109 Argillic horizon, starting at 25cm depth (scale in feet) and associated with a strong prismatic structure and extensive clayskins. This developed on silt loam in Oklahoma.

Plate 110 Natric horizon at 28-41cm depth (scale in feet) with prismatic and columnar structure below a mollic epipedon and albic horizon, and above a stratified substratum with salt efflorescence. Montana, USA.

Plate 111 Spodic horizon below an albic, having an upper black subhorizon (humus redeposition) and a lower redder one (iron).

105

106

107

108

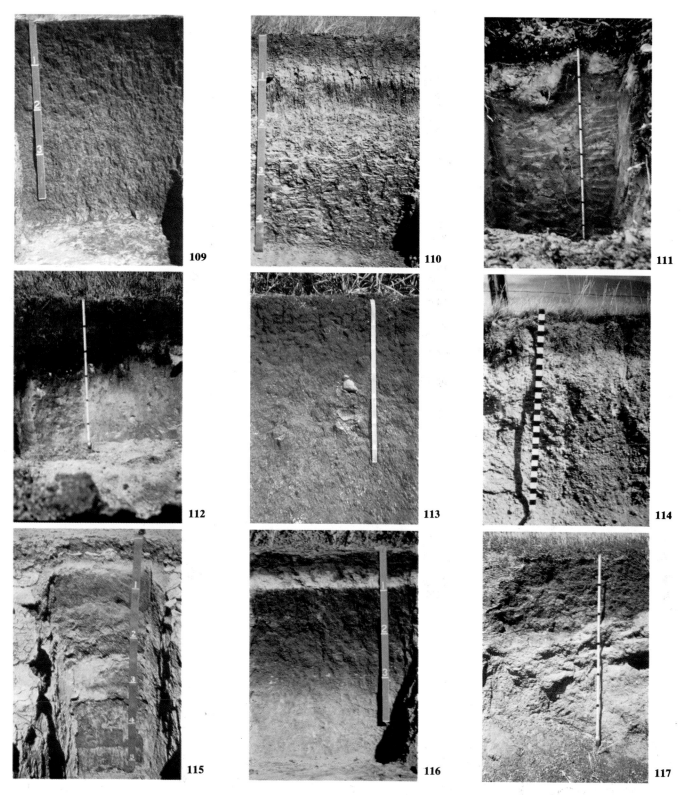

photographed in Minnesota. Irregular horizon boundaries may be related to the uprooting of trees. Ticks on the scale are every 15cm.

Plate 112 Cambic horizon occurs indistinctly below the mollic epipedon and above the parent glacial till in Manitoba, Canada, under grass. Ticks on the scale are every 15cm.

Plate 113 Oxic horizon between 36-170cm on a clay above lava flows on the island of Kaui (Hawaiian Islands). Scale in feet.

Plate 114 Calcic horizon below mollic epipedon and cambic horizon, developed on parent calcareous loamy and gravelly alluvium in Jefferson County, Montana. Scale in inches (1 in = 2.5cm).

Plate 115 Salic horizon, 10-36cm (scale in feet) in the silty clay loam of Clark County, Nevada. The gypsum content is high, and calcium carbonate is common throughout. When the soluble salts are removed from a sample its weight is reduced by half.

Plate 116 Albic horizon, 23-28cm (scale in feet), under mollic epipedon and above an argillic horizon on a silty loess in east-central Nebraska.

Plate 117 Duripan is a form of induration, or cementation, in soil layers: here it projects as a light-coloured zone 35-60cm deep (ticks on the scale are every 15cm). It has developed on an alluvium on granitic rocks near Riverside, California.
(All USDA Soil Conservation Service)

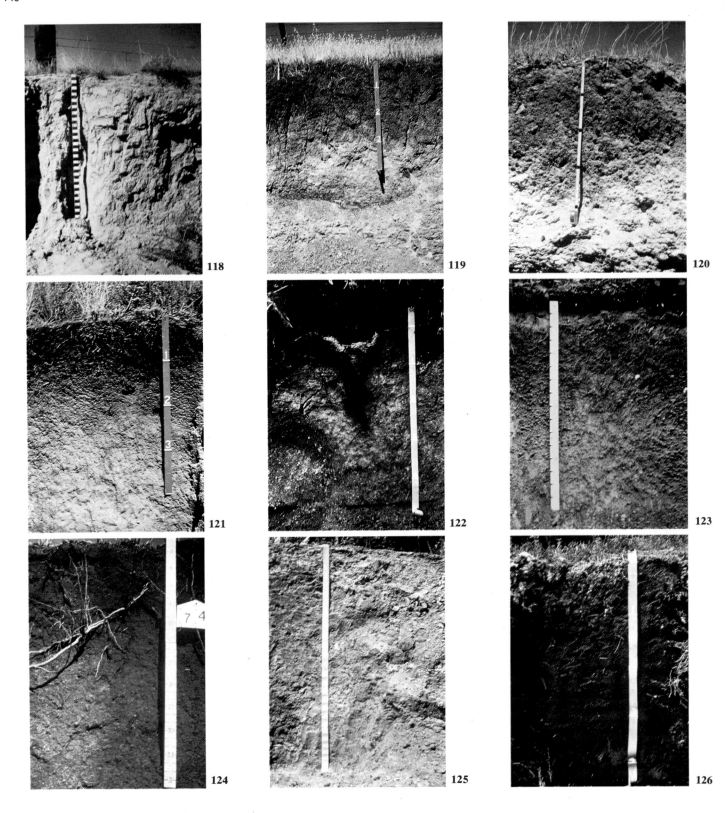

The US Soil Classification system: the major orders.

Plate 118 Entisols are weakly-developed soils. This one, on silt loess in south-eastern Wyoming, has a very thin ochric epipedon and lacks zonation beneath this. Scale in inches.

Plate 119 Vertisols develop on clays in dry regions: this one is found in the hilly country around San Francisco Bay, California. Carbonate concretions 1cm in diameter are scattered through the profile. Scale in feet. (Inceptisols, Plate 107, are developed over a short period of time, often on steep slopes.)

Plate 120 Aridosols are usually dry, like this one in Utah. A light-coloured ochric epipedon is underlain by reddish-brown prismatic argillic horizon and white massive calcic horizon. Scale ticks are every 15cm.

Plate 121 Mollisols have a mollic epipedon and a brownish cambic horizon over the greyish-brown parent silt loess. This one occurs in eastern Nebraska. Scale in feet.

127

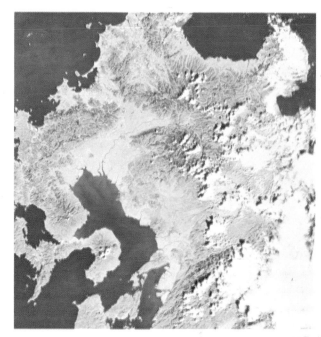

128

129

Seasonal biome types

Plate 127 A false-colour photograph of central California, taken from LANDSAT-1 on 26 July 1972. Areas of growing crops (red rectangles) can be distinguished from unused fields (blue/grey), and these from the less geometrical areas of woodland vegetation on the coastal ranges and bare rock and scrub inland. Cloud and fog covers the San Francisco/Golden Gate area. The central area is still partially flooded after heavy rains earlier in the year.

Plate 128/129 Japan and Florida. Two views from LANDSAT-1 taken in the summer/autumn of 1972. They show two contrasting areas on the eastern sides of continents. Both are false colour photographs, in which growing vegetation shows as pink/red; built-up areas as blue/grey; clouds as white. Compare the built-up and farmed areas, the cloud patterns over land and sea, and the depth of water offshore. Notice also the area of cypress swamp in central Florida and the wooded mountain slopes in the area of Japan around Nagasaki. (All NASA)

Plate 122 Spodosols have a thick (30cm) upper horizon, a thin white albic horizon, and a prominent spodic horizon with over 10 per cent organic matter in the upper part. This soil is on Kruzof Island, southern Alaska, and the parent material includes volcanic cinders and ash. Scale in feet.

Plate 123 Alfisols have argillic horizons and moderately high base contents. This specimen, developed on a late-glacial till in northern Minnesota, has a thin albic horizon. Others in warmer climates are redder. Scale divisions are every 7.5cm.

Plate 124 Ultisols have argillic horizons with a low base status (i.e. medium to strongly acid, low pH). This example is on clay loam in Georgia, and is 1m thick (scale in inches). The surface horizon has a granular texture, whilst the lower part is blocky.

Plate 125 Oxisols have a great variety of types: this one occurs on Kaui, Hawaii, and has hardened plinthite at the surface. The underlying oxic horizon extends for 2m and grades into the parent basalt lava.

Plate 126 Histosols are organic soils. All the material seen is peat, derived from the edge of 'muskeg' (sphagnum bog) in south-eastern Alaska. Scale in feet.

(All USDA Soil Conservation Service)

130

131

132

133

British woodlands have less natural variety than other temperate regions, but planting has increased this: many non-native trees will grow in British conditions. Notice soil/relief associations.

Plate 130 A stand of 150-year old beech trees (*Fagus sylvatica*) on Clay-with-flints on the Chiltern Hills.

Plate 131 A pine-birch-oak association on sandy soils near Bordon, Hampshire, with *Calluna* in the foreground colonising a burnt-over area.

Plate 132 Broadleaf and conifer growing together — an oak (*Quercus robur*) and a redwood (*Sequoia sempervirens*) showing a contrast in form.

Plate 133 A spruce plantation in Surrey.

Plate 134 A stand of large fir trees (*Abies grandis*).

Plate 135 A large elm tree (*Ulmus carpinofolia*), typical of many field hedgerows in England, but now dying out through disease. (All Forestry Commission)

134 **135**

Australasian woodlands include distinctive trees and associations.

Plate 136 Varieties of eucalyptids typical of Australia (*E. perrenianna* near camera, and *E. vernicosa*).

Plate 137 *Nothofagus obliqua* (southern beech) occurs in South Island, New Zealand, Tasmania, and also in southern Chile.
(Both Forestry Commission)

Arid biomes

Plate 138/139 Two views of desert vegetation in Nevada and the Sonoran Desert near Tucson, Arizona. Relate the covering of vegetation to soil conditions and relief.
(USDA Soil Conservation Service)

Plate 140 Apollo view (1970) of the Dead Sea (centre) and adjacent parts of the Middle East. What can be seen concerning the nature of surface features?

Plate 141 Southwestern USA from LANDSAT-1, 27 September 1972. The Colorado river crosses the area from north-east to south-west. The barren valley sides contrast with the irrigated farmland of patches along the valley floor, and the woodlands on the upper mountain slopes. (NASA)

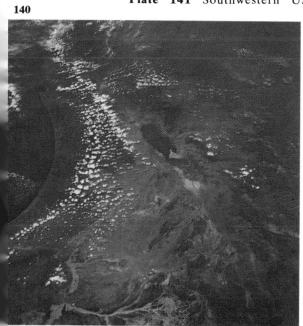

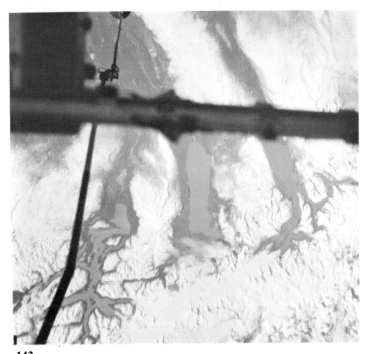

142

143

144

Cold lands and the oceans

Plate 142 The southern Andes in the fjord area of Chile: Skylab 3, mid-1973.

Plate 143 Northwest Territory of Canada by LANDSAT-1, summer 1972. An outlet of the Mackenzie river, Cape Bathurst and Arctic sea ice (largest mass 8/13km) can be seen. Meandering rivers cross the tundra, and a mass of lakes are present. Parallel islands in the Mackenzie are former terminal moraines. The pinks (rather than red hues) show the slow plant growth.

Plate 144 The Bahamas, LANDSAT-1, summer 1972. Grand Bahama, Great Abaca and Cay Berry islands (red colours) are seen with the surrounding reefs and shallow waters. Fine weather cumulus clouds in the sky. (All NASA)

Station	Latitude	Height above sea level (m)		J	F	M	A	M	J	J	A	S	O	N	D
Kano, Nigeria	12° 02'N	500	T	21.7	23.9	27.2	30.6	30.6	28.9	26.1	25	26.1	27.2	25	22.2
			R	0	0	3	10	68	111	205	310	142	12	0	0
Nairobi, Kenya	1° 17'S	1800	T	20	21.1	21.1	20.6	19.4	18.3	17.2	17.8	18.9	20	20	20
			R	35	40	114	210	129	45	12	45	25	53	114	68
Raleigh, NC, USA	35° 52'N	140	T	5	6.1	10	15	19.4	24.4	26.1	25	22.8	16.1	10	5.6
			R	83	83	91	88	86	103	139	126	114	72	74	83
Shanghai, China	31° 12'N	8	T	4.4	5	8.9	14.4	19.4	23.9	27.8	28.3	23.9	18.3	12.2	6.7
			R	48	58	83	93	93	180	146	141	129	71	50	35
Tokyo, Japan	35° 41'N	6	T	3.3	3.9	7.2	12.8	16.7	20.6	24.4	25.6	22.2	16.1	10.6	5.6
			R	58	76	109	131	150	170	141	180	256	200	89	55
Port Macquarie, NSW, Australia	31° 38'S	20	T	22.2	22.2	21.1	18.3	15.6	13.3	12.2	13.3	15	17.2	19.4	21.1
			R	139	177	162	165	141	119	109	83	96	88	93	126
Prince Albert, Saskatchewan, Canada	53° 13'N	450	T	−20	−16.7	−9.4	2.8	10	15	17.2	15.6	10	3.9	−6.7	−15
			R	17	15	22	22	37	72	55	50	52	20	22	20
Moscow, USSR	55° 46'N	160	T	−9.4	−8.3	−3.9	3.9	12.8	16.7	18.9	16.7	11.1	4.4	−2.8	−7.8
			R	37	35	28	48	55	73	76	73	48	68	43	40
Yakutsk, USSR	62° 01'N	115	T	−43.3	−36.1	−22.8	−8.3	5	15.6	18.3	15	6.1	−8.3	−28.9	−40
			R	5	5	3	5	12	28	32	40	22	12	10	8
Vladivostock, USSR	43° 07'N	30	T	−13.9	−10	−3.3	4.4	9.4	13.9	18.3	20.6	16.7	9.4	−0.6	−9.4
			R	8	10	20	30	53	74	83	121	114	48	30	15

FIGURE 12:9 *Continued. Mean climate statistics for regions with a wet season in the year.*

In North America the western cordilleras (Rocky Mountains, Coast Range, Sierra Nevada, etc) form a barrier to the passage of the cyclones at all seasons and result in a marked rain shadow area to the east. Cyclones begin to reform, however, over the prairies and are particularly intense when the cold polar air is brought into contact with the warm, moist air channelled up the Mississippi Basin in summer. Summer is the main rainy season in these regions. Like most summer rainfall in the middle latitudes it is often thundery in nature because the lapse rates are so high at this season and instability is easily induced. There may be long sunny periods between the heavy showers of rain. In winter the belt of disturbance and rains is forced southwards by the development of the midlatitude continental anticyclone with its pool of cold polar air and out-blowing winds.

13

Polar and mountain climates

There are two distinctive zones which do not fit into the general basis of classification used for world climates in this book. The climates of the ranges of young folded mountains like the Alps, Himalayas and Andes are modified so greatly by altitude, increased windiness and variable aspect that they must be considered as special cases. In addition the coldest parts of the Earth encircling the polar regions are dominated by low temperatures, extensive ice coverage and snow precipitation. In both zones, factors apart from the annual rainfall regime loom large.

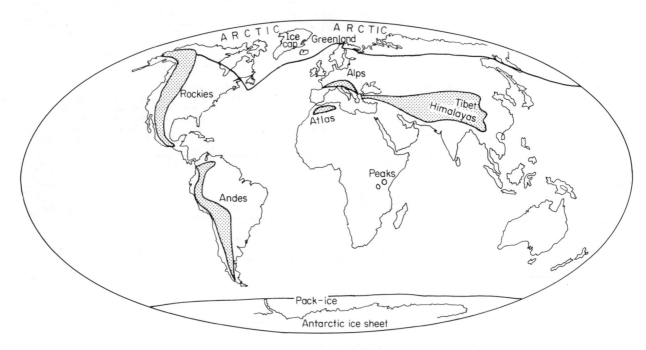

FIGURE 13:1 *The world distribution of mountain and polar climatic zones, and mean statistics of four stations.*

Station	Latitude	Height above sea leve; (m)		J	F	M	A	M	J	J	A	S	O	N	D
Mucuchies, Venezuela	8° 45'N	3250	T	10	11.1	11.7	11.7	11.7	11.1	10.6	10.6	11.1	11.1	10.6	10.6
			R	8	17	20	96	118	121	116	96	63	68	32	10
Point Barron, Alaska, USA	71° 23'N	8	T	−26.7	−27.2	−26.1	−17.8	−6.7	−1.1	1.7	4.4	3.9	0	−8.3	−23.9
			R	5	5	3	3	3	8	22	17	12	15	8	8
Chelyuskin Mys, USSR	77° 43'N	8	T	−25.6	−24.4	−28.3	−21.7	−9.4	−1.1	1.7	1.1	−1.7	−10	−16.9	−24.4
			R	3	3	3	3	3	17	28	28	10	8	5	3
Thule, Greenland	76° 30'N	8	T	−29.4	−29.4	−26.1	−17.2	−5	1.7	4.4	3.9	−2.2	−10	−17.8	−25
			R	3	3	5	3	5	8	20	17	8	3	8	5

Polar climates

Two major zones may be recognised within this grouping.

The **Arctic climate** occurs in a zone of intense atmospheric activity between the cool temperate climates and the ice cap. Weather records, however, tend to be restricted to the period since World War II. Before that time these regions were distinctly unattractive to human settlement and fewer than 50 weather stations existed, but strategic as well as scientific considerations have resulted in the establishment of another 200 since 1945.

The Arctic zone has long, dark winters in which there is a marked loss of radiant energy. Intense cold affects the area with mean temperatures falling to as low as $-30°C$ in January, and there are fewer weather system disturbances.

The summer is short, and the Sun's rays still received at a low angle, although there is almost 24 hours of daylight. Much solar energy is used in melting the winter snow cover and mean temperatures do not rise above $10°C$ in July. Temperature ranges around the Bering Sea and the northernmost parts of Scandinavia and eastern Greenland, where oceans penetrate, are much lower, and these areas receive more precipitation and fog. The more disturbed summer conditions, at the meeting of cold and warm air, also affect the coastal zones to the greatest extent. Elsewhere, continents and mountain barriers prevent warmth and moisture from reaching this zone and there are greater extremes of temperature and low precipitation totals (Figure 13.1). Such areas of long frozen winters, low precipitation totals, and intense atmospheric disturbance, experience snow falling in small amounts but lying for long periods. The snow is blown about by strong winds. The windiness emphasises the aridity of these regions caused by low precipitation figures and the frozen winters.

The **polar ice cap climatic zone** has the lowest average annual temperatures in the world: no mean monthly temperature is greater than $0°C$. Records for these areas are even more sparse than are those for the Arctic zone: whilst increasing attention has been given by expeditions, men do not remain for long enough to obtain records having the same status as those for other parts of the world.

In the northern hemisphere there is a contrast between the climate of the Greenland ice cap and the Arctic Ocean (Figure 13.2). Mean annual temperatures in central Greenland range from $-30°C$ to $-35°C$, but only from $-12°$ to $-4°C$ around the North Pole. The Arctic Ocean is covered by relatively thin pack-ice through which heat is transmitted from the underlying ocean waters especially in late summer when up to half the ice melts. At this season, with the temperatures rising to

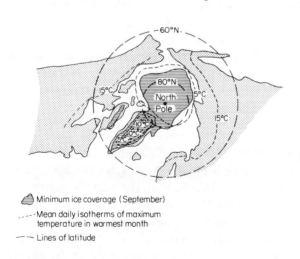

FIGURE 13:2 *The Greenland ice-cap and the Arctic in summer. Whilst the warmest day temperatures are well below 0°C on Greenland, they rise well above this over most of the Arctic Ocean at this time, leading to the melting of much pack-ice. (After Rumney, 1968.)*

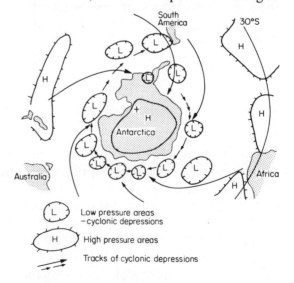

FIGURE 13:3 *The Antarctic climate. The distribution of cyclonic depressions and high pressure areas at 6 a.m. on 17 February, 1958. The arrows show the regular paths taken by depressions in that year. (After Rumney, 1968.)*

to over 0°C, there is some intrusion of cyclonic disturbances and precipitation may be in the form of rain. Temperatures fall in the long winter darkness when the pack ice increases as the ocean surface freezes fast: the air becomes cold, dense and extremely dry. On Greenland the ice cap reaches over 3000 m high and there is mountainous relief along the margins. In the interior, the air becomes very cold and dense over the ice and outward-moving air prevails for much of the year. But Greenland extends southwards into the belt of cyclonic disturbances, and these bring precipitation to the ice cap which may be as much as 3 m snow (i.e. equivalent to 310 mm rain). This is difficult to measure, since winds may whip up blizzards from the snow already lying on the ground. The disturbances bring rain to the coasts of Greenland in the summer months.

The climate of Antarctica, an area of upland plateau with two-thirds over 2000 m and six times as large as Greenland, has been studied by a handful of weather stations set up in the late 1950's. Temperatures as low as −87.4°C have been recorded in the interior, making this the world's coldest place. This immense mass of ice-covered land, surrounded by unfrozen oceans, has its own distinctive climate. Little warmth is received from the Sun even in the summer, and the climate is a more severe version of that experienced on the smaller Greenland ice cap. The cold dense air is associated with anticyclonic conditions on the ice cap, but the margins are affected by cyclonic disturbances intensified by the contrast in surface conditions provided by the surrounding ocean (Figure 13.3). It is virtually impossible to measure precipitation in the interior because of the high winds and ease with which snow is picked up again: there are few periods of calm. Measurements of ice levels and layers suggests that present precipitation totals are very small and that the ice is moving very sluggishly outwards, if at all.

The air movements aloft over Antarctica, as measured from cloud movements and balloons, seem to contrast with those at the surface. At higher altitudes there are regular incursions of somewhat warmer air, related to cyclonic developments to the north. These bring some snow to the central regions and are an important source of heat exchange.

The polar regions are thus areas of extreme cold and difficulty for man. They are also the source region for cold air masses, and much present climatological research is being concentrated on the effect which the variations in the melting of continental ice, or ice in the Arctic Ocean, have an atmospheric circulation. Thus the Arctic Ocean pack ice reflects a large proportion of the summer insolation, which would be absorbed by the ocean waters if the ice was removed. This would affect the world climates by reducing the temperature contrast between equator and poles, and would lead to decreasing atmospheric activity, greater warmth and less precipitation. The deserts would extend their area.

Climatologists now see the atmosphere as an integrated system, rather than a series of clear-cut and isolated regions. Most is still known about the features of middle latitude climates, but it is now seen that the keys to further understanding lie in the study of the zones at each end of the atmospheric circulation system — around the equator and poles.

Mountain climates

It is difficult to make generalisations about mountain climates. Mountain ranges extend through every major climatic region on the continental surfaces (Figure 9.2): the Andes, for instance, cross seasonal tropical, desert, equatorial, and temperate rainy areas. In addition, the variations in relief give rise to rapid alternations of climatic types: mountains, more than any other type of region, are subject to microclimatological changes. The major factor which mountainous regions have in common is that of altitude, and this affects the climate in a variety of ways.

1) The **temperature** decreases with increasing height, so that a range of temperature from the heat of equatorial lowlands to regions of permanent snow may be encountered as one ascends the Andes in Ecuador, or as one travels from Mombasa to the top of Mount Kenya. The snow line is the lowest limit of permanent snow on a mountain. Figure 13.4 shows that it is highest a little way from the equator and then declines towards the poles, reaching sea level at about 75° North and South. Nowhere in the British Isles is high enough to reach the snow-line, which averages about 1700 m in

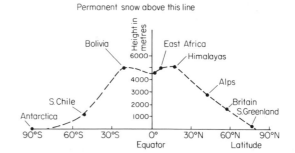

FIGURE 13:4 *The snow line. Why should it be lower at the equator than at 10-20° North and South?*

these latitudes, but high sheltered hollows retain snow for many years in areas like the Cairngorm Mountains of Scotland. The greatest number of changes due to altitude are found in the tropics, where there is the greatest difference between sea-level temperatures and the temperatures above the snow-line.

2) The **atmospheric pressure** also decreases with altitude. This makes it more difficult for people to obtain sufficient oxygen, but also means that the mass and density of the atmosphere are lower, allowing more heat to escape instead of being absorbed by the atmospheric gases. This is also helped by lower proportions of water vapour and atmospheric dust particles: at 2000 m the air can hold half as much water vapour as at ground level, and a quarter as much at 4000 m. Thus the decreasing temperature and pressure are connected closely. The rarified air at height gives rise to greater diurnal temperature ranges by allowing greater incoming radiation as well as greater outgoings. Figure 13.5 shows the effect of this as one ascends the Andes in southern Peru.

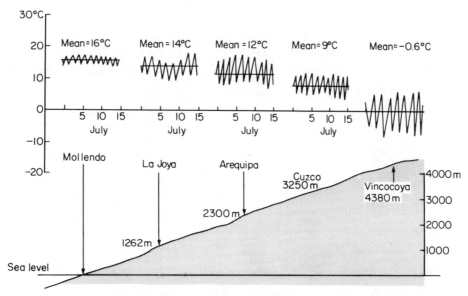

FIGURE 13:5 *Temperature conditions and altitude in Peru. What is the effect of increasing altitude on mean temperatures and on daily fluctuations of temperature? (After Rumney, 1968.)*

3) The **winds** increase in strength with height as the effect of surface friction is diminished (Figure 13.6). The winds at high altitude are not only more frequently strong, but are also stronger than the strongest winds experienced lower down, and blow less gustily.

4) **Rainfall** also tends to increase with altitude on the windward side of a mountain range. Air is forced to rise and the water vapour will condense to cloud and then may give rise to precipitation. Maximum precipitation occurs at approximately 3000 m, above which there is less, so that the upper parts of very high mountains are increasingly arid. In Britain there are higher areas on the west coasts, facing the humid air arriving from the Atlantic, but none rise above 1500 m, and the maximum rainfall is received some distance inland (e.g. at the foot of Snowdon in North Wales, or at

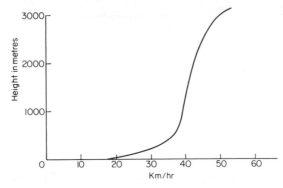

FIGURE 13:6 *The increase of wind speed with height on a mountain. (After Rumney, 1968.)*

Seathwaite in the Lake District). A reversal of this effect occurs on the lee slopes of mountain ranges, where the air is warmed and dried as it descends giving rise to increasingly arid conditions downslope (cf. Patagonia, chapter 11). The snow-line lies up to 1000 m lower on the windward side of the Andes than on the leeward.

Cloudiness is related closely to this pattern of events with high percentages on middle and lower windward slopes and little on the higher windward slopes or in the lee, although wave clouds are associated with the lee side and may form a canopy for several days (Figure 4.21). Cloud formation is also associated with the large diurnal range in mountain valleys, where inversions are established at night-time and mornings are often cloudy or foggy in the valley floors. At certain levels, regular hill-fog and high relative humidity is a characteristic of the mountain climate.

Whilst these effects of increasing altitude affect profoundly the climate of mountain chains as a whole, the individual valley environment is subjected to other influences which may override and at least interfere with these altitude-controlled patterns. The strongest factor is the interrelationship of the varying angle of the Sun's rays and the relief features of the valley. One side of an east-west valley will be mainly in the sunshine whilst the other will be mainly in the shade; there will be a daily regime of upslope (anabatic) and downslope (katabatic) winds (Figure 5.9); and there may be a variable pattern of valley-floor fog and cloud.

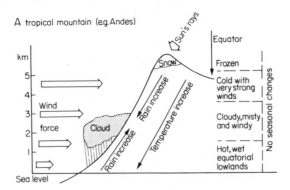
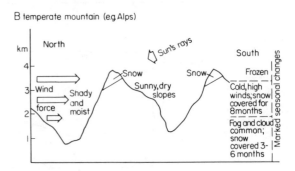

FIGURE 13:7 *What are the main differences between mountain climates in (A) tropical regions, and (B) temperate regions?*

The main feature of mountain climates is thus an extreme complexity of microclimates, together with increasing exposure to high winds and temperature variations. Figure 13.7 summarises some of the effects and variations experienced in the mountainous areas of the tropics and temperate latitudes. This serves to emphasise a final point. Whilst the average temperatures fall with altitude, as they do with latitude, the mountain region will retain a similar seasonal regime to the surrounding area. Thus the mountains of equatorial zones will not have summer and winter, whilst there will be a great seasonal contrast in midlatitude mountains like the Alps with mean annual temperature ranges of over 25°C in the valley floors and over 20°C at greater heights. The diurnal temperature range is much more important than the annual range in equatorial mountains.

14

Climates of the past — and future?

A consideration of the distribution of climates in the past is an important part of any account of the atmosphere and its workings. The evidence, however, is often of a different type from that which is used in meteorological and climatological studies. Several distinct lines of evidence can be followed, but they are valid in different ways.

1) **Meteorological evidence** covers a very short time: records of temperatures and rainfall go back no further than the mid-seventeenth century, and are sparse before the mid-nineteenth for even the European countries where observations began; elsewhere records may still cover less than 50 years.

2) Written **literary and historical records** enable certain details of climatic fluctuation to be reconstructed from travel accounts, high flood levels in certain rivers and good vintage years. Such records are again limited largely to countries with a history of civilisation and writings, but extend the evidence to a wider range of time and places.

3) Dating methods using the **radioactive iostope**, carbon-14, have proved useful especially over the postglacial period of the last 15 000 years. Such dates can be related to pollen sequences found in peat bogs.

4) The Pleistocene Ice Age provides a further range of evidence which can be used to reconstruct past climates. The most important evidence occurs in the microscopic **pollens** preserved in peat bogs, especially where these deposits occur between layers of boulder clay. River terrace deposits formed around the margins of ice sheets may contain **fossils** of animals which can be related to particular climatic types.

5) For more ancient climates (i.e. over 2 million years ago) it is necessary to go to the **geological evidence** provided by the limestones, sandstones and shales of the sedimentary rock sequence. It is fairly certain that a particular rock-type was formed in a desert, on a coral reef, beneath an ice sheet or in a humid tropical swamp, but such approaches necessitate talking in terms of fluctuations over a period of millions of years rather than a few tens of years. It is also important to bear in mind the theory which suggests that the continents have moved around on the Earth's surface: if this has been the case, then the search for past changes of climate will be made more difficult. Another source of evidence can be used, however. The ratios of oxygen isotopes in calcium carbonate sediments formed of planktonic shell matter reflects the surface ocean temperatures at the time of formation.

The historical past

The examination of rainfall or temperature records for the limited areas of the world which possess them over periods of 50 or 100 years reveals definite fluctuations which often show some regularity of form and repetition (Figure 14.1). Attempts have been made to link this type of variation with

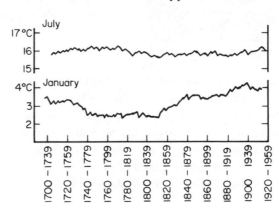

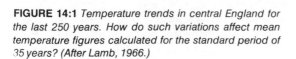

FIGURE 14:1 *Temperature trends in central England for the last 250 years. How do such variations affect mean temperature figures calculated for the standard period of 35 years? (After Lamb, 1966.)*

causal factors. Sunspots (bursts of solar energy) are thought to be responsible for some variation in the solar radiation (although it is most probably a matter of quality, increasing the small amounts of very short wavelength radiation — ultraviolet, x-rays and gamma rays — rather than the quantity). Sunspots occur in a regular pattern every 11 years (Figure 14.2), and this can be compared also with rainfall trends. The results show some degree of correlation, but the making of such simple and straightforward comparisons ignores factors which may be of greater or equal importance. The suggestion has also been made that the sunspot cycles showed greater correlation with climatic elements in the nineteenth century than they do now (see the Lake Victoria chart, Figure 14.2), although no explanation is available for this.

Records over a longer period of time are dependent on written historical records rather than meteorological observations (Figure 14.3). Deductions have been made from studies of tree rings in western North America, the extent of Arctic sea ice, the high levels of the river Nile and Hwang Ho flood waters, and the quality and quantity of European wine vintages (i.e. good quality and quantity together are associated with a lack of spring frosts and unusually long, sunny summers in Burgundy). It can be shown from a variety of records like these that the years between 800 and 1200 AD were relatively warm: Icelandic coasts were ice-free, sheep thrived in Greenland and grapes in England, the sea level was perhaps as much as 80 cm higher than today and the Viking wanderings in high latitudes were made easier by the mild conditions. From 1300 to 1600 AD the weather became colder in north-western Europe. Arctic sea ice blocked Icelandic coasts for half the year and reached the Shetland Isles, whilst Alpine glaciers reached their maximum extent, the sea level may have been 200 cm lower than today, Norwegian farmers were forced to abandon the more exposed lands and the river Nile experienced unusually high floods.

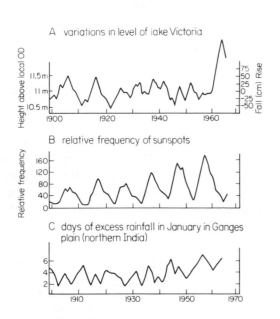

FIGURE 14:2 Can (A) and (C) be related to (B) in any way? What other factors may be involved? (After Flohn, 1969.)

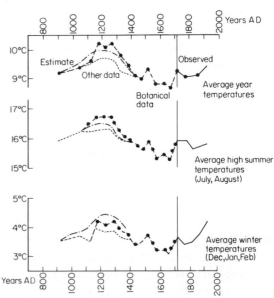

FIGURE 14:3 Temperature fluctuations in central England from 800 AD. Note the varying bases on which the evidence is reconstructed, and the scale of the changes involved. (After Lamb, 1966.)

Such records demonstrate a correlation between the extent of sea ice, the violence of the cyclonic storms and the severity of the winters experienced in north-western Europe. Figure 14.4 plots the variations in terms of summer and winter weather, traced across Europe. This diagram shows that phases of climate moved across Europe from west to east, reflecting the dominance of cyclonic or anticyclonic weather. Thus the main tracks of July depressions in the warmer phases (1100, 1940 AD) were through Iceland towards north Norway, and along the Mediterranean, whilst in the colder phases such as that around 1600 AD they crossed northern Britain.

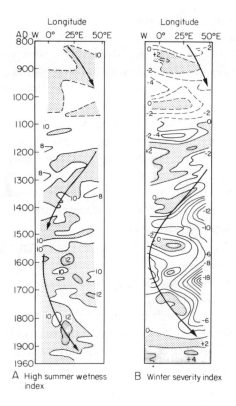

FIGURE 14:4 General trends in the weather of Europe at 50°N.
(A) The high summer wetness index, calculated by adding wet (i.e. more rain than average) July and August months per decade: unremarkable decades average 10; extreme values range from 4 to 17 in Europe.
(B) The winter severity index, calculated from the excess of mild or cold months, December, January and February: an excess of cold months is recorded as negative; unremarkable decades score 0; extreme decade values from = 10 to −20 in Europe.
What do the arrows indicate? (After Lamb, 1966.)

It is important to realise, however, that many of these fluctuations were merely of regional, rather than of global extent. Most records before 1800 come from the European area. It would be wrong for anyone in Britain to suggest that the world's climate was getting colder just because that country experienced a series of cooler-than-average summers or more severe winters. Changes from what is considered to be the routine pattern in one area are often balanced elsewhere. Thus the very cold winter of 1962-3 in north-west Europe was due to a persistent blocking anticyclone over eastern Europe; it was also associated with unusually high falls of rain in southern Europe and north Africa, plus a mild spell in eastern Asia. A form of variation and adjustment is at work in the atmosphere, and it seems to work in a cyclic fashion, repeating phases of weather over periods of a number of years. Smaller cycles of shorter duration are superimposed on longer cycles.

The cyclic pattern of climatic variation has practical results. Farmers tempted to settle in semi-arid regions during a series of wetter years have been broken by the succeeding dry years. Such droughts have occurred in the interior plains of the USA during the 1930's, in north-eastern Brazil, and in Kazakhstan (USSR) in the 1960's. The late 1960's and early 1970's saw an extension of the southern margin of the Sahara, bringing famine to countries like Mali and Tchad. A warming of the climate of southern Greenland over the years 1900-1945 led to the eskimos taking up sheep-farming and fishing instead of the traditional hunting, and enabled shipping to use Spitsbergen for seven months in the year, but more recently the climate of these areas has been worsening and causing hardship. The most obvious effects of these changes are felt near the limits of tolerance on the borders of lands which are too dry or too cold to attract settlement, or near the margins of economic crop cultivation. There is evidence, for instance, of widespread vine cultivation in south-east England in the warm phase from 1100-1300 AD; French growers objected to the competition, but after 1300 the worsening conditions led to the abandonment of English vineyards.

Is the climate of the Earth as a whole changing? This would mean that there would be some noticeable change in the total energy budget and therefore in solar radiation. The most recent historical evidence is conflicting, and most of the longer-term information is based merely on one region, north-west Europe. Thus, measurements between 1880 and 1941 indicated that the average temperature of the atmosphere and the surface ocean waters was rising at a rate of 1°C per 100 years, the sea-level was rising at 12 cm per 100 years, and the Alpine glaciers were shrinking at a rate which would suggest that the atmosphere was getting warmer. And yet it must be borne in mind that mountain glaciers like those in the Alps make up less than 1 per cent of the total land ice, and that the Greenland ice was not retreating, whilst that on Antarctica (90 per cent of the world total ice mass) was advancing. The Antarctic increase resulted in the freezing of a greater volume of water than was being released by the melting Alpine glaciers. It has been suggested that the ocean level rise could be due to density changes as the waters are warmed by solar energy and circulation, and that the glacial melting is a merely regional phenomenon. In short, the sea level and glacial fluctuations are not necessarily the ultimate criteria for assessing the nature and causes of climatic changes. The evidence examined here suggests that climatic changes in historical times are part of a cyclic pattern. Present knowledge is insufficient to account for such a phenomenon.

The geological past

Extensive areas of Europe and North America have landforms and rocks which record the fact that they have been covered by ice so recently that rivers have not had the opportunity to remould the landscape in a major way. The margins of deserts show that running water was once much more important in carving the landforms of these areas than it is today: fossil river valleys in Arabia are now filled with drifting sands. Such features indicate that the more recent geological past (i.e. the last few hundred thousand years) experienced climatic changes of more extreme magnitude than those recorded over the last few thousand years.

Period	Epoch	Stage	Pollen indication	Geological evidence in Britain	General situation
Quaternary	Holocene 10 000 years ago	Flandrian	Temperate		Post-glacial warming (see chapter 19)
	Pleistocene	Weichselian (= Devensian)	Cold	Glacial deposits, periglacial features	Last glaciation
		Ipswichian	Temperate		Interglacial
		Gippingian (= Wolstonian)	Cold	Glacial deposits, periglacial features	Glaciation
		Hoxnian	Temperate		Interglacial
		Lowestoftian (= Anglian)	Cold	Glacial deposits periglacial features	Glaciation
		Cromerian	Temperate		
		Beestonian	Cold	Periglacial conditions	Climate deteriorating in Britain, with earlier glacial periods elsewhere
		Pastonian	Temperate		
		Baventian	Cold	Periglacial conditions	
		Antian	Temperate		
		Thurnian	Cold		
		Ludhamian	Temperate		
	1.7 - 2 million years ago	Waltonian	Cold		
Tertiary	Pliocene				

FIGURE 14:5 *An outline of Pleistocene history. Notice the number of fluctuations in the climate within the period of time covered by the Pleistocene epoch. The British stages are given. (After Sparks and West, 1972.)*

The sediments deposited by the ice sheets, and the winds and rivers acting around their margins, left behind evidence concerning the environment of formation and the sequence of events which took place. The evidence these sediments now contain includes the fossil remains of animal life and microscopic pollen, which has been used as a good climatic indicator. It backs up the landform evidence concerning the Pleistocene Ice Age and the subsequent changes. Some of the main features of the history of the environment during this phase, and the implications to notice (Figure 14.5), are:
1) It involved a series (at least 4) of advances and retreats of the ice front, with consequent extension and reduction of the ice masses centred over north-eastern North America and the Baltic Sea.
2) As the ice advanced in the glacial periods, the sea level fell. During the last glacial period it was lowered to over 100 m below the present.
3) In the warmer interglacial periods the ice melted, and at times almost disappeared from the Earth. This melting was accompanied by rising sea-levels, which reached over 100 m above the present level in the most extensive periods of melting.
4) The world climatic belts were altered by the advance and retreat of the vast ice domes. As the ice sheets extended across northern Europe and the high mountain ranges of southern Europe, central Europe became tundra, the Mediterranean Sea area became a cool temperate wet-all-year climatic zone, the northern Shara experienced considerable winter rains and much of the equatorial region became arid.

The last ice retreat began 8000-10 000 years ago, and since interglacial periods have lasted up to 100 000 years the present phase may be regarded as an interglacial period. Even within the time since the last advance of the ice there have been important fluctuations. Some have been mentioned as being supported by historical evidence; others can be worked out from the succession of pollens occurring in peat deposits. The most conspicuous warm period was 4500-6000 years ago, just as the earliest human civilisations were beginning to develop. At this time the deserts became steppe grasslands, temperate forests extended north almost to the coasts of Siberia, and permafrost was reduced in the tundra soils (chapter 23).

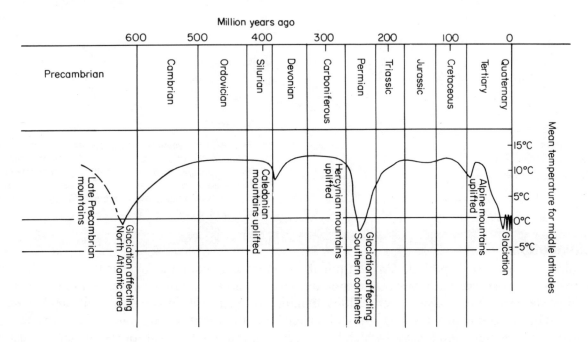

FIGURE 14:6 *Generalised temperature variations for the last 600 million years. Notice the correlation between the mountain-building phases and periods of glaciation. (After Dorf, in Kummel, 1961.)*

When enquiries are pursued back still further in time, the evidence becomes more vague, and the assumptions more debateable. Only the major features of climatic change can be discerned (Figure 14.6). The rocks of Britain contain evidence of glacial-type deposits over 600 million years old in the late Precambrian rocks of northwest Scotland (and these are similar to others of the same age all around the North Atlantic area). The overlying rocks record deposition in a warm climate over a period of 300 million years, culminating in subtropical coal swamps and in deserts with dune sands and evaporating lake deposits. Fossils of coral reefs and animals like the crocodile and rhinoceros suggest that these warm conditions still affected Britain until at least 25 million years ago, after which there was a gradual cooling towards another phase of glacial activity — the Pleistocene Ice Age. Other parts of the world show similar patterns (e.g. Figure 14.7), but the southern continents record yet another period of glaciation at about the time when the coal swamps were so prevalent through North America and northern Europe (i.e. the late Carboniferous period).

Certain patterns emerge, but they raise tremendous questions. Have the climatic belts changed due to change in solar radiation? Or has the geographical axis of rotation changed its orientation? Or have the continents changed their positions latitudinally?

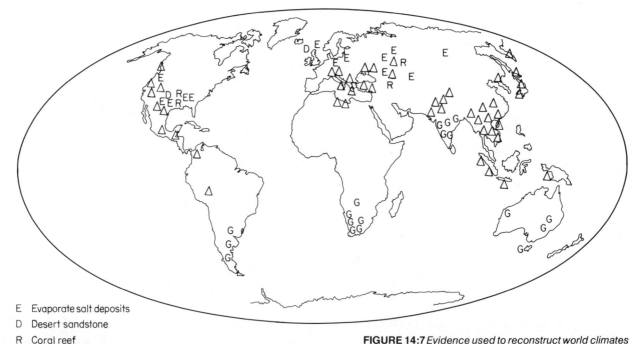

E Evaporate salt deposits
D Desert sandstone
R Coral reef
△ Warm-water organisms
G Glacial deposits, striae

FIGURE 14:7 *Evidence used to reconstruct world climates of 250 million years ago. Is it possible to reconstruct the ancient climatic belts without moving the continents? (Movements of the continents are also discussed in chapter 20 in relation to distributions of living things on the Earth.) (After Cailleux, 1968.)*

Explanations

Studies of this nature provide an additional dimension to climatic investigations. Today the atmosphere is observed at a 'snapshot' moment in time, and it is not at all certain that it is in an average state which has existed over periods of millions of years. Unfortunately so little is known in the field of palaeoclimatology (the study of ancient climates) that few definite conclusions can be drawn at the moment and theories which have been suggested to account for the evidence must be regarded as extremely tentative.

Some of the confusion can be overcome if the theories are related to a scale of activity. There are at least three levels.

a) The **shortest term changes** of up to a few hundred years can be investigated in terms of actual meteorological observations and historical records. They may be related to short-term cyclical variation, such as the sunspot cycle, or to such effects as the emission of vast quantities of volcanic dust which cuts down the penetration of solar radiation and thus lowers temperatures for a few years. Measurements on high mountains suggest that the solar constant increased between 1920 and 1940 (and this could account for increasing temperatures in that period), but the degree of error associated with such measurements was too great for them to be regarded as conclusive.

b) The **changes of medium term** include the Pleistocene Ice Age and its observed fluctuations. There is no evidence that these could have been caused by variations in solar radiation (although that does not entirely eliminate the factor), and it seems more reasonable to assume that the redistribution of the Earth's atmospheric energy budget due to large-scale effects such as the uplift of mountain ranges may be responsible. One theory suggests that the early Pleistocene uplift of the Central American isthmus caused tropical waters with their store of heat to be diverted northwards. This followed the earlier uplift of the Arctic margins, which had isolated that area and led to local cooling. Cold waters moved southwards and the meeting with warm waters led to increased atmospheric disturbance and heavy precipitation as snow over the higher parts of the adjacent land masses. Ice sheets built up and the process intensified until cooling had reached such an extent that precipitation was reduced and the decreasing supply of snow was not sufficient to maintain the massive ice sheets. They broke up, returning more water to the sea than the snow which was being supplied. The warming processes gained momentum and an interglacial period followed until the oceanic and atmospheric circulation re-established conditions for a further advance of the ice. This explanation does not answer all the difficulties associated with such a major event taking place at such a distance in time, but incorporates concepts of the right scale. Evidence from more ancient glaciations suggests that each has followed a period of mountain uplift, and there is at least some correlation between the presence of major ranges of mountains and an upsetting of the general patterns of atmospheric circulation.

c) The **longest term** is the rest of geological time: the Pleistocene Ice Age affected most of the last 2 million years, but there are nearly 4000 million years of geological time represented in the rock record. The continents have changed their positions with regard to the climatic zones and this affected the environmental patterns as a whole. It has been suggested from a study of palaeomagnetic evidence ('The Earth's Changing Surface') that Britain spent much of the Palaeozoic Era (600 to 280 million years ago) in the southern hemispheric tropics and moved northwards across the equator during the Carboniferous Period (Figure 14.8), during which time the coal seams were laid down.

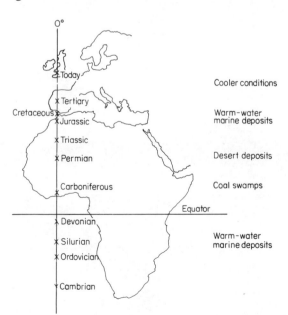

FIGURE 14:8 The positions of Britain during the last 600 million years according to palaeomagnetic evidence (see 'The Earth's Changing Surface' for the explanation). This evidence allows the ancient latitude to be plotted, but not the longitude.

Whilst moving through the northern hemisphere subtropics, it experienced a desert climate with easterly winds during the Permian and Triassic, and eventually gained its present position.

Thus if the question of scale is kept in mind it will be seen that a whole series of factors and causes have been responsible for the variations — and apparent variations — in the past climates of this planet. It must be clear that the evidence relates to changes in the atmosphere as a whole, or to changes in other aspects of the Earth environment.

Implications

A study of the past climates suggests that, whilst there have been major variations at certain stages in the Earth's history, these have all been within certain limits which can be understood in terms of presently observed cyclic fluctuations in Earth processes. There has not been a great modification of the external influences, and a consideration of how finely balanced are these external influences (principally the Sun) should make mankind thankful: a little less solar radiation and this planet would freeze; a little more and it would become parched.

What does the future hold? Some studies have suggested the way ahead for the longer term forecasting of our weather (chapter 7), and the cyclical patterns of events mentioned also assist such methods of forecasting. Man is concerned with the future fluctuations in tens of years. It is superfluous, for instance, to suggest that continental movements will affect the climate of Australia in 50 million years' time! It must also be remembered that man is taking a major part in influencing his environment (chapter 7). It is unlikely that he will be able to alter the weather to suit his precise requirements for some time ahead because of the scale of the energy changes involved. And yet man is filling the atmosphere with fumes, the result of which is unknown. It has been claimed that the increasing carbon dioxide content will raise the average temperatures. So far the effects are small, but no one knows what would be sufficient to set off a catastrophic series of events. As the recent studies have shown, more knowledge concerning the basic working of the atmosphere is needed before useful statements can be made on this subject. It would seem that the energy budget of the atmosphere (chapter 3) is fundamental, and that significant alteration of an aspect of this budget could be of greatest importance. And yet the sheer quantities of energy involved are so great that even their measurement is not at present within man's grasp.

15

Life on the Earth

The biosphere is the zone in which living things are found (Figure 15.1). It is confined to the outer surface of the Earth, including the full depths of the oceans and the lower layers of the atmosphere. Living things may exist a few metres or so into the soil, rock or the ocean-floor bottom sediments whilst some have also been found in crude oil extracted from rocks several kilometres deep.

The biosphere is a very thin layer, at the most a few metres deep, on the continental surface; even in the oceans it is only a few kilometres deep. Yet within this shallow zone there is an immense variety of living organisms. One million animal species, and a quarter of a million plant species have been described so far by biologists, but this by no means includes all those which actually exist: estimates vary between 2 and 4 million species. And it must be remembered that the 3700 million human beings at present inhabiting the Earth belong to the single species, *homo sapiens*, since they form a potentially interbreeding group. All the varieties of skin colour, hair type, facial structure, blood groups and bodily shape characteristic of mankind today are included within the limits of a single species, the basic unit of biological classifications.

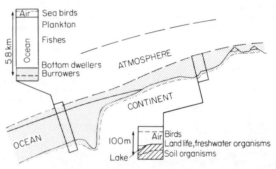

FIGURE 15:1 *The biosphere (i.e. the dotted area on this diagram). Compare the narrowness of this zone with the radius of the Earth, or the depth of the atmosphere. Compare the characteristics of the oceanic and continental realms within the biosphere.*

All the forms of living things can be classified and are divided broadly into two Kingdoms — the *Animalia* and *Plantae* (Figure 15.2). The major groups within the animal kingdom are known as phyla, and within the plant kingdom as divisions. These are subdivided into classes, orders, families, genera and species. Plants have the property of being able to convert inorganic matter such as soil minerals, water and atmospheric gases into organic matter by the process of photosynthesis. Animals eat plants or other animals. Most forms of plant life in the sea constitute the microscopic phytoplankton, and only a few larger forms occur there. These are the seaweeds which are also algae. On the land, however, the plants possess more varied structures and forms. Land plants include the fungi and mosses, the ferns, club mosses and horsetails; the conifers and other related gymnosperms whose reproductive bodies contain unprotected seeds; and the flowering plants (angiosperms) which make up 95 per cent of all land plants today, protecting their young seeds within carpels that eventually develop into fruits. Plants vary in size from the microscopic single celled algae (less than 10μ) to the giant redwood trees of north-west California, over 110 m high.

There is also a variety of microscopic animal forms: the zooplankton parallels, and feeds on, the photosynthesising phytoplankton in the surface waters of the oceans, but is composed of more varied forms. Whilst marine phytoplankton is made up largely of single-celled diatoms and flagellates, the zooplankton includes single-celled protozoans plus tiny crustacea, molluscs and the larvae of many groups. Larger animals can be divided into the chordates (i.e. those with a backbone — fishes, amphibians, reptiles, birds and mammals), and the non-chordates. The non-chordates include sponges, corals, worms, molluscan forms (mussels, oysters, snails, squids), arthropods (insects and crustacea), and echinoderms (sea urchins and starfish). Some of the non-chordates are

A

Phylum	Known species Today	Fossil	Main forms	Habitats
Protozoa	30 000	9 000	Unicellular, mostly microscopic. Move by pseudopodium (*Amoeba*), cilia (*Paramecium*) or flagella.	Fresh and salt water. Some in soil. Many parasites (e.g. malaria).
Mesozoa	50			
Parazoa	4 200	1 760	Sponges: collection of cells in common skeletal structure.	Mostly shallow seas: few freshwater or deep ocean. Filter feeders.
Cnidaria (Coelenterata)	9 600	4 500	Cells specialised in groups (in tissue). Hydroid attached (corals), medusae (jellyfish).	Mostly salt water: shallow benthos, plankton. Some freshwater (*Hydra*). Carniverous.
Ctenophora	80	–	'Comb jellies'	
Platyhelminthes	15 000	×	Unsegmented flat worms: free-living planarias, parasitic flukes and tapeworms.	Many freshwater. Wide range of parasites.
Nemertina	550	×		
Aschelminthes	12 000	×	Includes nematodes (round worms). Free-living and parasitic (hook worms).	Free-living common in soil and freshwater. Wide range of parasites.
Acanthocephala	300	–		
Entoprocta	60	–		
Bryozoa (Polyzoa)	4 000	3 000	Encrusting colonial organisms, often very small.	Marine filter feeders.
Phoronida	15			
Brachiopoda	260	15 000	Two-valved, mostly attached to sea floor.	Marine or brackish water. Filter feeders.
Mollusca	100 000	40 000	Soft body: delicate organs protected by mantle. Often with one (snail) or two (bivalve) shells. Some complex (squid).	Mainly marine. Some snails and bivalves freshwater. Some snails, slugs on land.
Siphunculoidea	275	–		
Echiuroidea	80	–		
Annelida	7 000	×	Segmented worms. Marine forms have eyes and appendages.	Marine, freshwater, land (earthworms), parasites (leech).
Arthropoda	765 000	16 400	Jointed, paired appendages + external skeleton. Class Trilobita. Extinct group. Class Insecta. Largest number of species. 3 pairs walking legs. Many with wings. Class Crustacea. 5 or more pairs legs. (lobster, crab, crayfish). Class Arachnida. 4 pairs walking legs. (spiders, scorpions).	Marine. Mostly land; some in freshwater, soil: parasites. Mostly marine: planktonic to large benthos. Mostly land.
Chaetognatha	50	×	Arrow worms.	
Pogonophora	43	–		
Echinodermata	5 700	14 000	Bilateral or pentameral symmetry. 'Spiny-skin' animals — sea urchins, starfish, sea lilies, sea cucumbers.	All marine, found at all depths. Herbivores and carnivores.
Chordata	45 000	24 000	Protochordates: notochard, but no vertebral column. The rest are vertebrates: Class Agnatha (lamprey): jawless fishes. Class Chondrichthyes: sharks, rays. Class Osteichthyes: true (bony) fishes. Class Amphibia: frogs, salamanders. Class Reptilia: lizards, snakes, crocodiles. Class Aves: birds. Class Mammalia: hair-covered, warm-blooded	Mostly marine. Now marine; formerly also freshwater. Marine. Marine, freshwater. Land, reproducing in water. Land. Land. Land; some marine.

Division	Known species		Main forms	Habitats
	Today	Fossil		
Cyanophyta			Blue-green algae: unicellular, filimentous. No nucleus; reproduces by imprecise division.	Mostly marine; some freshwater, hot springs or meltwater; deserts — rocks — wood . . . wide range of land conditions.
Euglenophyta Chlorophyta	20 000	—	Green algae. Most varied algae: unicellular or filamentous colonies. Cells with nuclei.	
Chrysophyta Pyrrophyta			Yellow-green algae, diatoms. Dinoflagellates.	
Phaeophyta			Brown algae No unicellular or small forms: includes large seaweeds.	Mostly marine; a few freshwater
Rhodophyta			Red algae: anchored to sea floor.	Mostly marine; several freshwater
Schizomycophyta	4 500	×	Bacteria. Tiny unicellular forms with poorly organised cell nucleus. Spheres (cocci), rods (bacilli) and spirals (spirilla).	Some autotrophs; others parasitic (typhus, pneumonia, tuberculosis). In soil, water
Myxomycophyta		—	Slime moulds. Multinucleate mass of protoplasm; produces spores (*Penicillium*).	Mostly land, in moist places
Eumycophyta			True fungi. Includes some moulds, yeasts, mildews as well as club fungi (rusts, mushrooms). Diverse.	Land, in moist places (N.B. Lichens are algal-fungal associations)
Bryophyta	25 000	—	Liverworts, mosses. Small plants.	Mostly land, some aquatic; some in marshes (peat moss-sphagnum)
Tracheophyta	10 000 250 000	— —	Subphylum Psilopsida. Primitive vascular plants. Subphylum Lycopsida. Club mosses. Subphylum Sphenopsida. Horsetails. Subphylum Pteropsida. Ferns gymnosperms (conifers). angiosperms (flowering plants).	Mostly land; some aquatic

FIGURE 15:2 *The main groups of living things on the Earth.*
(A) The Animal Kingdom
(B) The Plant Kingdom

N.B. × = extinct worms, total 1000 known species.
— = no numbers available for extinct species.
(Numbers of living species after Buffaloe, 1968; numbers of extinct species after Easton, 1960.)

fixed to the sea floor, and for this reason may have plant-like names (sea anemone, sea lily, sea cucumber). Other non-chordates move around in search of their plant or animal food. Like the plants there are varieties which live in water (salt or fresh) and those which live on land.

Life in the past

Fossils in the rocks provide a record of life in the past. It is a partial record, but its study provides some idea of the development of life on the Earth and shows that the diverse picture found today has existed for many millions of years.

Fossils are produced normally when an animal or plant dies and is entombed rapidly in a sediment which in turn slowly becomes a rock by means of compaction and cementation. Fossilisation is thus a special process and the likelihood of any one organism being fossilised is remote. Marine life is more favoured than land life, since few sediments accumulate on the land and most land animals or plants are decomposed rapidly by the weather, bacteria and a variety of scavengers which are particularly active on land. Hard parts like bones and shells are less liable to rapid destruction and so stand a greater chance of fossilisation than soft flesh or plant material.

The fossil record is thus partial (Figure 15.3): it contains lots of bivalve shells, coiled shells, stony coral skeletons, calcified exoskeletons and plant leaves, but rarely any of the delicate forms such as

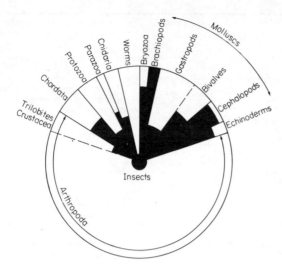

FIGURE 15:3 *Living and fossil species of animals. The shaded areas show the representation of a group in the fossil record; the divisions of the circle are proportionate to the numbers of species in each phylum. Notice the groups which are important as fossils, but not as living forms today, and vice versa. (After Easton, 1960.)*

jellyfish, worms, slugs or insects, and very few remains of land vertebrates — in spite of their often massive bones.

A detailed look at the fossil record shows that some of the groups which were prominent in the past are also important today, but that many are much less so. There must have been many changes of emphasis in the past, and some of these can be traced through the geological record (Figure 15.4).

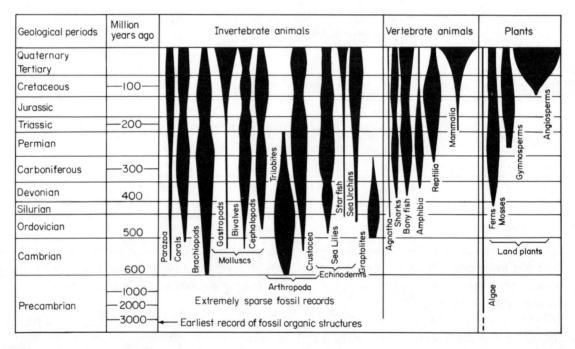

FIGURE 15:4 *The diversity of life in the past: the fossil record of the major groups of animals and plants. Many groups are not included because they are known from few fossils: they comprise large groups like the insects together with soft bodied forms like the worms. Compare the representation of major animal and plant groups in, say, the Silurian, Carboniferous, Jurassic and early Tertiary.*

Some groups, like **the chordates**, have a record of generally increasing diversity with time. Thus the fossils of primitive, jawless and armoured, fish-like creatures found in the Silurian and Devonian rocks (440-350 million years ago) are found in association with varieties possessing jaws, followed by the cartilaginous sharks and the true, or bony, fishes. Changes in the forms of the fish skeletons may have given rise to the first land-dwelling amphibians by the late Devonian period (350 million years ago), and in rocks 300 million years old there is a record of the first reptiles. All these groups continued to diversify, although the amphibians became less important in competition with the reptiles, and about 100 million years ago the reptiles themselves underwent massive extinctions. The reduction in the varieties of reptiles was accompanied or followed by expansions in other groups. The warm-blooded mammals and birds, which had existed insignificantly alongside the reptiles for over 100 million years diversified rapidly between 70 and 60 million years ago and have now become the dominant large animals.

There is thus a conflict between the increasing numbers of species in some groups and the extinctions in others. In general the chordates have become more diverse because new groups have evolved alongside remnants of the longer-established forms and a similar pattern of evolutionary change can be charted for **the plants**. After a long history of primitive marine forms, lasting from at least 3000 million years ago when the first algal-like structures occur in the rocks, the first land plants evolved just before the first land animals. Their fossils occur in rocks formed just over 400 million years ago, some 50 million years ahead of the earliest record of land animals. The land plants were apparently confined at first to the lowlying, moist locations until more efficient methods of seed dispersal and protection developed, and it was not until approximately 100 million years ago that the flowering plants evolved as a distinct group. Grasses do not appear in the record until 30 million years ago. All the main groups of plants are still in existence, and there seems to have been an overall increase in variety through the course of time.

The **non-chordate animals** provide a less obvious record of increasing diversity, since most of the phyla have representatives in the Cambrian system (i.e. rocks formed between 600 and 500 million years ago). Hardly any animals have a record before this period of time, and the previous history of these groups is conjectural. There may have been gradual evolution in Precambrian time — though there is no record of many-celled (metazoan) forms preserved — or there may have been a 'sudden' rise of metazoan forms at about 600 million years ago.

The **origin of life** on the Earth is a question to which such studies lead, but although the oldest known rocks are 3900 million years old, those formed before 600 million years ago contain very few fossils. The origin of life will always remain a matter for speculation. It is possible that the organic 'building blocks', the compounds of carbon, hydrogen, oxygen and nitrogen, could have been synthesised from components of the atmosphere (possibly water vapour, methane, carbon dioxide, carbon monoxide, ammonia and nitrogen) using energy supplied by ultra-violet solar radiation. This type of synthesis seems to have occurred outside the Earth, since meteorites of rock over 4000 million years old contain complex organic molecules. There are bacteria-like structures in the 3000 million year old Figtree Series of rocks in South Africa, whilst the Gunflint Chert of Ontario contains fossil impressions resembling the structural features of blue-green algae. This latter formation is 2000 million years old, and by that date photosynthesis and nitrogen-fixation had evidently begun. Oxygen was liberated into the atmosphere by photosynthesis, and further fossil evidence suggests that by 1400-1200 million years ago modern cell structures had become established with reproductive processes involving the blending of genetic material in chromosomes. Fossils of the first metazoan (many-celled) creatures with hard skeletons are found in rocks 600 million years old. Fragments of human fossils appeared in the record only 2½ million years ago. The general development of life on the Earth may be viewed in summary (Figure 15.5).

The Earth presents a complex record related to the evolution of living organisms, which is unparalleled, so far as is known, elsewhere. Life on the Earth has a long history of interaction with the physical environment: the rock-atmosphere-ocean-organism system has developed as a unit and the interdependence of these components is basic to an understanding of each, even though they may be isolated for the purpose of specialist study.
1) The depth of the atmosphere and the properties of its gases filter out the more harmful fractions of the solar radiation and yet allow sufficient energy to reach the surface so that important metabolic reactions can take place.

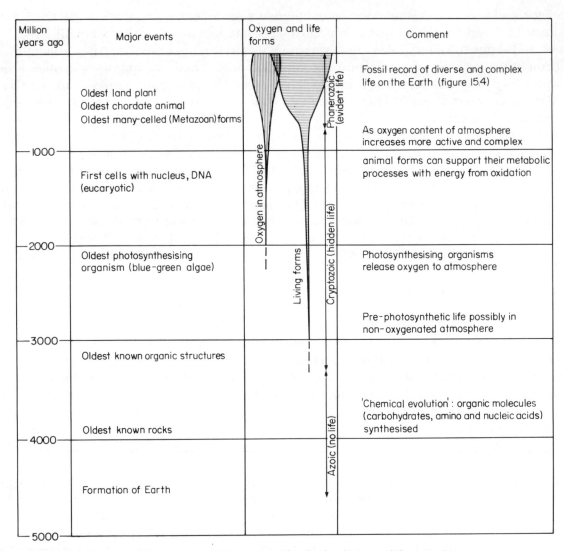

FIGURE 15:5 *Some evidence and recent ideas concerning the development of life on the Earth.*

2) The Earth moves round the Sun at a distance so that the heat energy received gives rise to temperatures in which water can exist in the liquid form and in which living systems can thrive. Light energy makes possible the basic process of photosynthesis.

3) Solar energy leads to a cycling of the water from the ocean into the atmosphere and thence on to the land surfaces before returning to the oceans. Living forms can exist on the land as well as in the sea because of this process.

4) The distribution of land and sea across the Earth's surface, combined with the relief variations within continents and oceans, produce a great variety of environments which in turn have supported a wide diversity of life.

5) The atmosphere and surface rocks supply the chemical elements necessary for the material requirements of the Earth-based living things.

All these factors are finely balanced, and small changes in some of them could lead to disastrous effects for organisms on the Earth. Thus an increase of carbon dioxide due to the burning of coal or oil could lead to a rise in atmospheric temperatures, causing the melting of ice sheets and so raising the sea level and endangering large areas of major cities which lie close to sea level. Apart from some cosmic disaster — such as the collision with another body in space — life on the Earth, which has such a long history, could continue to exist for hundreds of millions of years yet, but its quality will be impaired progressively if man does not keep a careful watch on his environment. For the moment the biosphere is a region of plentiful water, mineral and gaseous resources, and has a constantly replenished source of energy arriving daily from the Sun. The chemical and physical basis of life on the Earth is well provided for, although the balance between the continuance and destruction of this living zone is always rather fine.

This study of the biosphere will approach it from two main points of view. The first can be summarised as the **ecological view**, which studies the interrelationships of the living organisms with each other and the physical environment; and the second as the **biogeographical view**, which investigates the distributional inequalities amongst organisms and seeks explanations for them. Before these approaches are pursued it is important to look at the basic controls affecting the biosphere — the supplies of energy and matter.

Energy in the biosphere

The basic controlling factor in any system — including the relief of the continents, the movements in ocean waters and atmosphere, and the life of the biosphere — is the energy input and utilisation. Without a supply of energy, no 'work' (i.e. growth, movement, etc.) can be done.

The energy source for life on the Earth is radiant energy from the Sun. The supply is thus external to the Earth, but is renewed daily (chapter 1). Plants are the first link in the chain by which this energy is passed to the entire living realm. Only about one per cent of the incoming solar radiation is accepted by plants (Figure 15.6), although this is a vast quantity when considered on a world-wide scale. Plants absorb only the visible part of the radiation spectrum, which makes up less than half of the total reaching the Earth's surface. The proportion will vary with the season of the year and from cloudy to sunny days. Greater amounts of energy are received in the tropical regions (Figure 3.9) than in the polar areas. In addition, a proportion of the visible light will be reflected from plant surfaces, and the quantity involved in the production of plant matter will depend on other factors such as the right proportions of certain chemical elements in the plant structure.

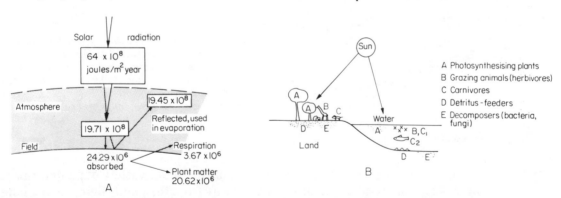

FIGURE 15:6 *Energy supply for the biosphere.*
(A) The energy which passes to plant matter for transmission to other forms of life is approximately 1 per cent of that reaching the Earth's surface from the Sun.

(B) Generalised patterns of energy reception on the land and in the water. Trace the transfer of energy through the different constituents of the biosphere in each realm. (After Watts, 1971.)

The process by which the light energy is absorbed by plants and used in the formation of organic material is known as **photosynthesis**. This involves two basic reactions.
1) The **light reaction** (which takes place only in light) occurs only in the presence of pigmented molecules, known collectively as chlorophyll, which are involved in the conversion of light energy to chemical energy. Water is necessary in this process, from which hydrogen can be incorporated into reducing molecules in the new chemical substance storing this energy: adenosine triphosphate (ATP). The oxygen from water is released freely into the atmosphere.
2) The **dark reaction** (taking place at all times when the product of the light reaction — ATP — is available) results in the transformation of the ATP, in reaction with carbon dioxide, to carbon-rich food compounds, like glucose. Other organic compounds — a variety of carbohydrates, fats and

proteins — are formed in further reactions which incorporate mineral nutrients (nitrogen, sulphur, potassium, phosphorus). The combination of these organic compounds forms protoplasm, material which has the property of being alive.

Photosynthesis thus requires the presence of a light source, chlorophyll-pigmented plants, phosphate-rich energy-storing compounds, water, carbon dioxide and mineral salts from the soil. The system which provides for life on the Earth is complex, depending on a balance of factors.

Plants manufacture the organic materials which are the food supplies for other organisms. They are known as the **primary producers**, or **autotrophs** ('self-feeders'). Most plants are phototrophs, using light energy, but some autotrophs obtain their energy by oxiding inorganic matter and are known as chemotrophs. These include some bacteria and blue-green algae, which are important inhabitants of the soil.

Autotrophs are the basis of life in all major environments — dry land, freshwater and seawater. In the oceans the autotrophs are dominated by microscopic phytoplankton, restricted to a surface layer by the penetration of light (the photic zone), and include single-celled algae like the diatoms and flagellates. Although they are so tiny, they may multiply at such a rate that vast areas of seawater may be coloured by them during reproduction. The largest marine plants are the multicellular algae — the seaweeds — which are again limited to the zone of light penetration. These are also concentrated near the water surface, either in floating masses (e.g. in the Sargasso Sea) or attached to rock around the shores just below low tide. Their colours indicate the depth to which appropriate light wavelengths extend (Figure 15.7).

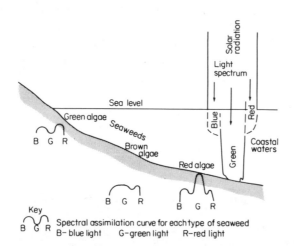

FIGURE 15:7 Light penetration in north European coastal waters related to seaweed colouration. How do the different seaweeds assimilate the sector of the spectrum reaching them? N.B. light is not the only factor involved in this colour zonation, and green, red and brown algae may occur together in a rock pool. (After Lewis, 1964.)

Bodies of freshwater, such as lakes, rivers and ponds, have a wider variety of plant forms than the seas. Most of these forms have become adapted secondarily from the plants living on the land: there are varieties which have roots in the pond floor (water lilies), or float freely (water hyacinths). The growth of aquatic plants is linked closely to the period of daylight and the supply of nutrients. Thus the temperate seas become the scene for explosive phytoplankton growth in spring when daylight hours become longer and in autumn when strong currents stir up the nutrients (chapter 24).

Land plants are more complex in structure and functions than aquatic forms. The climatic conditions and the supply of mineral nutrients vary much more than in the sea. Thus the Sahara desert has an ample supply of sunlight, the temperatures are within the range for encouraging plant growth, and certain areas possess the correct soil nutrients. The lack of water over most of this desert prohibits extensive plant life and restricts vegetation to a few extremely specialised forms which can withstand the conditions (chapter 23): where water is available in the oases, living forms flourish. There are also regions where it is too cold for plant growth during all, or part, of the year; or where winds are too strong for taller plants to exist; and where soils are deficient in mineral nutrients.

All other organisms feed in some way on organic matter which has been manufactured by the autotrophs: they are known as **heterotrophs** ('different feeders'), and may feed on plants, other animals (either living or decomposed), or on organic compounds in solution, derived from once-living plants and animals. The energy stored in plant matter following photosynthesis is transferred to other creatures who eat the plants. There is a fairly simple progression: plant (primary producer) eaten by herbivore animal (primary consumer) eaten by carnivore animal (secondary consumer) eaten by another carnivore animal (tertiary consumer). This is known as a **food chain**. Thus a sheep (herbivore), grazing on grass may be attacked and eaten by a wolf (carnivore); insect larvae feeding on the stem of a plant may be eaten by a bird; crustacea in the zooplankton will feed off the phytoplankton and will be eaten by herring; small molluscs will graze on seaweed in shallow water, and will be attacked by other molluscs or fishes. In practice the food chain is too simple a concept, since many animals will have a varied diet (Figure 15.8), and the complex interrelationships are shown more correctly in terms of a **food web**. The food chain illustrates the way in which energy is transferred through the biosphere. This type of food chain is known as a 'grazing pathway', but

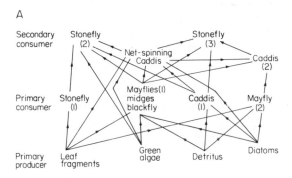
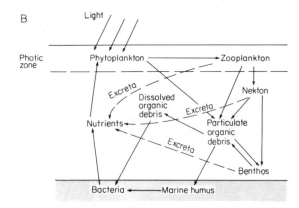
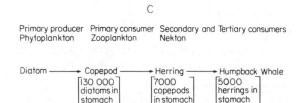

FIGURE 15:8 *Food webs and chains.*
(A) Detail of a food web in a stream in south Wales. Notice the varied diets of some (compare Mayfly (2) and Net-spinning Caddis), and the restricted diet of Stonefly (1). (After Watts, 1971.)
(B) The general food web in the sea.
(C) A single marine food chain, giving some idea of the vast numbers of microscopic plankton required: the whale's stomach contained the equivalent of 8×10^{12} diatoms!

energy is also transmitted through the 'detritus pathway'. The latter method is slower, and is helped by burrowing grazers, but may be responsible for a high proportion of energy transfer in a particular area. The detritus pathway includes small animals feeding on decomposing plant and animal matter, together with bacteria and fungi (Figure 15.9).

It is not easy to demonstrate the progress of energy transmission in a food chain, since it requires analysis of the gut contents of a range of consumers. A large proportion of the energy transferred is 'lost' to the system at each level (Figure 15.10), being transferred to heat energy and passed up in the processes of muscular work and respiration by which organisms acquire and then break down the food supplied to them in a reaction with oxygen. Something like 90 per cent is expended in this way at each stage in the food chain: only 10 per cent is stored for the next link in the chain to eat. This 'loss' limits the number of links in the chain, and also means that the biosphere needs a constant renewal of the energy supply.

Energy is used by both autotrophs (A) and heterotrophs (B) (Figure 15.11). Both are important in terms of farming, since it is possible to assess the productivity of crops and herbivore meat-producers, and to identify the points where improvements can be made. This has been done for rice and wheat yields where strains can be bred to absorb more light energy, or to put less material in useless stalks, etc, but of course there are also economic, technical and social factors affecting farming which may mask the effect of such improvements: too much potential food is lost in

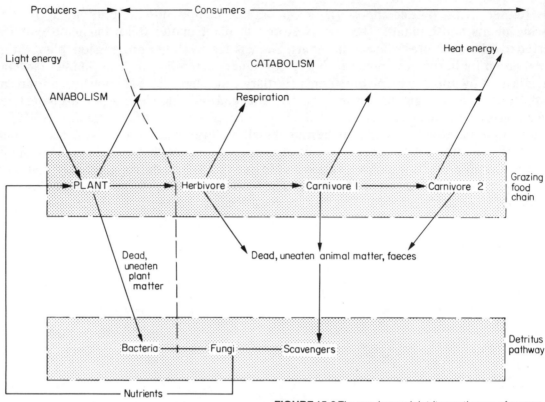

FIGURE 15:9 *The grazing and detritus pathways of energy distribution.*

storage after it has been harvested, for instance. It is realised that young animals (e.g. steers, chickens, lambs) use a far higher proportion of their food intake for making flesh than older animals, and so animals are killed for meat before they mature.

These considerations are important in view of the increasing world population. Some interesting changes in diet might occur if prejudices concerning food could be overcome: palatability interests the inhabitants of the wealthier nations more than energy-transferring efficiency. Thus, rabbits put on flesh weight four times as rapidly as beef cattle, and certain waterfleas are even more efficient!

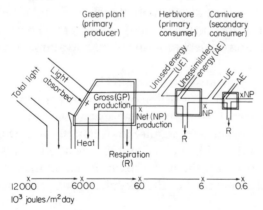

FIGURE 15:10 *Energy transfer through a food chain. At each level only 10 per cent is passed on. (After Odum, 1963.)*

Fish flour, made from uncommercial fish, contains 80 per cent protein (cf. wheat flour 12 per cent), and could be the cheapest form of food for large numbers of people. Whilst European cattle have been introduced to the tropics, they have suffered too readily from disease, and it has been suggested that it would be more sensible to domesticate the zebra or antelope, which are adapted to the conditions. From another point of view it might seem that the 90 per cent energy loss involved in producing herbivore meat might indicate that man should become exclusively herbivorous, but

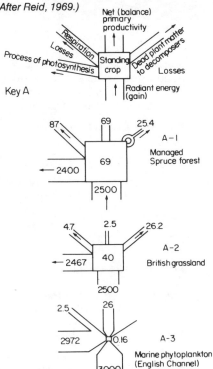
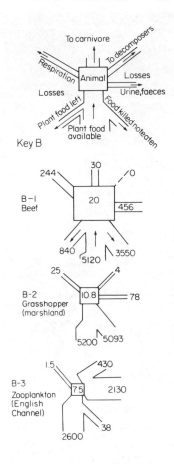

FIGURE 15:11 *Energy flow in primary producers (A) and in primary consumers (B). The units used are known as Standard National Units (SNU): 1 SNU = $10^9 \times 4.184$ joules/hec year. (After Reid, 1969.)*

proteins are more plentiful in meat than in plants. Certain plants do contain more than others, but this is a matter for future developments.

Carnivores are at the end of the food chain. They produce little flesh which is edible, and so are not eaten much by men: fish are the only group in this category eaten in large quantities. Carnivores assimilate more of their food intake into body muscles and flesh than herbivores, and are an important factor in the general ecosystem balance. In places where they have been exterminated by man, the herbivores have been allowed to grow in numbers, leading to overgrazing and destruction of the soil. In New Zealand, this occurred as deer were introduced to an area without large carnivores.

The productivity based on this transfer of energy varies greatly from one part of the world to another: areas in the tropics receive more intense sunlight over a much longer period; the sea reflects more sunlight than the land; and forest tends to be more productive than cultivated land, since crops normally take part in photosynthesis for a smaller proportion of the year. Major zones of the world may be delineated in terms of primary productivity — i.e. the productivity of plants, which is basic to the overall productivity (Figure 15.12).

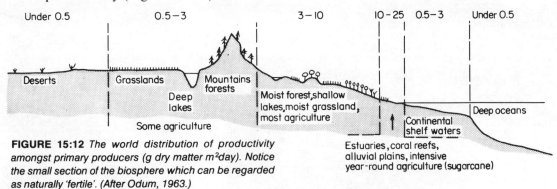

FIGURE 15:12 *The world distribution of productivity amongst primary producers (g dry matter m²day). Notice the small section of the biosphere which can be regarded as naturally 'fertile'. (After Odum, 1963.)*

145

The Arctic. An April 1970 NIMBUS satellite view of snow-covered Scandinavia, with sea ice between 60°N and the Pole. Convection cloud forms as cold air from the ice blows over the sea. Notice the extent of ice in the Baltic Sea at this time of the year.

Plate 145 (Crown Copyright)

Life on the Earth and Moon

Plate 146 The absence of any biosphere on the Moon is emphasised by the complex life-support systems required by the Apollo astronauts, such as Irwin of the Apollo 15 mission, seen here against the background of Mount Hadley on 31 July 1971. (NASA)

Plate 147 Earth vegetation in relation to mineral nutrients. On the left is a 24-year old Sitka spruce (*Picea sitchensis*), which has grown in a nutrient-deficient soil. A similar tree on the right had been fed 340g of potassic superphosphate four years before the photograph was taken. (Forestry Commission)

146

147

148

149

150

Wind and plants. The effects of wind on plant life.

Plate 148 Scots pines (*Pinus silvestris*), average 16m high, blown down after a gale affecting Speymouth Forest, Scotland. (Forestry Commission)

Plate 149 Wind causing soil to move in a sheet-like manner covering the sorghum field on the right in Colorado, USA.

Plate 150 The combined effect of windbreaks and strip-cropping in a former Dust Bowl area of South Dakota near Hecla. (149 & 150 USDA Soil Conservation Service)

Matter in the biosphere

Plants and animals are composed of a wide range of chemical elements (Figure 15.13). These occur in both living and non-living matter: living things are thus dependent on the non-living matter. Oxygen, carbon and hydrogen are present in the largest quantities in all organic molecules, and other elements are associated with particular compounds or functions. Nitrogen and sulphur occur in proteins; a magnesium atom occurs at the centre of each chlorophyll molecule; phosphorus is essential to the storage of chemical bond energy in ATP; other elements occur in enzymes, which are protein catalysts speeding biochemical reactions.

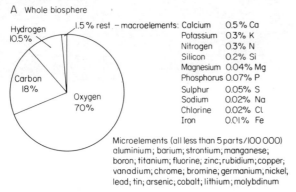
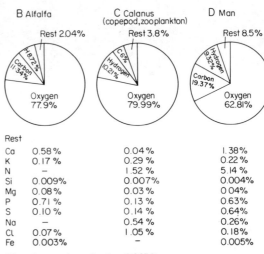

FIGURE 15:13 *The chemical composition of living matter by weight, expressed as a percentage: the average for the whole biosphere (A) can be compared with individual organisms (B)-(D). Three elements are always dominant (oxygen, carbon and hydrogen), but these and others very markedly in proportion. (After data in Watts, 1971.)*

The elements which are important in the building up of organic matter (protoplasm) are known as **nutrients**. They are involved in a series of circulating systems, a contrast to the supply of energy to the system, although once again the plants have a fundamental role to play in making the elements available to living organisms (Figure 15.14). Nearly all the nutrients pass through rocks, the atmosphere, living creatures and water in systems like those known as the 'rock cycle' and the 'hydrological cycle' (Figure 4.1). In fact the several cycles overlap at various points. Many nutrients are cycled back to the Earth's surface in rainwater (Figure 15.15). Man adds other substances to these cycles in the form of weed-killers, pesticides and waste effluent: these may pass through animals with tragic results (Figure 15.16), or may affect the biosphere system at large. The major

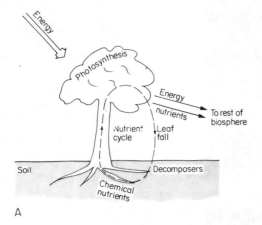
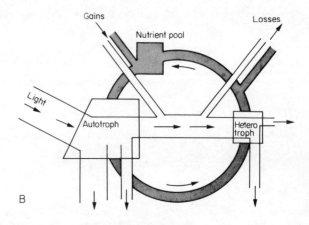

FIGURE 15:14 *Energy and nutrients in the biosphere. (A) The primary place taken by plants. Only plants can transform solar energy into a form which is available to organisms; only plants can take up chemical nutrients and incorporate them into organic material. (B) The one-way passage of energy compared with the re-cycling of nutrients (c.f. Figure 15.10). (After Odum, 1963.)*

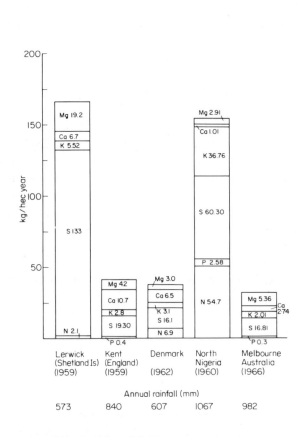

FIGURE 15:15 *Nutrients collected from rainwater. (After data given in Watts, 1971.)*

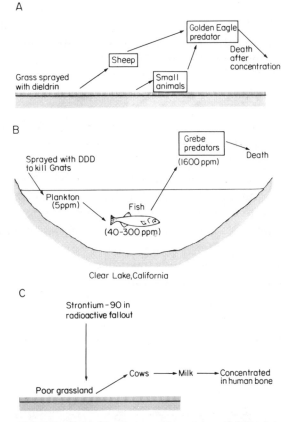

FIGURE 15:16 *Three examples of harmful (toxic) chemicals being taken up by organisms and passed through food chains so that they become concentrated in the end members.*

nutrient roles and cycles may be summarised diagrammatically (Figure 15.17), and some results are worth studying.

1) Much more water is taken up by the plant than is used. Thus a crop weighing 20 tonnes at harvest may include 15 tonnes of water: i.e. only 5 tonnes of dry weight crop, of which 3 tonnes are composed largely of matter transformed and fixed from water. Whilst this crop is growing, 2000 tonnes of water pass from the soil, into and through the plant: only 0.15 per cent of this is used, and 40 per cent of the solar energy reaching the plant is taken up in the evaporation of transpired water. Potential evapotranspiration in an area is related to the climate (chapter 4) and is most important in terms of plant physiology. The plant must have the capacity for allowing sufficient moisture to pass through it, or for protecting itself against the loss of too much.

2) Oxygen both supports life, and arises from it! The original Earth atmosphere may have contained no free oxygen, and it was only after the development of photosynthesising organisms from anaerobic (i.e. living in unoxygenated conditions) heterotrophs that we find oxidised iron compounds occurring in rocks, and the beginnings of the higher forms of life; advanced metazoans could develop only after the ocean surface spread of phytoplankton had released sufficient oxygen to provide an ozone screen in the upper atmosphere against ultra-violet radiation. It is thought that oxygen in the atmosphere then increased to the point when extensive forests developed on the Earth's surface. Man could now be reversing this process by cutting down the forests, building over vegetated land, and providing a thin film of oil over the ocean surface. At the same time he is increasing the carbon dioxide content of the atmosphere, which encourages plants to grow more rapidly and to give up more oxygen, and is advancing cultivation into desert areas, which produced little oxygen in the past.

3) In the natural state there is a balanced system, whereby the loss and gain of nitrogen in the soil and available to plants are roughly equal. Industrial processes introduced in 1914 make it possible for man to synthesise nitrogen from the air to make compounds which can be used in fertilizers. This is obviously upsetting the balance, and considerable extra quantities are being washed out of the soil and deposited in ocean sediments, instead of being re-cycled to the atmosphere. Questions concerning the amount of nitrogen compounds which the denitrifying bacteria can accept and re-cycle are unanswered.

4) Phosphorus is not re-cycled like the other materials mentioned. Very small amounts are washed out of phosphate rocks, decaying bones, and excreta, and are used in plants as phosphate compounds. Some is lost via rivers to the sea, where it is incorporated in sediments, but none occurs in the atmosphere, so it follows only a one-way path from the land to the sea, except for the occasional uplift of the sea-floor deposits. Thus phosphorus is normally the limiting element in an ecosystem: doubling of the content in a pond doubles the production of plant matter. When this happens — and it is becoming common in smaller lakes and ponds forming drainage sinks for fertilizers, detergents and treated sewage — nitrogen becomes the restricting element for growth. Blue-green algae then take over, since they fix atmospheric nitrogen, thriving on the increased phosphorus. Their rapid increase gives rise to masses of decaying organic matter, which fouls the

FIGURE 15:17A

Recycled compound or element	Role in biosphere
Water	Water is a major constituent of protoplasm: e.g. makes up 99% some marine invertebrates; 85% some mammals; 66% most vertebrates; 50% wood. Water supplies hydrogen for organic compounds in photosynthesis, and oxygen to the air. Water is a solvent for the mineral nutrients which are only available to plants in solution. When it freezes ice floats to the top of a pond or lake: life continues beneath.
Oxygen	Oxygen is very active chemically. It combines with most elements in nature and the process of oxidation speeds metabolic processes. Oxygen forms an average of 70% atoms in living matter, and is a basic building block of carbohydrates, fats and proteins.
Hydrogen	Hydrogen is an important constituent of living matter: on average 10.5%
Carbon	Carbon is an important constituent of living matter: on average 18%
Nitrogen	Nitrogen is an important constituent of protein molecules
Sulphur	Sulphur is an important, though small constituent of some proteins
Phosphorus	Phosphorus is a constituent of protein molecules: these include ATP, the chemical energy-storing compound so important in photosynthesis. It is normally the element which limits ecosystem growth.

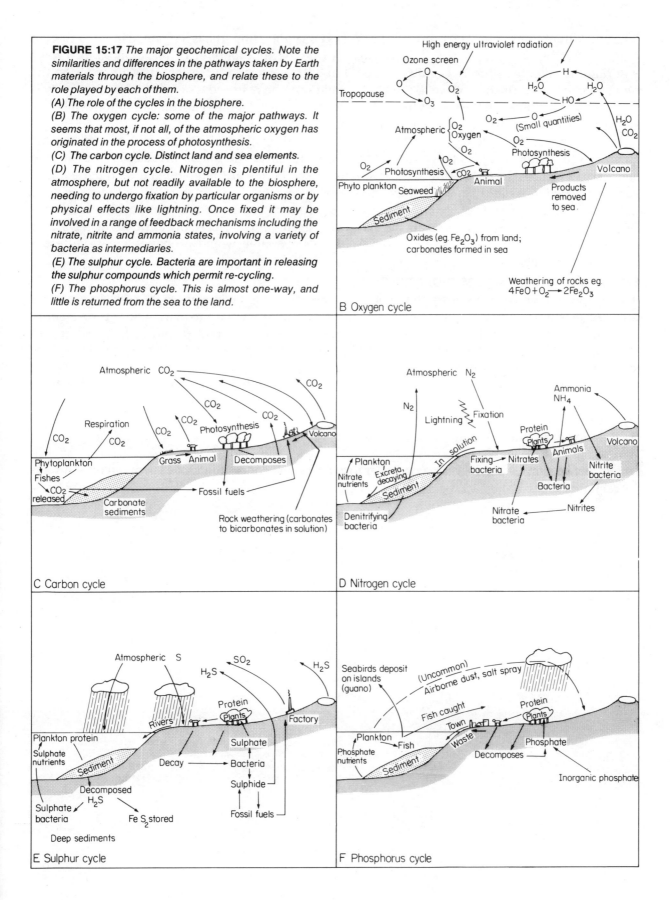

FIGURE 15:17 The major geochemical cycles. Note the similarities and differences in the pathways taken by Earth materials through the biosphere, and relate these to the role played by each of them.
(A) The role of the cycles in the biosphere.
(B) The oxygen cycle: some of the major pathways. It seems that most, if not all, of the atmospheric oxygen has originated in the process of photosynthesis.
(C) The carbon cycle. Distinct land and sea elements.
(D) The nitrogen cycle. Nitrogen is plentiful in the atmosphere, but not readily available to the biosphere, needing to undergo fixation by particular organisms or by physical effects like lightning. Once fixed it may be involved in a range of feedback mechanisms including the nitrate, nitrite and ammonia states, involving a variety of bacteria as intermediaries.
(E) The sulphur cycle. Bacteria are important in releasing the sulphur compounds which permit re-cycling.
(F) The phosphorus cycle. This is almost one-way, and little is returned from the sea to the land.

lake and kills off other life. The process is known as **eutrophication**. It has happened in Lake Erie, in North America, but since it is relatively cheap to restrict the amount of phosphorus in sewage entering the lake, the process can be stopped. Lake Erie is fortunate in that its waters are on the move, and it can be flushed out in 3-5 years: it could soon be restored to a living lake. Other lakes, such as Lake Michigan in the same Great Lakes area, are in less immediate danger, but the flushing time is longer (100 years), so the long-term result could be worse.

The biosphere thus demands certain chemical nutrients, which it incorporates in organic chemicals for a short space of time. Most of these are present in the soil, or in the ocean waters, and some come directly from the atmosphere. Many of the cycles are inter-related, and agricultural research which looks at them separately will not be successful. In the natural state there seems to have been some regulation of the circulation of these vital nutrients, and the composition and forms of plants and animals are related to the particular availability of certain nutrients in the Earth environment: another planet with different materials available would probably have a different aspect to its biosphere.

Constraints affecting the biosphere

The energy and nutrient flows are the major controls affecting the biosphere, but distributions are affected also by a number of constraints, varying in intensity from place to place. Older texts identify these as the major factors affecting plant and animal life on the planet, but in fact they are of secondary importance to the controls exerted by energy and matter supply. The constraints will be studied in four major groupings:

1) The 'atmospheric and oceanic characteristics', which include a range of effects resulting from characteristics of sunlight, temperature, rainfall, windiness and the ocean environment (chapter 16).
2) The soil (chapter 17).
3) Relationships amongst plants and animals (chapter 18).
4) The influence of man (chapter 18).

The factors are isolated in this way in order to examine the contribution of each, but it must always be remembered that they act together in nature.

16

Life in atmosphere and ocean

Perhaps the greatest contrast in physical constraints is between those experienced by plants and animals living on the land and those in water (either freshwater lakes and rivers, or saltwater seas and oceans). These two realms provide such a contrast in physical conditions (Figure 16.1) that forms of life in each are quite different. In both, however, the controlling factors are similar: there is an energy flow from solar radiation through plants to herbivorous, then to carnivorous animals and detritus-feeders; and there is a constant re-cycling of the chemical nutrients.

Characteristic	Land	Oceans (lakes, rivers)
Area covered	148 300 000 km² (29% Earth surface)	361 700 000 km² (oceans) (71% Earth surface)
Altitude/ depth of life zone	Highest mountains 8848m above sea level, but life restricted to narrow zone at surface: highest trees 100m, average 20m	Deepest oceans 11 516m; average depth 5000m with life throughout
Density	Atmospheric density falls with height above sea level. Plants and animals require more rigid support than in water	Water density increases with depth. A buoyant medium
Light	A good supply at the base of the atmosphere, but does not penetrate soil; partially lost in woodland	Penetrates only a limited surface zone: the 'photic zone'
Temperature	Rapid changes: organisms have range of tolerance	No rapid changes: organisms have narrower tolerances
Water	Periodic precipitation: obtained by plants via soil, or by animals from rivers, etc. Organisms require special means of storage	Present all around
Wind	Important factor in transpiration; may prohibit growth where very strong (e.g. on coasts or at high altitude)	Affects only surface zone
Nutrient supply	From soil, atmosphere	From water (gases or mineral ions in solution) or bottom sediment. Ocean water is a saline solution

FIGURE 16:1 *The major physical constraints affecting life in the biosphere. N.B. there are also differences between salt- and fresh-water conditions, particularly in terms of salinity and often depth, but these are of less significance than the major distinctions made here.*

Density

The most obvious difference in the physical nature of the two environments is the density contrast experienced by living things in each. Life on the land exists mainly above the surface of the soil at the base of the atmosphere, density 0.0012 g/cm³. The atmosphere acts as a filter on the solar radiation spectrum, and is a realm of turbulent movement causing weather changes. The low density of the atmosphere means that living forms receive no physical support from it and therefore are concentrated at the base: birds and flying insects have to return to the ground at frequent intervals. The density of seawater (1.025 g/cm³) is almost identical to that of living protoplasm, so that living forms have little difficulty in maintaining themselves at a particular level, and they are found distributed throughout the ocean depths. At the same time pressure on marine organisms increases with depth, but since they are found living in the deepest waters it must be possible for them to adapt to such pressures. Human beings find descent into water difficult, since the pressure increases at 1 kg/cm² for every 10 m descent, and this affects the gases dissolved in the blood, but marine animals overcome this in various ways, chiefly by equalising the water pressures inside and outside of their bodies. Thus whales can dive from the surface to 300-600 m depth and return within minutes, and many of the most active marine organisms can do this. Most fishes have a gas-filled swimbladder, which enables them to adjust to different pressures, and if they are hauled up in nets too rapidly from depth, the bladder is forced out through the mouth and the fishes become distorted. Problems of reduced pressure are experienced by men ascending high mountains into rarified air, and temporary sickness may occur.

Light

Light energy is the basis of life in the biosphere: variations in the reception of light give rise to differences in organic activity. The contrast between land and sea environments affects the proportion of light used, the types of life forms present, and their distribution. Marine plants accept a smaller proportion of the light reaching them than land plants because of the greater reflectivity of the sea surface at most angular positions of the Sun. Although marine life is distributed throughout the ocean depth, sunlight is available only in the uppermost layers (Figure 16.2). The maximum light penetration in the sea occurs where the Sun is overhead, and in the polar regions light enables photosynthesis to take place only in three months of the year. Thus marine plants are largely confined to the microscopic phytoplankton floating in the surface waters, except for the landward margins of the oceans where seaweeds attached to the floor become important. Light is important also to the marine animal life, stimulating it to move towards food sources and being concerned with the regulation of many life processes in daily or seasonal rhythms.

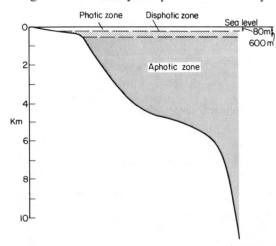

FIGURE 16:2 *Light penetration in the oceans. Light is sufficient for photosynthesis only in the photic zone; dim light may penetrate under good conditions to 600 m (the disphotic zone). More light will penetrate deeper in tropical regions because the Sun is more nearly overhead and reflection from the surface is less. (After Gross, 1972.)*

Plants on the land have developed erect, bushy habits in order to accept as much sunlight as possible (the leaf surface area is often 20 times the area of ground covered by the plant), and strong woody stems and branches support this foliage in the low density medium of the atmosphere. Different plants have different requirements for normal growth and development. Each has a minimum light intensity below which growth will not occur, and each has an optimum intensity for maximum growth. Most land plants have higher optimum light intensities than the phytoplankton: some, such as trees, cereal crops, grasses and most 'weeds' grow best in bright sunlight (**'heliophytic'**), whilst others, such as mosses, ferns, and some herbaceous and shrubby plants grow best in the shade (**'sciophytic'**). Most plants are tolerant of a range of light conditions and many trees (e.g. beech, spruce, fir, yew) have sciophytic seedlings (though willow, poplar, birch and Scots pine do not).

The length of daylight varies seasonally with latitude, and this may affect life processes, as it affects the basic process of photosynthesis in the oceans. There are short-day species of plant which respond to their critical day length by flowering and producing seeds (e.g. chrysanthemums, tobaccos and spring bulbs), whilst long-day species require more than a critical period of daylight (e.g. beet, lettuce, wheat and potato). Animal life also may be sensitive to light: some species are active only at night, and some produce their own light in the dark ocean deeps or during the night.

Water

Water must be present before light-activated processes will take place. It is primarily a protoplasmic constituent, and few living cells survive if the water content falls below 30 per cent of normal. Water availability is no problem in aquatic environments, but is often a major limiting factor for the development of living organisms on the land. Whilst the marine plant simply has to absorb water from its surroundings, most land plants have to absorb it from the soil through roots. Land plants are therefore dependent on local precipitation, water movement in the soil, and on evaporation rates. The land plant will wilt when a state of 'physiological drought' occurs: this may happen when the rate of water absorption by the roots is insufficient to offset losses by transpiration — i.e. when the soil is dry after a period without rain, or when it is frozen. Most land plants lose too much water during the day, but make up losses at night when transpiration virtually ceases.

In any land plant as much as 98 per cent of the water taken up by the roots will pass through and be transpired. Needs vary. Thus a wheat crop will take 507 parts of water per part of dry matter; alfalfa 1068; millet 275. Whilst a maize plant will take up 2.4 litres per day, an oak tree will demand 675 litres. Size is the main factor in the latter case.

The relationship of water supply to land plant form is so important that plants have been classified on this basis:

> **hydrophytes**, growing in water;
> **helophytes**, growing in waterlogged soils (seasonally or permanently);
> **mesophytes**, transitional forms;
> **xerophytes**, adapted to drought in various ways.

Few plants, except for primitive algae, lichens and mosses (which lie dormant until water arrives) tolerate reduction in cell water content, so that special means have to be adapted to acquire more or retain what is held. Thus the xerophytes will prevent water-loss in transpiration by means of hardened, cuticularized leaves (**sclerophyllous**), by protecting the stomata, or by reducing the size of leaves (**microphyllous**); although the leaf size may play a small part in reducing water loss. Others will have extensive root systems penetrating to the deepest retreats of the water table; and others will conserve water in underground tubers or thickened stems (*Cacti*).

Water is also important to animals, but they do not have to obtain it from plants and most can move to a source of supply; water holes in the East African savannas draw a wealth of herbivorous animal life each evening and in this way supply the carnivores with their food. Desert animals live in holes where loss of body fluid is reduced. Many insects, like the locust, which breed on the margins of deserts, do so only after rains, and scientists studying their movements refer constantly to the records of rain in the breeding areas (chapter 19).

Wind

Wind is important in different ways to the organisms of land and sea. On the land, strong winds will tend to increase transpiration, so that plants in coastal and high altitude situations are often stunted. Wind distributes pollen and seeds, but can also uproot trees or blow away dry topsoil.

Winds blowing across the ocean affect the surface movements of waters. At speeds exceeding 12 km/hour a circulation is set up in the surface waters, breaking up slicks of organic matter into parallel 'windrows' (Figure 16.3). Offshore winds also bring up nutrient-rich waters to the ocean surface in areas like the coasts of Chile and Peru (chapter 11), and winds generally help to promote turbulence and consequent mixing of the waters.

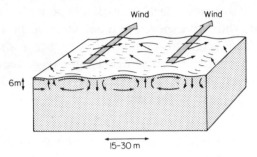

FIGURE 16:3 *Windrows. A steady wind (over 12 km/hour) sets up a surface circulation in the open ocean: oily films and phytoplankton are swept into zones of convergence. Zooplankton may also accumulate in these zones: there may be concentrations 100 times greater than elsewhere. (After Gross, 1972.)*

Temperature

Temperature conditions govern the rates of biochemical reactions and thus the metabolic activity of organisms. If other factors are regarded as constant, chemical reactions approximately double in speed for every rise of 10°C. Living things are composed of extremely complex molecules, so that too great a temperature rise (to over 40°C) leads to some of these molecules breaking down; enzymes are irreversibly inactivated at temperatures over 45°C. Mammal blood is maintained at about 37°C, and that of birds a little higher still, near the top of the possible scale and enabling the greatest degree of metabolic activity, but a rise of a few degrees during illness can be critical. Living processes also cease below 0°C. A few specialised organisms can still retain their living properties at temperatures as low as −75°C (though they are in an inactive state), or as high as 92°C in the waters of hot springs: both are rare, special adaptations.

Temperature differences are the major factors governing the distribution of life in the sea. Whilst temperature differences are important on the land, the presence or absence of water becomes a limiting factor more frequently. And yet changes of temperature experienced in water bodies are much less sudden or extreme than those on the land (chapter 3).

Temperature differences in the oceans are most effective in the shallower zones: plants and animals living along the shores, on the floor of shallow continental shelves, or in the surface waters, are affected. Apart from the marine mammals (whales, sea lions, porpoises and dolphins) the animals in the sea are cold-blooded — i.e. their body temperature closely resembles that of the environment — and most live within a water mass having a low range of temperature. A series of major latitudinal zones can be recognised in the oceans (Figure 16.4). Pronounced seasonal cooling will lead to a period of quiescence when the creatures of temperate zones will require little food or oxygen, and breeding is often restricted to a season of warmer temperature. The greatest seasonal variations occur in middle latitudes, which is the scene of the most common migrations: polar species migrate there to breed in winter, whilst subtropical species come in summer.

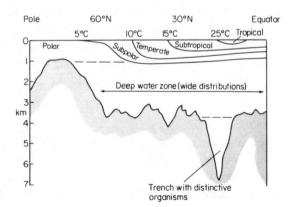

FIGURE 16:4 *The effect of temperature on life-zones in the oceans. Surface forms can be zoned latitudinally, but at depth the forms have wider distributions and from the small amount of information available concerning what happens at great depths it seems that some forms may be found from the equator to the poles. The trenches, on the other hand, are isolated from each other with extremely cold, dense water in them, and each contains a distinctive group of organisms.*

At depths below 2000 metres, temperatures in the oceans are uniformly low. Organisms living on or above the absyssal plains may have a worldwide distribution, although those living at even greater depths in the trenches tend to be restricted to a particular locality.

There is a particularly interesting correlation associated with a study of life in the warm and cold sectors of the oceans. The tropical oceans, with relatively warm waters throughout the year, contain large numbers of different species of animals and plants (Figure 16.5), although each species may include relatively few individuals. In the colder oceans, and in those with fluctuating seasonal

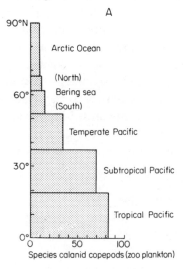

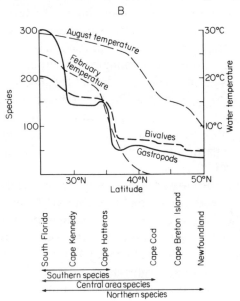

FIGURE 16:5 *Species variety and latitude. Two aspects of ocean life: compare this picture with the species composition of temperate and equatorial forests.*
(A) The numbers of species of a common member of the zooplankton from the equator to the North Pole. These organisms live near the ocean surface. (After Gross, 1972.)

(B) Numbers of species of two groups of largely bottom-living molluscs along the eastern coast of North America. The warmer water species have northern limits, but northern species extend to the southernmost points. (After Stehli, in Laporte, 1968.)

temperatures, there are relatively few species present, but large numbers of individuals within each species, and the individuals commonly grow to larger sizes. There is a temptation to suggest that life is easier in the warm tropical waters, and that fewer species have been able to adapt to the cold conditions. The increased metabolic rate in the warm areas might account for earlier sexual maturity, shorter life spans and more rapid changes in genetic material, giving rise to a greater variety of smaller forms. Two facts must be remembered in making assertions of this nature. Firstly, it is found that the total biomass (i.e. the mass of all living things) is far greater in cold temperate regions (Figure 16.6). The upwelling waters bring nutrients to the surface and result in a high

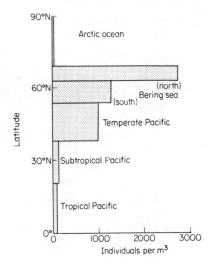

FIGURE 16:6 *Biomass and latitude. Compare this diagram with Figure 16.5(A). (After Gross, 1972.)*

production of phytoplankton which carries through to other sectors of the food chain. Thus life is not so difficult in these regions as may have been thought, and they form the world's greatest fisheries (chapter 24). Secondly it must also be remembered that the tropics have been subjected to less drastic temperature changes in the past than have the middle and higher latitude oceans. The Pleistocene ice sheets would have affected life in the oceans on the poleward side of latitudes 30°N and 30°S. The communities in these areas today could be regarded as 'immature', having evolved over a much shorter length of time than those in the tropics, and are thus less diverse.

On the land, local changes of temperature are much greater than those in the sea — and these changes affect shallow bodies of freshwater as well. The changes may be from day-to-night, or from season-to-season. The world can be zoned latitudinally in a similar way to the oceans, although the basis is slightly different, since seasonal differences become more important. Plants can be divided into groups needing a particular range of temperature to complete the life cycle:

megatherms are plants of tropical and subtropical regions requiring at least four months over 20°C;

mesotherms require temperatures between 10°C and 20°C;

microtherms are plants able to live in areas with 8-11 months under 10°C;

hekisotherms are those living in polar belts where the temperature never rises above 10°C.

Each species of plant is adapted to one or other set of conditions. Each has a minimum temperature, below which it is inactive (e.g. from −2°C to 5°C in temperate cereal crops; but as much as 15°C to 18°C in melons, sorghums and date palms); an optimum temperature at which all functions occur most efficiently; and a maximum temperature, beyond which metabolism ceases. Figure 16.7 shows how these affect the growth of certain crops. When temperatures fall below the minimum for growth, plants will become inactive and, if the climatic regime imposes such a period, the life cycle will be adapted to it and protection supplied. Thus the cycle of growth will be completed in a favourable season, and in the unfavourable season (i.e. too cold, but similar effects are apparent on

Crop	Cardinal points (°C)			Number of days needed for germination at certain temperatures			
	Minimum	Optimum	Maximum	4.38°C	10.25°C	15.75°C	19°C
Wheat	3-4.5	25	30-32	6	3	2	1.75
Barley	1-2	25	30	4	2.5	1	1
Maize	8-10	32-35	40-44	—	11.25	3.25	3
Sorghum	8-10	32-35	40	—	11.5	4.75	4
Rice	10-12	30-32	36-38	—	—	—	—
Tobacco	13-14	28	35	—	—	9	6.25
Sugarbeet	4-5	25	28-30	22	3.75	3.75	3.75
Alfalfa	1	30	37	6	3.75	2.75	2

FIGURE 16:7 *The cardinal points of germination of some crops. The minimum cardinal point is the temperature below which metabolic activity does not occur; rates of activity (e.g. growth) will increase up to the optimum temperature, and will then decline to the maximum (beyond which activity again ceases). (After Tivy, 1971.)*

land when it is too dry) transpiration will be reduced by the shedding of leaves and the protection of young leaves in buds, or by the storage of plant matter in seeds, tubers or bulbs. Deciduous trees do not necessarily represent an adaptation to seasonal temperature changes: many tropical plants also have distinctive rhythmic cycles. In addition many temperate plants either have a short growth cycle (e.g. beech trees complete theirs in 2-3 weeks), or require the 'stimulation' of winter cold (e.g. many British trees and shrubs need temperatures of under 9°C for periods of 200-3000 hours before re-starting a new life cycle; and cereal crops fruit earlier if exposed to low temperatures).

A concept which has been much used by geographers is that of the '**growing season**'. This is, however, defined in a variety of ways:
1) number of days above a minimum — e.g. 6°C (42°F) in temperate latitudes (tropical regions have higher threshholds);
2) number of frost-free days;
3) accumulated temperature, summing the mean daily temperatures (day-degrees) or multiplying the day-degrees by the average hours of daylight. This alternative shows up the quantity of heat available: thus Chicago and Aberdeen both have 7 months above the 6°C threshold, but Chicago (maximum mean monthly temperature 23°C) has 2328 day-degrees in this period whilst Aberdeen (maximum mean monthly temperature 14°C) has only 1095 day-degrees. Such approaches also tend to obscure the fact that on the land a variety of factors affect plant growth. In the tropical parts of the world temperatures are suitable for growth throughout the year, but seasons are imposed by rainfall distribution. In such areas the relationship between precipitation and evaporation is more closely related to the growing season than the temperature (although this is one of the factors affecting the evaporation rate). A better concept is the 'physiological growing season' based on a period of time when a whole range of factors is correct, but knowledge of all the factors is still rudimentary, and records of most of them are no more than of local occurrence.

Animals are affected by temperature differences, as well as plants. Cold blooded reptiles and frogs have a rest period when the environmental temperatures fall below a level at which body metabolism ceases, and prefer shady conditions during the hottest weather. Mammals, however, can live in a much wider range of temperature conditions, both as a group and as individuals. Varieties which live in colder climates protect themselves with longer body hair: the polar bear is an obvious case, but the lions introduced to Safari Parks in Britain have also grown longer hair which helps them withstand the winters.

Temperature changes affect the wider aspects of life on land. Soil-formation is determined by the rate of weathering and the activity of micro-organisms: both are more rapid in warmer (and wetter) conditions. The soil water is also affected by temperature: the greater solubility of the nutrient minerals at higher temperatures leads to more rapid plant growth, whilst if the water in the soil is frozen it is denied to the plant.

Tolerance

Every living organism in the biosphere is thus affected by a combination of factors. Too much, or too little of only one of these may destroy the organism or reduce its capacity for growth: the weakest link in the chain may be the limiting factor. Thus plants may have a wide tolerance of temperature, but a low tolerance of soil acidity. Each organism will have a range of conditions in which it can exist actively, including the optimum conditions, and a range that can be tolerated. It seems that the range of tolerance is related closely to the reproductive period. It is also clear that such ranges vary with other factors in the complex system of the biosphere: the complete balance of the life system can be upset by man's interference. Thus the introduction of excess phosphate to Lake Erie led to eutrophication as the limiting factor was altered; and the disposal of heated wastes around nuclear power stations has led to subtropical migrant fish being tempted to stay on through a winter season to be killed off in a period when there has been a break in the effluent.

151

152

153

154

Soils. Some aspects of soil study.

Plate 151 The soil forms a thin layer in which grasses can grow. It has developed on a 50cm layer of glacial debris in Scotts Bluff County, Nebraska, USA (the base of this is shown by the arrow). This in turn lies on a siltstone rock.

Plate 152 Soil structure. A prismatic structure has developed in the subsoil in Kimball County, Nebraska, based on a gravelly parent material. Such structures are often produced by the drying out of the soil.

Plate 153 Parent material. This soil profile, developed on a river terrace in Kansas, shows a series of sections resulting from vertical accretion on the site: each dark layer is a former topsoil.

Plate 154 Organic soil. The black soils of the so-called 'Muck Area' occupying an old glacial spillway between Jennison and Zealand in Michigan, and contrasting in land use with the surrounding hills. Celery and onions are grown on the rich soils. (All USDA Soil Conservation Service)

LIFE IN THE ATMOSPHERE AND OCEAN

Land classification. These photographs are related to the US Department of Agriculture system of land classification.

Plate 155 Class I land in Missouri. Deep, level silt loam soil producing a good crop of alfalfa.

Plate 156 Class III land in Minnesota. A fine sandy loam which is rather too well-drained and has problems of drought and wind erosion.

Plate 157 Class V land in Georgia. Pitcher plants grow between sparse pines on wet land.

Plate 158 Class VII land near Santa Barbara in California, with steep slopes on clay loam soil. The natural cover is grass and oak, but this area is extremely susceptible to sheet and gully erosion if over-grazed.

Plate 159 A variety of land types in California. What are the features of each group? (N.B. Beans are growing at I; artichokes at II and III; peas at IV; pasture at VI.) (All USDA Soil Conservation Service)

17

Life and the soil

The soil is a most important part of the land biosphere: it is the source of water and mineral nutrients for plants living in it; it is also the zone where plant material decays and is re-cycled into the nutrient 'bank'; and it is a home for many tiny animals, bacteria and fungi which form an essential part of the 'detritus pathway'. There is no equivalent medium in the oceanic environment, since the ocean waters themselves supply many of the requirements afforded by the soil on the land.

Soil is basically broken-down rock material. This may be found directly on top of the parent rock, or at some distance from it after transport. Thus some valley floors in south-east England have soils formed on material which was blown there by winds affecting the tundra regions in front of the Pleistocene ice sheets, whilst farther north the varied boulder clays provide another form of transported parent material. This basis of mineral matter is acted upon by the atmospheric processes, which bring about chemical and physical changes, dominated by the passage of water through the soil. Organisms living in and on the soil also modify its nature, so that the original mineral matter becomes mixed with organic waste and decomposition products, and both are re-arranged by movements taking place within the soil environment. After a time the soil often takes on a layered arrangement, related partly to the parent material, but more closely to the internal movement of water, which is associated with local conditions of climate, relief and vegetation cover.

Soil is thus something more than dirt or 'the stuff that plants grow in'. What may seem to be lifeless and uninteresting (Sir John Russel wrote: 'a clod of earth seems at first sight to be the embodiment of the stillness of death') is in reality a whole world of life with its own physical and chemical characteristics. Soil is also more than the 1917 definition of Ramann: 'Soil is formed of rocks that have been reduced to small fragments and have been more or less changed chemically, together with the remains of plants or animals that live on or in it'. A more modern approach to the study of soils is reflected in: 'The soil is a natural body of animal, mineral and organic constituents differentiated into horizons of variable depth, which differ from the material below in morphology, physical make-up, chemical properties and composition, and biological characteristics' (Joffe, 1949).

This last definition leads to a distinction between the true soil, or topsoil — which is the growing medium for plants and is a definite environment within the biosphere — and the subsoil, which is merely broken-down rock. The subsoil is, in effect, a transitional stage between the true soil and the underlying rock.

This definition of soil is that applied by the geographer, biologist or farmer. The engineer, on the other hand, will include any loose rock debris as 'soil': it may include river and glacial sand and gravel, beach deposits, and all rock materials of low strength.

Soil science, or **pedology**, has become established in the last few years as one of the most vital of the applied sciences which will contribute to man's continuing existence on his increasingly crowded planet. Soil is important in at least two ways:

1) It is **the basic medium for food and timber production**. Each soil has certain characteristics — which can be defined in terms of the mineral nutrients required by plants, its resistance or otherwise to water movement through it (closely related to grain-size), and its internal structure, which is an outcome of its developmental history. The soil scientist is able to examine a soil and assess what may be called its 'fertility' (i.e. its ability to support the growing of crops) or 'capability'. He will be able to save the farmer or forester long periods of trial and error and suggest what improvements are possible within limits set by the basic nature of the soil and the terrain in which it occurs. The Soil Survey of Great Britain has published a land capability classification, and has produced some maps on this basis which can be related to the Soil Survey and Geological Survey maps. Much present

farming practice is injurious to the natural soil, since constant ploughing and the addition of chemical fertiliser, together with the removal of almost all the organic material by, say, a wheat crop, breaks down the distinctive structure of the soil and returns it to the status of parent mineral material from the living soil environment.

2) It is also **the foundation for buildings and roads**. Thus land-use planning should begin with a consideration of the soil quality. Which areas should be kept for farming? Which are best for housing, or factories, or forestry, or recreational uses?

The soil is a natural resource, which cannot be replaced if it is destroyed. As such it is closely inter-related with all the factors affecting the biosphere — climate, vegetation, animals, underlying rock type and relief — all of which are in a state of delicate balance. It is thus vital to understand more about soils, their nature and origin, and to treat them with far greater respect. It is significant, however, that pedology is a recently-developed science. The Soil Survey of England and Wales was first recognised in 1939, and systematic mapping began in 1946. Most countries now have organisations for this purpose, although the late start means that too many decisions regarding land-use have been taken without reference to them and the damage has been done.

Soil characteristics

A detailed soil study — either by examining a section exposed at the top of a quarry, or better by digging a pit in a field — may begin with a field exposure (Figure 17.1). Several observations can be made which give an insight into the way in which the soil was formed.

1) **Three main divisions** can be identified (Figure 17.1). A pit is likely to be restricted to the upper two layers, but a depth of 2 metres is often sufficient to show these. The depth of each layer should be measured, and a scale section drawn in a note-book.

2) **Colours** tell a lot about the nature of the soil, and there is a chart of standard colours (the Munsell colour chart). Thus dark brown or black soils have a high content of humus (i.e. the products of decaying organic debris), whilst light grey soils have little. Iron compounds give yellow or red colours to the soil. Poorly drained soils are often bluish or spotted.

3) Samples of soil should be taken from the various layers and kept for **simple laboratory analysis**. One test is carried out by measuring a portion of each sample and pouring it into water: a proportion will float to the top (the undecayed organic litter), whilst most will sink.

4) A few fine-mesh sieves will enable the dried samples to be analysed in terms of **soil texture**, or grain size. The soil can then be classified (Figure 17.2). It is difficult to measure the clay sizes, since large numbers often cling together, even after drying.

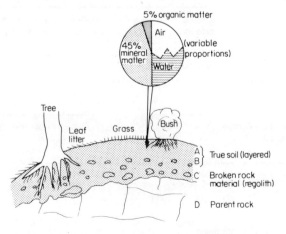

FIGURE 17:1 A soil section as seen in a quarry wall. The true soil may show a vertical layering, and will contain most of the plant roots. Beneath this will be weathered rock fragments and solid rock. The inset shows the average general composition of soil by volume. (Inset after Bridges, 1970.)

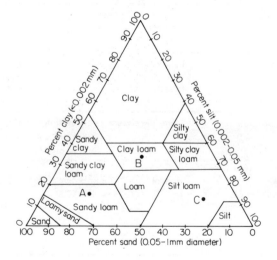

FIGURE 17:2 Soil texture. A ternary diagram showing the range of terms used after a soil has been dried and sieved. For A, B, C see Figure 17.3. (After US Department of Agriculture in Strahler, 1969.)

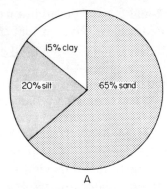
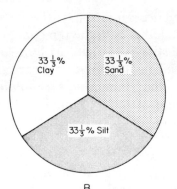

A — Sandy loam: sand fraction obvious; moulds readily when moist, but normally does not stick to fingers

B — Clay loam: distinctly sticky when moist; sand fraction difficult to detect

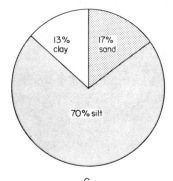

C — Silt loam: moderately plastic, but not sticky; smooth, soapy feel of silt main feature

FIGURE 17:3 *Three examples of common soils are shown in terms of fractional constituents and a description of the hand specimen found in the field. (After US Department of Agriculture, in Strahler, 1969.)*

5) The **pH value** of the soil is another test which is often carried out, using a simple field kit. The pH value is essentially a measurement of the concentration of hydrogen ions (H^+) in the soil (Figure 17.4). A sandy soil with poor heath vegetation growing on it will have a low pH value (5-5.5). Such a soil is said to be acidic because compounds of calcium, magnesium and potassium (sometimes known as 'bases'), which are essential nutrients for plant growth, are easily dissolved away by water

pH	4.0	5.0	6.0	7.0	8.0	9.0	10.0	11.0
Acidity	Very strongly acid	Strongly acid	Moderately acid	Slightly acid	Neutral	Weakly alkaline	Alkaline	Strongly alkaline	Excessively alkaline
Lime requirements	Lime needed (not for acid-tolerant crops)			Lime generally not needed		No lime needed		
Occurrence	Rare	Frequent	Very common amongst humid farmed areas	Common in subhumid and arid areas				Limited areas in deserts
Soil groups	Podzol	Grey-brown podzol Tundra soils		Brown forest Prairie soils Latosols Tropical black earth		Chestnut and brown soils		Black alkali

FIGURE 17:4 *Soil acidity and alkalinity, based on the hydrogen ion concentration (i.e. the pH value). (After Strahler, 1969.)*

percolating through the soil. A soil on chalk or limestone has a high pH value (7.5-8) because few calcium ions have been replaced by hydrogen ions. The pH value is quite a good measurement of soil fertility. If it is low, little will grow unless lime is added; if it is too high, the concentration of calcium will mean that it may be lacking in hydrogen or nitrate ions. A balance is clearly best.

The measurement of pH value assesses only one aspect of soil chemistry, but has the advantage of rapid field application. An alternative way of expressing the same general relationship is to say that a soil has a high or low 'base status' (i.e. high or low proportions of calcium, potassium and magnesium).

6) Examine a sample under a microscope. What **forms of life** inhabit the soil? Clearly much experience is required to identify the animals, but some idea of the sorts of organism living in the soil can be gained.

A soil study of this nature investigates the constituents of the soil. Which of these result directly from the initial breakdown of the parent rock material? Which are due to subsequent chemical activity? Which are due to organic activity? Most of the lines of investigation followed concern the solid material, mineral and organic. Water and air are also important constituents (Figure 17.1). They vary in relative proportions. Thus after rain there will be more water and it will be moving downwards until the soil becomes saturated (i.e. all pore spaces are filled), whilst during a hot sunny day there will be more air, and water will be drawn up to the surface and evaporated (gradually emptying the pore spaces). Soil water is a complex chemical solution, including a wide variety of soluble salts — bicarbonates, sulphates, chlorides, nitrates, phosphates and silicates. Soil air tends to have a higher proportion of carbon dioxide than the air above soil level.

The combination of soil texture and water/air content results in a range of **soil structures**. The particles of soil are held together by tiny clay particles and organic particles having colloidal properties, which carry out this function in the presence of water. As well as holding the soil together, the colloidal substances hold in the mineral nutrients. The units of soil structure are known as **peds** and the spaces between are important for the circulation of water and living organisms in the soil. There are five main categories: structureless, platy, crumb (granular, small aggregates), blocky (irregular bounding faces on larger masses), and prismatic (large column-like masses).

Thus a soil with a high proportion of clay-colloidal matter will be formed of large cohesive structures and will contain a high proportion of mineral nutrients. Whilst this makes it basically fertile, it will also be difficult for the farmer to plough it because it is heavy and sticky. Ploughing, and the use of chemical fertilizer tends to break down the soil structure, and creates another problem since increasing quantities of mineral nutrients will have to be added in order to compensate for the easier loss to solution in percolating waters.

The arrangement of the soil in layers of varying textures and colours is known as the **soil profile** (Figure 17.5), and this has come to be regarded as a feature of mature soils and a suitable basis for classifying soils.

Soil formation

The characteristics of a soil are thus due to a combination of factors: parent material, climate, relief, organisms and time. Many are interdependent, so that climate, for instance, is often related to relief features in mountainous areas, and affects the types and numbers of organisms present.

1) The **parent materials**. The immediate parent material is the *C* horizon of the profile (Figure 17.5) — either the local rock broken down, or the transported 'drift' material lying on top. Weathering causes the rocks to break down into a mixture of soluble and insoluble materials. The soluble matter is mostly in the form of carbonates and bicarbonates of the mineral bases calcium, magnesium and potassium. The effect of weathering on a rock like granite is as follows:

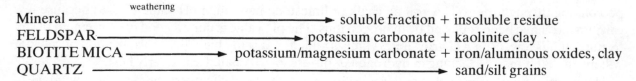

Some soluble matter is removed in stream water; the insoluble residue forms a framework for the development of a soil. The silt and sand grades produce the 'soil skeleton', whilst the clay is important in retaining the soluble plant nutrients since water cannot move quickly through its fine pore spaces. Soils on the same rock-types vary, however, and similar soils can occur on a range of different rocks. There must be other factors which control the development of soils and may even override the contribution of the parent material.

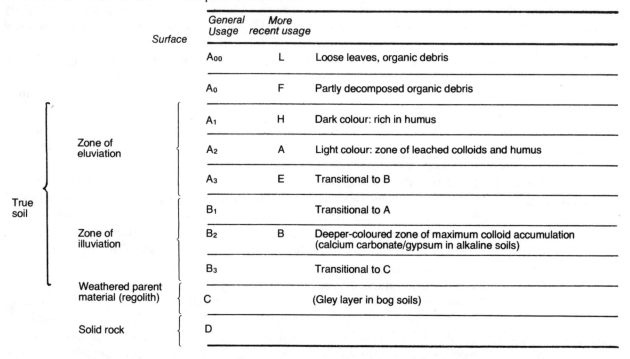

FIGURE 17:5 *Horizons in a generalised soil profile: not all of these are always present in a particular soil. (After US Department of Agriculture, in Strahler, 1969.)*

2) **Climate** is probably the most important factor acting on the weathered rock material. It affects particularly the moisture content of the soil (e.g. rainfall, evaporation and transpiration and humidity) and the temperatures in which soil-forming processes can take place. Rainfall varies from place to place in total annual quantity, in its distribution through the year, and in its intensity. Only a proportion enters the soil (after run-off and evaporation have removed much of what falls). Heavy rainfall is associated with downward movement of water and the soluble materials in the soil, and this process is known as **leaching**; an excess of evaporation over precipitation leads to water, and the salts it contains, rising by capillary forces to the surface. Where leaching is important, in humid regions, a **zone of eluviation** (i.e. 'washed out') develops in the A horizon of the profile, and a **zone of illuviation** (i.e. 'washed in') in the B horizon (Figure 17.5). In the tropics, silica is affected by this activity instead of the iron and aluminium, which remain in the surface soils. Dry areas experience the drawing-up of calcium carbonate to the surface where it may be deposited as caliche (found widely in northern Chile), and $CaCO_3$ may accumulate at intermediate depths in transitional climates, such as those of the drier margins of the prairie and steppe grasslands. In the USA a line can be drawn along the 600 mm isohyet to divide the **pedocals** (soils which contain calcium carbonate) in the drier west, from the **pedalfers** (more leached soils) in the wetter east. The temperature of the soil

and the presence of soil moisture together affect the rate of chemical weathering activity and also of bacterial activity: clay proportions increase greatly in regions of higher temperatures and humidities, and the bacteria decompose the organic debris back to mineral nutrients so rapidly that no forest floor litter can accumulate.

3) **Organisms** living in the soil also affect its gradual modification and evolution. Plants draw the ions of the bases calcium, magnesium and potassium out of the soil, but return them after decomposition. Trees may thrive on soils with little calcium, but grasses need calcium carbonate. Dead plants contribute to the humus content of the soil, and the process of humification releases organic acids which often speed up the decomposition of the mineral matter. Hydrogen ions from the acids replace the leached calcium, magnesium and potassium in cool, humid regions, making the surface zones of soils 'acidic' and of low fertility. If the acids are released into an already acidic soil the bacterial activity is reduced. Coniferous forests supply a continual rain of hard, needle-like leaves which break down slowly and release few bases to the soil, whereas temperate deciduous leaves are incorporated more easily and contribute more bases. The humus enters into an intimate mixture with the clay fraction of the soil, which contains the mineral nutrients (Figure 17.6). If the mineral nutrients are lost rapidly by this process the soil has a low base-status and high H^+ ion content (the H^+ ions from rainwater replacing the bases in the surface zone), measurable as the pH value: a solution with one part in ten thousand of H^+ ions has a pH value of 1; with one part in one hundred thousand the pH is 2; and so on; a soil with a high pH value thus has a low hydrogen concentration, but a higher concentration of bases. This characteristic is also important in the soil structure, since clay particles will often bond together if the pH value is 7 or more, but will fall apart and are commonly leached in acid soils.

Bacteria also occur in the soil, producing the humus: this occurs very rapidly in warm, moist climates, allowing little time for humification or the release of organic acids. The differences between the tropical, temperate and cold climates (and bacteria are hardly able to affect the organic debris in the last) can be explained on this basis alone (Figure 17.6). Bacteria are also involved in the nitrogen and sulphur cycles (chapter 15). Animals in the soil are effective in turning over and burrowing in the material, which is redistributed in the process: a well-burrowed soil will have less vertical differentiation than one which contains few soil organisms. Small mites also assist in the breakdown of plant tissue by reducing it to very small particles and by digesting it.

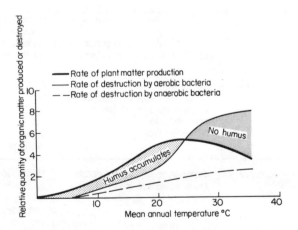

FIGURE 17:6 *The relative rates of production and destruction of organic matter in soils related to bacterial activity in different climatic zones. (After Senstivs, in Strahler, 1969.)*

4) **Relief** is a further factor (Figure 17.7), since true soils with a full profile can develop only on fairly flat surfaces where erosion is slow. Flatter upland surfaces have well-leached, thick soils, since uplands also attract heavier rainfall; flat bottomlands are poorly drained and have thick gley soils. In addition, relief affects aspect and sunniness, and this is an important factor in soil warmth, soil moisture and hence in the activity of the soil organisms.

5) As with any process involving evolution, or development, the production of a true soil requires **time**. Soils are less well-developed on parent material which has been deposited recently in a valley

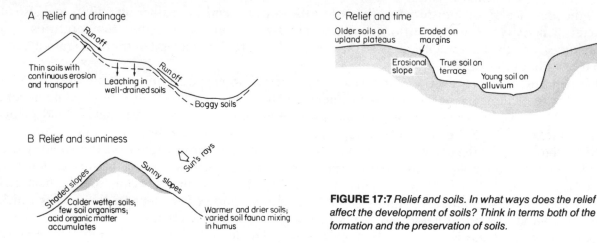

FIGURE 17:7 *Relief and soils. In what ways does the relief affect the development of soils? Think in terms both of the formation and the preservation of soils.*

floor than they are on a well-established mantle of rock debris. The world's oldest soils are probably to be found on the flat upland surfaces of the low latitude areas which have not been disturbed by the effects of glaciation during the last million years. In most areas, several thousand years are needed to produce a mature soil (i.e. with a fully-developed soil profile), and the oldest soils in Britain north of the English Midlands can be no older than the last glacial episode, some 10 000 years ago. Thus the thoughtless stripping of vegetation in many parts of the world leading to soil erosion in time of heavy rains has resulted in an irrevocable loss. Figure 17.8 demonstrates how soils and vegetation communities develop together over a period of time.

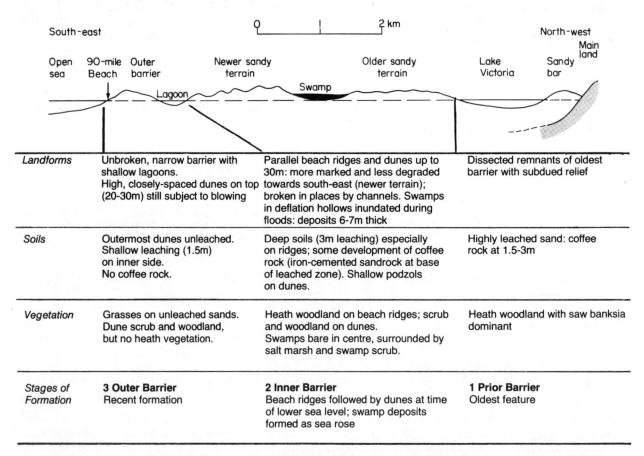

FIGURE 17:8 *The East Gippsland sand bars, Victoria, Australia. This shows an example of the development of relief, soils and vegetation together over a period of time since the later stages of the Ice Age (i.e. possibly over the last 20 000 years). (Data from Bird, in Steers, 1971.)*

These individual factors work together in a range of soil-forming processes, or trends.

1 Podzolisation occurs in the soils of cool, humid regions where the leaching process is dominant. True podzols are limited to the coniferous forest and heath vegetation communities, but a wide range of leached soils occurs under deciduous forest and pastureland in these climates.

True **podzols** (Figure 17.9) are associated with the acid humus produced by heath and coniferous plants and known as **mor**: such plants do not take up much calcium, magnesium or potassium and therefore do not return these to the soil, so that the base status is low and the pH value acidic. The mor is produced by the very slow decomposition of organic matter. Rainwater passing through this layer becomes very acid by taking up the excess of H^+ ions and is able to dissolve iron and alumino-silicate minerals. A bleached, eluvial zone of quartz sand, known as the **albic zone**, is underlain by a darker, illuvial (**spodic**) horizon, where the leached material (i.e. humus and iron salts) accumulates. The reason for this accumulation is not fully understood, but it may give rise to a layer which the water cannot penetrate, known as 'hardpan'.

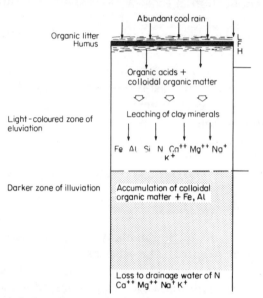
A Podzolisation process

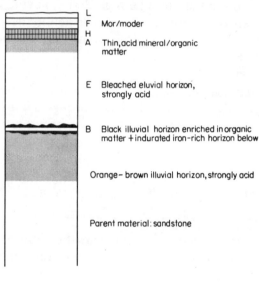
B Humus-iron podzol profile, Surrey (Folkestone sands)

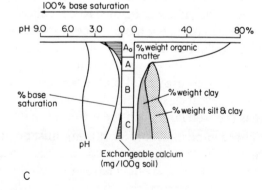
C

FIGURE 17:9 *Podzols.*
(A) The process of podzol formation.
(B) A typical podzol profile.
(C) The profile diagram indicating the nature of vertical differences in composition.
((A) and (B) after Strahler, 1969; (C) after Cruikshank, 1972.)

Brown earths are less intensely leached soils which develop with more rapidly and fully decayed humus known as **mull** (or the intermediate **moder**) beneath a variety of deciduous woodland and grassland communities. This is less acidic and surface leaf debris breaks down each year. Strong leaching still occurs, carrying the surface clays down to the *B* horizon, but the minerals themselves are not broken down chemically. Clay skins cover the peds at this level, an **argillic** horizon. The greater quantity of organic humus in the soil encourages the presence of organisms, which turn over the soil and restrict the vertical differentiation in the soil profile.

2 Calcification is characteristic of dry regions. The low rainfall causes little downward movement and evaporation follows, drawing up and precipitating calcium and some magnesium ions. Grasses produce a mull humus in large quantities in regions of slightly higher rainfall totals, returning the fertile base constituents to the surface soil and forming a deep, rich A horizon.

3 Ferralization (formerly called laterization) occurs in the humid tropics where heavy rain and uninterrupted warmth give rise to a deeply weathered layer. Dense forests produce a heavy leaf fall, but bacteria break this down at once: the circulation of mineral nutrients is particularly rapid. Leaching is heavy, but silica is removed rather than iron or aluminium, which accumulate at the surface. The removal of the silica leaves a porous and firm soil, rather than a soft and sticky one.

In regions of more marked dry season, the alternation of wet and dry conditions gives rise to increased movement in the soil, and the formation of iron crusts. Soils throughout the tropics tend to be low in fertility due to the lack of lasting base and humus content. It must be said, however, that tropical soils have not received the attention accorded to temperate soils. No simple picture of this nature can therefore be regarded as having a wide application in those areas: what is mentioned is merely one of many processes in action there.

4 Salinization takes place in arid areas where drainage is impeded. In well-drained sectors the soluble salts remain in the soil, or are re-distributed to lower areas. Salts accumulate where evaporation is not compensated by salt removal, and where salts are drawn up to the surface with the water in capilliary action. Surface accumulation in the form of encrustations form in **solonchak** soils. Sodium ions tend to disperse through a humic horizon, giving a **solonetz** soil.

5 Gleying is due to the presence of water in the soil for long periods, which results in anaerobic (non-oxygenated) conditions: plant growth is inhibited and the absence of oxygen leads to the chemical reduction of iron minerals with a consequent change to blue colours. This is most common in low-lying, boggy areas in moist, cool climates — tundra and a range of humid temperate areas. Peat forms on top of a structureless, sticky clay.

Soil classification

A whole range of different classifications of soil has been made, and in many cases these have been related partly to the local view of the formulating soil scientist. The Russian soil scientists, who played a large part in the development of pedology in the early twentieth century, based their classifications on the associations of soil-types with the climatic and vegetation zones, which were so clearly demonstrated in the open expanses of their country. Early British classifications of the same period were based on differences in the parent material and its geological relationships rather than on the world climatic regions. Recent classifications have referred more closely to the basic processes of soil formation, but have varied because they have often been designed for different ends. Some are based on the degree of leaching, drainage characteristics and type of humus; others on one or two of these characteristics, or on whether the A, B, and C horizons of the profile are present: other classifications are based on the soil texture and whether there are changes between the topsoil and subsoil. Local classifications can have their own emphases: thus the classification of soils in irrigated areas is based largely on the differences in salt content.

Three important considerations must be borne in mind in devising a world soil classification today. The first is, that like all worldwide classifications of natural phenomena, only the most distinctive forms can be picked out. There are a whole range of intermediate and transitional soil types, and one attempts to characterise the varieties which demonstrate the really basic features of soil types. Secondly, the traditional classifications were based on the concept of a fully-developed (i.e. mature, or zonal) soil found beneath woodland, whereas the greater part of the world's land surface is now occupied by soils which have been modified by farming — ploughing and the addition of fertilizer. And thirdly, much work has been carried out in the last few years and advances in soil science technology have been made. A system which follows up these points is that devised by the American Soil Survey in 1960 and called the 'Seventh Approximation': in other words it is regarded as a stage in the journey towards a fully adequate system. The Seventh Approximation adopts new names

A Horizon Characteristics

	Mollic	Anthropic	Umbric		Plaggen			Histic		Ochric
Surface Horizon (Epipedon)	Dark, thick; over 50% exchange capacity saturated by base cations	Similar to mollic plus high phosphate after long farming.	Dark, but with less than 50% exchange capacity saturated by base cations		Man-made, over 50cm thick with some characteristics of original soil.			Thin, seasonally saturated with water; high organic carbon.		Light colour; low in organic carbon; thinner than histic.

	Argillic	Agric	Natric	Spodic	Cambic	Oxic	Calcic	Gypsic	Salic	Albic	Duripan	Indurated Fragipan	Plinthite
Subsurface Horizons	Significant clay accumulation. Illuvial.	Compost after farming adds clay/humus.	Argillic + columnar structure >15% sat. sol. Na+.	Accumulated free sesquioxides and/or organic carbon.	Altered, liberating iron, forming clay; + loss of original structure.	Low content weatherable matter: clay is kaolinite.	Enriched with CaCO₃ in secondary concretions >15cm thick	Enriched with CaSO₄ >15cm thick	Enriched with more soluble salts >15cm thick	Clay + free iron oxides removed	Cemented by silica or aluminium silicate	Loamy + platy structure; hard when dry	Highly weathered, poor in humus; hardens

B Ten Soil Orders

Orders	Suborders		Orders	Suborders
Entisols Weakly developed	Aquent Arent Fluvent Psamment Orthent	Water saturated seasonally (gleyed) Strongly disturbed by man On alluvial deposits Sandy/loamy texture to 50cm depth Others inc. arid soils, lithosols	**Spodosols** Spodic horizon (podzois)	Aquod Ferrod Humod Orthod	With gleying Little humus in spodic horizon: Fe instead Little iron in spodic horizon: humus, aluminium Spodic horizon with both iron and humus
Vertisols Cracking clays of areas with dry season	Udert Ustert Xerert Torrert	Usually moist Dry for short periods, unmottled Dry for long periods Usually dry	**Alfisols** Argillic horizon + mod-high base content	Aqualf Boralf Udalf Ustalf Xeralf	With gleying Others in cold climates Others in humid climates Others in subhumid climates Others in subarid climates
Inceptisols Moderately, often quickly developed	Aquept Andept Tropept Umbrept	With gleying after water saturation High content of clay; often on volcanic ash Tropical areas Umbric epipedon	**Ultisols** Argillic horizon + low base status	Aquult Humult Udult Ustult Xerult	With gleying With humus-rich A horizon In humid climates Others in subhumid climate Others in subarid climate
Aridisols Usually dry; ochric epipedon + calcic layer	Argid Orthid	Argillic or natric horizon Cambic; duripan; calcic/gypsic/salic/horizon	**Oxisols** Oxic horizon or plinthite near surface	Aquox Humox Orthox Ustox	With gleying Humus-rich A horizon Others in humid climates Others in drier climate
Mollisols Dark surface, high base status	Alboll Aquoll Rendoll Boroll Udoll Ustoll Xeroll	Albic, orgillic horizons Gleying Parent materials > 40% CaCO₃ Others in cold climates Others in humid climates Others in subhumid climates Others in subarid climates	**Histosols** Organic soils	(not finalised)	

FIGURE 17:10 *The American Soil Survey staff classification, known as the 7th Approximation (1960). Notice how this is related to other classifications, and to man's effect on the soils. (Data from Bridges, 1970.)*

(such as epipedon for surface horizons darkened by organic matter), a range of terms for subsurface horizons, and three terms for indurated horizons (Figure 17.10). All the horizons are combined in a three-dimensional unit of soil, known as the **pedon**. Ten soil orders are recognised and these can be divided and subdivided.

This classification has many merits as a scientific and rigorous basis unrelated to the earlier, often confused, terminology, but it has not been adopted everywhere, and it seems that it may remain as simply one more interesting soil classification. It illustrates an important point concerned with the use of classifications. Any classification should be made because it is going to be a useful codification of knowledge: it is not made up as a sort of intellectual game. Thus the soil scientist will wish to have a basis of classification which matches recent developments in his work. At times there will be the need for a completely new look and even a new set of terms, and this may mean that his classification becomes less useful to others who require a less specialized approach.

A simpler classification is adopted here, related to the climatic classification in chapters 9-14. The supply of water to the soil is perhaps the most important climatic influence on soil formation, and is also the major limiting factor in the development of the land biosphere. The temperature factor, which is the next most important, is regarded as the major subsidiary climatic factor in these classifications.

This classification considers mainly the mature, or **zonal soils**, which characterise the fullest play of the various soil-forming factors. If these have not developed it is because of some inhibiting factor. There are wet soils where drainage (and thus leaching or upward movement) is impeded; dry soils due to aridity or extremely permeable rock material; soils on steep slopes where transport of weathered debris is rapid; and soils in lowlands where deposition is rapid. Soils where drainage conditions are unusual (i.e. impeded or too rapid) are classed as **intrazonal**; those where no soil profile is differentiated as **azonal**.

When soil classification is brought to the more local level, climatic conditions tend to be less overwhelmingly important. Soil types can be seen to be related to the landforms, underlying parent material and their moisture content. Thus one hill-valley cross-profile may show a transition from freely-drained hill-top areas with deep soils (e.g. brown earth), to a shallow slope soil, and a gleying soil on the valley floor. These can be grouped together as a **catena** (Figure 17.11). Catenas are made

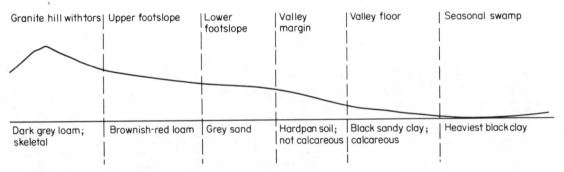

FIGURE 17:11 *A catena: i.e. soil types which occur regularly with particular landforms. This is a useful concept when mapping large areas, like the one depicted in East Africa. (After Bridges, 1970.)*

up of a number of smaller units, known as **soil series**, which are the units of mapping on a detailed scale. This system is adopted by the British Soil Surveys. The classification on which the following descriptions of world soils are based is zonal (Figure 17.12).

The major world soil groups
Soils of the humid tropics
The humid tropical regions of the world are found close to the equator (chapter 10). They receive over 2000 mm of rainfall each year, and temperatures average 25°C with little seasonal variation. Moisture and temperature conditions are therefore conducive to high rates of chemical activity, and

Zonal (layered soils)	Climate wet all year	Tropical	Few true soils (Laterite — not a true soil)
		Temperate	Podzol Brown earth
	Climate seasonally wet	'Mediterranean'	Brown soils Cinnamon soils
		Tropical	Ferrisol Ferruginous soils Vertisol
		Temperate	Chernozem Chestnut brown
	Climate arid		Grey desert soils (Few true soils)
	Cold all year	Tundra	Few true soils
Intrazonal			Rendzina, Ranker Gley Solonchaz, Solonetz
Azonal			Lithosols Regosols Alluvial soils

FIGURE 17:12 *A simple classification of world soils.*

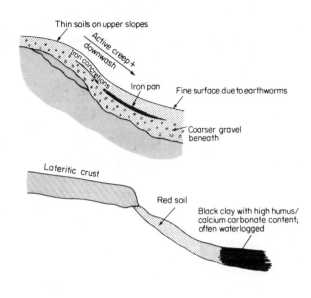

FIGURE 17:13 *Soils of the humid tropics. Heavy rainfall and rapid weathering scarcely allow the development of proper soil profiles. (After Selby, 1971.)*

to the rapid growth of luxuriant vegetation. Massive, dense forests supply a continual rain of organic material to the forest floor, where bacteria are extremely active and water movement is continuous. In addition many of these areas have experienced the present soil-forming conditions over a much longer period than the more temperate parts of the world, where the Pleistocene glaciations involved climatic fluctuation and the deposition of new parent materials. Weathered rock material may be 30 m deep, and may often be difficult to distinguish from the true soil (i.e. the section affected by soil forming processes at the moment).

In these regions where rainfall totals are high, and dry seasons are too short to make an impact, the main soil varieties are often determined by the position of the soil with respect to the drainage conditions. If the soil is well-drained, leaching is intense, abetted by complex chemical reactions under high temperatures. The litter supplies a high base content to the surface soil horizons, which remain neutral to only moderately acid despite the intense leaching. Silica becomes more soluble than iron or aluminium, which are left behind, colouring the surface soil red. If drainage is impeded there may be zones of silica enrichment (mostly in the drier parts) or of iron enrichment (Figure 17.13). There is, however, much variation from this general pattern, depending on the nature of the underlying rock (particularly whether or not it supplies base ions to the soil), the length of any drier seasons experienced, and the position of the water table. Drier conditions lead to surface cracking and the accumulation of salt in certain horizons. The extremely wet conditions in which mangrove swamps form along coasts and river banks give rise to gleyed soils in which nodules of iron pyrites form. Thick accumulations of peat are unusual in the humid tropics unless tree debris builds up at a rate sufficient to overcome the rate of bacterial breakdown.

Laterite is a term used often in connection with tropical soils, but it is best kept strictly for a hardened, concentrated horizon of iron and aluminium oxides, and avoided as a soil name. Laterite may be formed of nodules, or it may have a vesicular (cellular) or even slaglike nature. The laterisation process is associated with exposure of an iron-accumulation horizon to the atmosphere, leading to irreversible hardening after erosion has stripped the overlying leached soil horizons (Figure 17.14). Such horizons are exposed commonly on ancient erosion surfaces, and may even be re-weathered to become the parent material of a new soil. Laterite can be dried out and used as

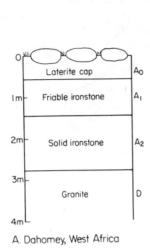
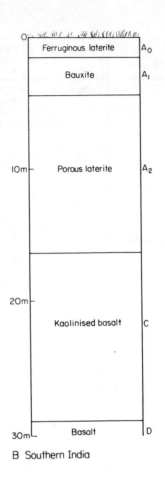

FIGURE 17:14 *Lateritic 'soils'. Iron and aluminium accumulate at the top of the soil, but any organic matter or mineral nutrients have been lost through the porous structure of the laterite. (After McNeil, Ehrlich, 1971.)*

building blocks, but, unless it can be broken up in some way, it is a very poor basis for farming. In many tropical areas the removal of forest for farming has resulted in rapid loss of topsoil and exposure of laterite crust: thus at Iata, a Brazilian experimental station in the Amazon Basin, the soil disintegrated after only the second planting and soon baked hard.

Soils of humid temperate regions

As in the tropics, leaching is dominant in the soil-forming processes of humid regions in middle latitudes, but once again there is a wide range of final products. The rainfall which engenders the leaching also gives rise to the dominant forest-type vegetation in these areas. Two main groups of soils occur beneath the forests: the podzols, which are found mainly in the northern coniferous forest belt and on infertile, sandy and gravelly areas in warmer climes, and the brown earths, normally associated with deciduous forests. It must be emphasised, however, that there are many intermediate varieties, as well as a whole range of intrazonal and azonal soils in these regions: all are characterised by a degree of leaching.

The distinctive feature of **podzols** is the light-grey horizon just below the surface (Figure 17.9). They are the dominant soils of the zone between 50° North and the Arctic Circle, where fluvio-glacial sands and gravels are widespread amidst heavier boulder clay soils and bare rock, and podzolization proceeds rapidly. Outside of this zone they are limited to areas where porous parent materials (especially sands) and conifer/heath vegetation predominate. They even occur on well-drained river terrace deposits in the tropics.

In the main zone there are long, freezing winters and brief, cool summers. Precipitation is not excessive (500 to 1000 mm), but evaporation is low and snow-melt in spring supplies further moisture which encourages leaching. Organic material from the conifer/heath plants is low in bases and plant nutrients: the soil becomes acid, and few soil organisms can live in it. Mor humus, with slow breakdown by chemical, rather than bacterial, activity is present. The leaching process carries iron and aluminium compounds out of the surface horizons, leaving the silica to predominate in the

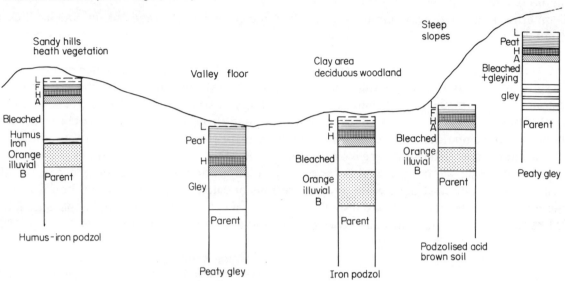

FIGURE 17:15 *A variety of podzol soils found in various situations in Britain. (After Bridges, 1970.)*

ashy-grey *A* horizon. Variations occur according to the original parent material, vegetation cover, relief and drainage characteristics (Figure 17.15).

Red and yellow coloured podzols are common in the warmer, east coasts of the temperate zone: these are transitional between podzols and ferralites due to year-round rainfall in the warm conditions of south-eastern USA, Brazil, southern China and eastern Australia.

There is also a transition between podzol and brown earth soils. Brown soils can be more acid than usual, and podzols can be reduced to the slightest grey horizon. **Brown earths** are also essentially leached soils, but lack the distinctively coloured podzol layers. They develop on a wide range of parent materials, varying from clay to sand in texture; they are commonly found in areas which were covered originally by deciduous forest and are thus often developed on clay or loam parent materials in England. The climate of these regions is not so extreme as that of the main podzol zone: winters are shorter and rainfall is more evenly distributed throughout the year. Leaching is thus more moderate, and the litter produced from the deciduous trees provides more nutrients encouraging a greater development of the soil fauna and more efficient breakdown of the humus, which is then distributed and mixed with the soil minerals by earth-worms. Such soils are more fertile and less stratified than podzols, and have been largely cleared of the original forest and used for farming in western Europe. The leaching process in a brown earth removes carbonates from the surface, at times giving a lower pH value (4.5 to 6.5) than with podzols. Humus is normally of the mull type, and the clay fraction may form a constant proportion throughout the profile.

Varieties of brown earth are dependent on the humus produced and the parent material:

	Humus	*Parent material*	
Acid brown soils	Mull/Moder	Sand/silt, under 20% clay	Clay enriched in *B* horizon
Leached brown soil	Mull	Clay/silt	
Ferritic brown soil	Mull	Iron sands	
Gleyed brown earth	Mull	Poorly drained alluvial material	
Brown warp soils	Mull	Better drained alluvial material	
Rendzina (azonal)	Mull	Limestones	

Soils of seasonally wet regions

These regions can be divided into three groups.

1) Areas with wet winters: the 'Mediterranean' regions, in which a variety of soils including red (terra rossa), brown and cinnamon coloured soils are found.

2) Tropical areas with wet summers: the zones of savanna grassland and seasonal forest, where ferruginous soils are the most common variety.
3) Temperate areas with wet summers, also often the province of grassland, in which black and chestnut brown soils occur. The marked seasonal regime of alternate drought and rainy conditions affects the development of soils in these regions.

The **Mediterranean-type regions** are characterised by cyclonic winter rain which causes leaching, and a summer drought leading to concentration of calcium carbonate near the surface in areas where limestone rocks form the parent material. A wide range of soils develops under these conditions, from the brown earths in areas of very short summer drought where the leaching process is dominant, to the cinnamon soils where the drought lasts for five or six months. Around the margins of the Mediterranean Sea itself there is an unusual predominance of limestone rocks (Figure 17.16). Regions with longer summer droughts have soils which grade into the varieties characteristic of deserts: moisture penetration is slight and concretions of calcium carbonate occur below a shallow humic zone as long as there is at least a small proportion of calcium carbonate in the parent material. Soils like this have a light, yellowish-brown colour, and are termed **cinnamon soils**.

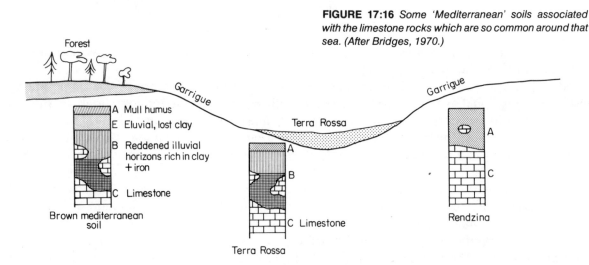

FIGURE 17:16 Some 'Mediterranean' soils associated with the limestone rocks which are so common around that sea. (After Bridges, 1970.)

Tropical areas between the evergreen rain forests close to the equator and the subtropical deserts have a summer maximum rainfall, and are often clothed with savanna grassland. Under these conditions **ferruginous soils** are found. Horizon development is more marked than in the soils beneath the tropical forests, but the soils are not so deep in these regions of seasonal rainfall (cf. Figure 17.17 and Figure 17.13). The **vertisols** of these regions suffer alternate shrinkage and expansion, which has the effect of redistributing the soil materials.

Temperate areas with summer rainfall maxima are found in the interiors of continents in grassland areas to the south of the boreal forest and deciduous forest belts — particularly the prairies and steppes. These regions have a low rainfall, which comes largely in summer, and cold winters. Chestnut brown soils are the most typical, and grade into podzols or brown earths on the moister margins. **Chernozems** are the most characteristic soils and their greatest development is in south-central USSR; other smaller areas occur in the prairies of North America and in parts of the pampas of Argentina. The most common parent material beneath the chernozems is loess, a loamy deposit formed by wind action winnowing out the finer particles from the bare soils in front of the Pleistocene ice sheets and transporting them to regions farther from the ice. The soils in south-central USSR and the northern prairies are frozen in winter, and covered by moderate falls of snow; snow melt in spring provides little soil moisture, and summer rain never gives rise to intense water percolation through the soil. Grasses, but not taller plants, will grow, though there may be patches of herbaceous plants, and trees increase towards the forest margin. The grasses produce a

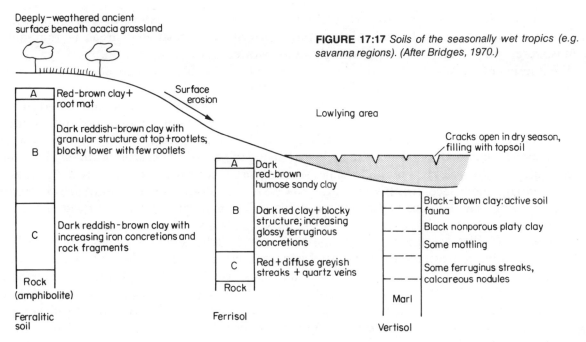

FIGURE 17:17 *Soils of the seasonally wet tropics (e.g. savanna regions). (After Bridges, 1970.)*

thick mat of roots in the upper soil, but decomposition is retarded by late summer drought and winter frosts. The humus is thus not lost from the surface, and is distributed through the upper soil layers by the rich fauna. The humus content of the *A* horizon may range from 10 per cent near the surface to 2 per cent at a depth of 80-100 cm. Most leaching occurs in spring and early summer, but a later reversal of water movement concentrates calcium carbonate in the lower soil horizons. Such soils are extremely fertile because of the humus content near the surface, but this fertility can soon be lost by ploughing. The North American equivalent is slightly degraded due to greater leaching and less calcium carbonate accumulation.

The **chestnut brown soils** of the drier margins develop beneath bunch grasses and salt bush. Organic matter is in shorter supply due to slower growth resulting from lower rainfall and higher transpiration rates. Surface accumulation of humus is in a shallow zone (up to 25 cm) and calcium carbonate accumulates nearer the surface. At times salts may accumulate to give rise to sodium-enriched horizons (**solonetz soils**).

Soils of arid areas

A large sector of the Earth's land surface is desert, where there is little rainfall and unusual weathering conditions. Plant and animal life are highly specialised and the soils are also distinctive. A coarse product of weathering is redistributed by occasional floods and the wind, and during this process soil parent materials are sorted into gravel, sand and the finer materials. Moisture is insufficient for leaching, the humidity is low and daytime temperatures high. Upward water movement is dominant, and calcium salts (carbonates and sulphates) together with some iron or silicate compounds are deposited on the surfaces of rocks and large stones as the water is evaporated into the atmosphere. The calcium salts give rise to a whitish surface concretion known as calcrete, whilst the iron and silicate materials give a dark, shiny surface ('desert varnish'). The small supplies of organic material are soon blown away by the wind. There is little profile development: desert soils have often been called 'non-soils' or youthful soils with limited leaching — e.g. the **grey desert soil**. In the lower lying parts of deserts where the water table occasionally rises to a position close to the surface, salts accumulate in the soil due to the evaporation of lakes formed after wet periods. periods.

Soils of cold areas

The coldest parts of the world outside of the permanent ice caps have a tundra climate, and the soils are frozen for a large part of the year. Only 2-4 months have above-freezing temperatures, and precipitation is low: these areas are virtually frozen deserts. The most extensive areas of this type are in northern Siberia, Canada and Alaska, but there are also restricted areas in the southern hemisphere, and on high mountains throughout the world.

Weathering processes are slow, concentrated in the period of alternating freezing and thawing of the soil. The subsoil does not melt, even in summer, and is known as the **permafrost zone** which may extend down to over 300 metres: only the top few centimetres are modified by soil-forming processes. Soil thaw in summer means that the water in the soil cannot percolate downwards through the impermeable permafrost, and the surface layer becomes mobile, flowing gently downslope (solifluction). Finer material collects in hollows, leaving the coarser debris on the hills. These tundra areas occur mostly poleward of the treeline, since permafrost near the surface prevents trees from getting enough root space. Any plants which grow do so very slowly and have short growing seasons, so that they are small in size and provide little litter. Organic matter decays slowly in the frozen conditions, and little water movement occurs in the soil for three-quarters of the year.

Meltwaters leach the surface soils in the early summer, and the depth affected is determined by the upper level of permafrost and the amount of water already in the soil. The movements within the soil due to solifluction cause local piles of soil materials, which may break through the covering of mosses and spill out on the surface. This sort of event involving movement and churning up of the rock debris militates against the production of soil horizons (Figure 17.18). True soils develop only where there is better drainage, as on the margin of a river terrace or a hill-top area over permeable rock; elsewhere **gleys** are common.

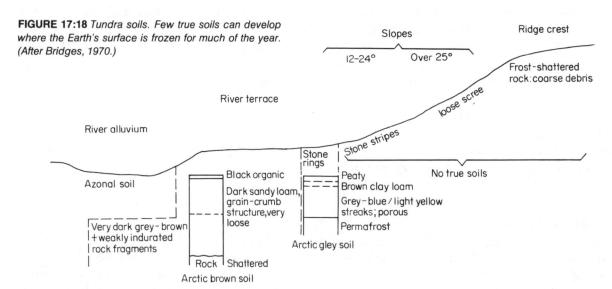

FIGURE 17:18 Tundra soils. Few true soils can develop where the Earth's surface is frozen for much of the year. (After Bridges, 1970.)

The study of tundra soils has gained importance as regions like Alaska, northern Canada and northern Siberia have become involved in schemes of defence, and in the problems of transporting oil in heated pipelines. It has been realised that many of the soils in areas like Britain, which were on the margins of the later extensions of the Pleistocene ice sheets, were formed in similar ways. Soils in former tundra zones demonstrate the typical churning, polygonal structures and ice-wedges found in tundra soils today, although the contemporary humid temperate processes are modifying them. A close study of soils can often provide evidence for past climatic changes.

Intrazonal soils

Within any of the soil groups local conditions of rock-type, water content and mineral content may override the climatic factors.

Limestone rocks have particularly distinctive soils, known as **rendzinas**, which grade into brown calcareous soils in temperate regions. The calcium content of the soils inhibits water movement, so that leaching effects are minimal. It also encourages vegetation which leaves a base-rich litter, and hence a varied soil fauna. The soils are thus fertile, red or brown in colour, and the main drawback is that they are dry because of the underlying permeable rock. Brown calcareous soils form on limestones with a large insoluble residue, such as the Jurassic limestones of the Cotswolds: they are deeper, subject to more leaching, and have definite horizons. Thin, unstructured soils on non-calcareous rocks are known as **rankers**.

Poor drainage also affects the soil development in most parts of the world: **gley soils** are the result, and are often part of a catena of otherwise zonal soils (Figure 17.11). If a soil is saturated with water the air is driven out and anaerobic (non-oxygenated) conditions are established. Iron compounds are chemically reduced from the brown ferric to the green/blue ferrous state, becoming more soluble and easily removed. The remaining minerals give the soil a grey colour. Gley soils develop where soils extend below the water table, or where the soil and subsoil are very slowly permeable due to fine texture or an impermeable layer near the surface.

Soluble salts accumulate in drier soils, and have a bad effect on the soil fertility. This is a problem in irrigated areas, where salts are brought to the surface by the artificially high water table position. Such soils occur within most of the major soil groups. The main salts are the sulphates, chlorides and carbonates of sodium and magnesium. These salts may form a surface crust (white alkali soils, or **solonchaks**), and leaching may cause an accumulation at a particular depth: if sodium is involved it prevents the formation of horizons and results in a black alkali soil, or **solonetz**.

Azonal soils

Azonal soils can also occur in any major soil region, but have no characteristics of a developing soil. This may be due to the recent accumulation of the parent material, or where erosion is too rapid for weathered material to remain for long. **Lithosols** are mountain soils on steep slopes. **Regosols** occur in recently-formed volcanic or dune-sand material. **Alluvial soils** occur in river valley floors. **Organic soils** develop on peat bogs.

A   B

Man's effect on soils and vegetation

Plate 160 Pelorotani Mountains in north-east Sicily. Eroded highlands and a valley floor choked with debris washed from the slopes in the background contrast with the hillsides terraced for afforestation in the foreground. (Forestry Commission)

Plate 161 Ploughing some soils may lead the formation of a hard layer just below ploughing depth. This 'plough pan' has caused a cotton plant in Arkansas to root sideways instead of downwards.

Plate 162 A Dust Bowl farm in Colorado, abandoned in the 1930's. Footprints were buried in 5 minutes by blowing sand. Later fences were removed so that the sand could move away, and the land was planted with sudan grass and cane.

Plate 163 High surface concentrations of soluble salts in formerly irrigated land of the Imperial Valley, California.

Plate 164 Contour-ploughing has prevented much soil erosion in dry areas like Nebraska and even on areas of lower slopes, as in Ohio it helps to retain surface water around the plant roots. (161-164 USDA Soil Conservation Service)

Plate 165 Rabbits were excluded from this area on a thin rendzina chalk soil in southern England. After 20 years the short, rabbit grazed grass has given way to Calluna, and beech trees have begun to grow through the gorse patch behind. (Forestry Commission)

LIFE AND THE SOIL

162

163

A

B

A

B

164

165

18

Competition and living together

Life in the Earth biosphere is dependent upon incoming energy from the Sun, and the circulating supply of mineral nutrients in the ocean and on the land (chapter 15). Its distribution is affected also by such factors as temperature, water supply, light and soil quality (chapters 16, 17). Living organisms also influence each other: they may compete for occupation of a particular area, or they may depend on one another for survival in the same space. The physical environment permits a range of possibilities; the organic environment includes all the organisms in a particular area and their relationships to each other, and operates to select a particular group of plants and animals best suited to the physical conditions and each other. As with the other factors affecting plant and animal distributions it is difficult to isolate these 'biotic' elements. There is a close-knit interdependence between organisms, the soil, ocean waters and atmosphere, and the inter-relationships between organisms are particularly diverse and complex. On the land it is generally true to say that plants modify the environment, creating local niches in which the animals can feed and live. In aquatic conditions, however, the watery environment is so inflexible to change and plant structures are so small — except around the shore margins — that there is no comparable situation.

Land plants modify the physical environment by creating shade and shelter: light intensity, air temperature and wind force are cut down (cf. microclimate of woodland areas, chapter 8). This introduces a competition factor, in which some plants are better able to function in a particular set of environmental factors. Since light is the basic energy source for plant growth, the ability to capture it gives trees an advantage in other ways. They are better able to use the water and nutrient supplies when photosynthesis can take place over as large an area as possible. Trees with denser foliage like the beech take a larger proportion of the incoming light than those with more open leaf arrangements (e.g. oak, ash). This results in a poorly developed group of low-growing herbaceous plants living in the shade beneath such trees. Competition is greatest in the early stages of growth: it has been shown that only 1 out of 2000 beech seedlings reach maturity, and superphosphate fertilizer is applied to the roots of young trees transplanted in forestry schemes in order to give them advantage over the surrounding and competing plants like bracken and heather. Plants will naturally grow best in what are called their optimum conditions, but where several plants compete for that environment a number will be unable to establish themselves there and will be forced into areas less suitable to their best growth. Thus a plant like sorrel (*Lumex acetosella*) is commonly found on soils with an acidic reaction, but it will grow more vigorously on non-acid soils if its natural competitors there are removed.

There is also **competition in the animal world**. This is to be separated from the food chain interrelationships: the lion and antelope are not competing for the same resources. Within a particular area there will be competition for food resources on a number of levels, and perhaps the most intense variety will occur between members of the same species having similar diets. Thus species of birds in the back gardens of town dwellings (Figure 18.1), the large carnivores of the African savannas, and the baboon communities of tropical forests will establish territories which they guard jealously. The variety of animal life within an area is decided partly by the number of different types of food which are available.

This introduces another aspect of the inter-relationships between living things in the biosphere. **Many live together** without affecting each other adversely, and even provide a beneficial situation for others. Thus insects play a vital part in pollination, sometimes very specifically: red clover needs the bumble bee. A variety of animals assist in the dispersal of plant seeds. Many animals and plants have an interdependent association with each other, known as **symbiosis**. Algae live with corals in

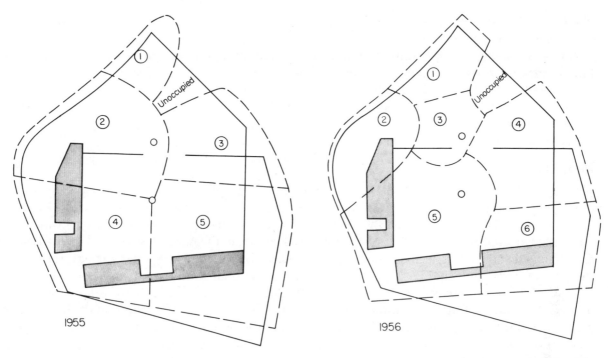

FIGURE 18:1 *Territories of male song thrushes in the Oxford Botanic Gardens at the start of the breeding seasons.* (After D.W. Snow, in Watts, 1971.)

reefs, whilst nitrogen-fixing bacteria live in the roots of leguminous plants. Some may associate together without a direct affect on either: the remora fish accompanies a shark to the place where it kills, and then feeds off the remains — a process known as **commensalism**. Others give definite help to each other (**mutualism**), as when elephants and rhinos give birds a lift to the meal, whilst the birds are removing insects which irritate the skins of the larger animals carrying them. Lichens are a close amalgamation of algae and fungi. **Parasitism** benefits only one party, although the most effective parasite will not kill his host.

Migration and dispersal

It is not sufficient, however, to look at the momentary picture of life in a particular area, because most living forms have the capacity to move or to be transported from one region to another at some stage in their lives. This means that no group of animals and plants in an area can be regarded as a completely closed system for the purposes of observation: leaves, pollen and seed are blown or dispersed in other ways beyond its boundaries, whilst animals move in and out.

Plant distributions can be altered by the dispersal of seeds. This occurs in a variety of ways, assisted by features of the seeds themselves, which enable them to stay up in the air whilst being blown by the wind, or to become attached to animal fur or bird feather, or to float in water and thus be transported by river, ocean current or ice floe. Charles Darwin recorded the fact that natives of Pacific coral islands obtained stones for their tools from the roots of drifted trees, and that seeds from the soil trapped in the same roots had germinated — although most had been killed by immersion in seawater. Birds, with their extensive seasonal migrations, can be effective agents of seed dispersal.

Such considerations might suggest that the same plants could grow everywhere in the world. That this is not the case is due to the fact that certain barriers prevent complete dispersal. The most obvious physical barriers are the oceans (for land plants) and the continents (for sea plants), and high mountain ranges often prevent dispersal across them. There is another sort of 'barrier', however, which prevents the plants from becoming established in a new area once its seeds have arrived there. This involves all the physical and biotic factors which inhibit the growth of a particular plant: the climatic, ocean water or soil conditions may not be suitable, and the competition from other plants may be too intense.

Animals can migrate by means of their normal locomotion, though they are also subjected to the limitations of the barriers mentioned above. Some are limited in their capabilities by small size and slowness of movement, and some living in the sea become fixed to the sea floor — although these often have a mobile larval stage which encourages their wider distribution. Marine animals meet a further type of barrier — that of the deepest waters: bottom-living forms from the continental shelf areas on one side of an ocean cannot migrate to the far side.

Man in the biosphere

Man is the 'highest' of all the 'higher' animals in the biosphere and has had a greater effect in modifying life on the Earth than any other creature. Man has indeed become 'lord of creation', although he has not always made the best use of the resources provided. In the past this was due largely to ignorance and to the seemingly unlimited bounty of the natural world, but the situation has now changed and there is an increasing emphasis placed on the conservation of the natural world and the most effective utilisation of the system which provides man with food.

Little of the biosphere can be regarded as unaffected by man's activities in the relatively short time he has inhabited the Earth. The total effect has grown to enormous proportions in the last century as western civilisation with its mechanical means of modifying the natural environment has spread to all parts of the globe, but man began to alter the scene as soon as he acquired the use of fire and was able to fell trees. At the same time it would be wrong to give the impression that the effect has been only harmful. Man has made many areas of the world far more productive than they were in the natural state. Dry regions have been irrigated; wet marshes have been drained and new land has been added by reclamations from beneath lake or seawater; poor soils have been made fertile by the addition of mineral nutrients and organic matter; and pests have been controlled. On the debit side of the balance sheet the construction of irrigation schemes in lowlying areas has often resulted in the accumulation of too much salt in the soil; removal of forest has led to soil erosion; and many groups of animals and birds have become, or are becoming, extinct due to overkilling by man because of their nuisance value, their use as a source of meat, fur or feather, or merely as sport.

Man and the soil

Man's progress in farming techniques, and in extending the farming area, has affected the natural vegetation over large areas of the world. Since the vegetation grows in close association with the underlying soil, involving the recycling of nutrients, removal of forest leads to changes in the soil. Many of these changes cannot be reversed within the time available to man in a single generation. In the early days of man's advances, however, it was easier for him to obtain large crops by means of simple irrigation systems in dry regions. The Nile and Tigris-Euphrates areas are the best-known of these, and supported the earliest-known civilisations, but others also developed in lowlying valleys throughout Asia and in South America. Many eventually failed because the raised water table led to waterlogged soils and brought up salts to the surface, making the soils sterile. Such a situation has also affected the Indus Plains of West Pakistan, irrigated by British-built canals in the late nineteenth century. By 1961 waterlogging and saline soils were rendering 23 000 hectares useless each year. A solution has been found whereby individual farmers sink their own tube wells and pump out water for irrigation: this has the effect of lowering the water table and leaching out the salt accumulation. Salty, waterlogged land is being reclaimed here, but this is one of the few situations in which soil, once rendered useless, can be used again within a short period of time.

As soon as man developed domesticated animals and crops he needed to remove the forest and plough up the natural grasses. The animals were harnessed in order to drag ploughs, whilst trees were cut down for fuel and so that more land could be farmed. The process has advanced at an increasing rate. Thus North America, virtually untouched by these processes and still largely forest-covered in the eastern half at the beginning of the nineteenth century, now possesses few natural forests. Ploughing has had the effect of mixing the top layers of soil, and of exposing them to the wind and rain. Even where there is a pastoral economy, the pressure of population often leads to overgrazing. Thus in Rasjathan (India) the semi-arid conditions support sparse drought-resistant shrubs, but overgrazing by goats has destroyed these plants, and winds have blown away the soil.

Such dramatic examples of soil destruction are not the only effects of farming. Every farmed area of the world has experienced increased erosion, although the results may not be so immediately obvious. Some areas have adopted farming practices which are not in great conflict with the local climate: thus the effects of soil erosion in lowland Britain, where the rainfall occurs as drizzle and the slopes are not too steep, are much less than in the lands bordering the Mediterranean Sea, where the heavier storms have a greater effect on the soils of steep slopes. Even in England, however, the wind has taken a toll by removing much fine soil and peaty material from the ploughed surface of the lowlying Fenlands around the Wash. The hilly and mountainous lands of eastern Spain, southern Italy and Greece have lost a large proportion of their original soils, which either have been washed into the sea, or have extended the areas of alluvial coastal plain and delta, largely since Classical times. Ruinous effects resulted from the transfer of the farming methods developed in the cool, humid conditions of northern Europe to North America and tropical regions. Tobacco-planting in the south-eastern states of the USA had exhausted the soils over wide areas by the mid-nineteenth century, and upland areas became dissected and gullied. The grasslands of the middle west were ploughed and the winds carried away the fine particles from the broken sod. In these areas the periods of wetter years were followed by periods of drier years when the winds blew away vast quantities of topsoil and disaster struck the 'dust-bowl' areas (Figure 18.2). This happened on a large scale in the 1930's on the high plains of the USA, but continues to be a problem, and more recently 40 million hectares of virgin lands in southern Siberia have been ploughed with the same fate. Such areas of low rainfall require special methods of farming, in which moisture is conserved by leaving the fields fallow for one year in two, run-off is deterred by contour ploughing, and drying winds are moderated by well-sited wind-breaks. Topsoil cannot be produced by nature at a speed equal to the effects of erosion.

In the humid tropics rainfall becomes heavier, the weathered layer is often deeper, and the soils are in an even more delicate state of balance with the environmental conditions. It is suicidal (literally) to remove the natural forest cover and plough the soil in traditional northern European farming style. Farming developments in such lands must be based on tree-crops, although the complexity of tropical conditions, and the slow increase of knowledge concerning these areas, do not yet permit final judgment concerning the best methods. In Dahomey, West Africa, even the replacement of forest by tree plantations led to conversion of the soils to brick-hard laterite over a

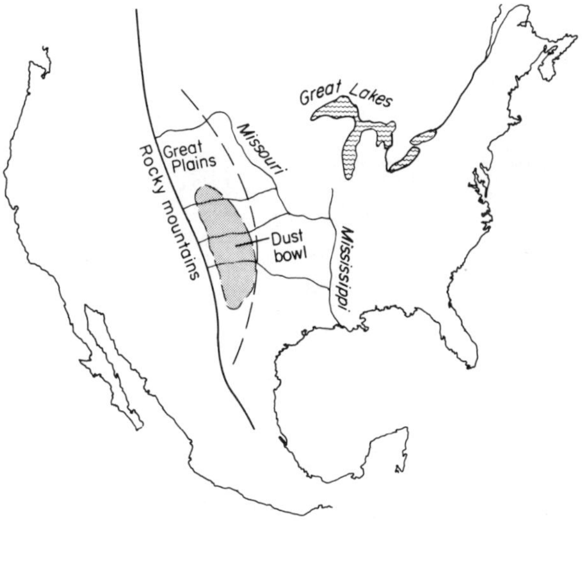

FIGURE 18:2 *The Dust Bowl. The high Great Plains at the foot of the Rockies are in a semi-arid rain shadow zone. A series of dry years in the 1930's caused the loss of injudiciously-ploughed topsoil, especially in Kansas and south-eastern Colorado.*

period of 60 years. Small clearings for native shifting agriculture lead to exhaustion of the soil nutrients within two years, but the soils can recover if they are returned to natural forest growth and left for over 40 years. Unfortunately increasing pressure on the use of land does not permit this.

Man, fire and grazing

A growing body of ecologists are coming to the conclusion that the natural vegetation throughout the world was probably dominated by trees before man was able to affect it (Figure 18.3). Whilst local areas within the widespread grasslands in temperate areas (prairies, steppes, pampas, veld), and the tropical savannas, may be due to purely natural factors, particularly relief and soil conditions, the great majority of grassland has become dominant only because tree growth has been suppressed. Savanna grasslands, for instance, occur in regions of marked winter drought, but distinctive forests, adapted to the seasonal regime, also occur in similar climates. The soil-type, moreover, is related most closely to the immediate vegetation cover, and to its supply of humus and nutrients, rather than directly to the climate: the supply of nutrients from the grass cover is different to that from the leaf-fall.

In North America the boundary between woodland and prairie, as discovered by the early European settlers, was often abrupt, whereas one might expect a transitional (ecotone) type of vegetation community. It is known that the Indians began fires, and that these swept across the prairies in uncontrolled fashion until they were doused in more humid regions. Fire destroys tree saplings, but grasses soon spring to life again. The vast herds of bison roaming the prairies would also prevent the re-establishment of trees by their grazing habits. It is interesting to record that when the large herds of bison were killed off and the fires restricted in the early days of European occupation, trees began to grow in the wetter prairie regions — until they were cut down for farmland. The evidence relating both to the burning by Indians and the revived woodland is now lost, and evidence of a similar type was lost even earlier in the Russian steppes. Burning by Indians is also thought to be responsible for the unusual thickets of Ponderosa pine in south-eastern USA (Figure 18.4).

Savannas are most widespread in Africa, significantly the continent which has seen the longest occupation by man. The dry season makes firing easy and few woody species survive. Those which

FIGURE 18:3 *The effect of man's activities in removing woodland and adversely altering the whole picture of the biosphere in the Mediterranean area. This is not something new!* (After Watts, 1971.)

Culture	Farming activity	Vegetation cover
Pre-Greek	Grains, herd animals in balanced farming	Woodland
Roman	Intensive grain farming: less balanced	Farming
Berber	Merino sheep: land deteriorates	Scrub
Recent	Sheep and goats, scattered food growing: deterioration continues	Herbs and grasses

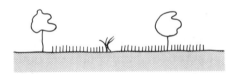

Open parkland: after years of Indian periodic (3-10 years) burning (over 10 000 years)

A

FIGURE 18:4 *Two systems of human economy which have a marked effect on the natural scene.*
(A) The original forest has given way to open parkland.
(B) The Ponderosa Pine thickets have become cluttered with undergrowth and dead plant matter, thus constituting a major fire hazard.

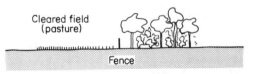

Last 100 years: no deliberate burning or unfenced pasture

B

do are resistant to the fires, with thick bark (e.g. palms and baobabs). As in North America, grazing may affect the situation, and forest/savanna boundaries are sharp. Experiments have been carried out, protecting areas of savanna from the effects of fire, and these have led to the increase of woody species, but an area denuded of its trees is short of seeds for the regeneration of woodland, so this takes some time. In addition the soils are altered by the change to grassland and become unsuitable for forest: they are too shallow and too hardened for penetration by tree roots.

Some savannas may be a result of natural factors, but where this is the case it is related to soil — the edaphic factor. 'Water savannas' occur in waterlogged areas, where sedges, mosses and marsh plants constitute a distinctive community. Grassland is also found in upland areas like north-eastern Brazil where long-established erosion surfaces are covered by soils poor in nutrients, and some have concluded that this is the natural vegetation, but it is found that fires occur here as well.

If savannas and temperate grasslands are largely the result of man's intervention (though the fires may also have been induced by natural forces like lightning), other implications follow for the development of the whole biosphere in these regions, since man has possessed the means for such intervention only for a few thousand years. The increase of herbaceous grasses would have encouraged the predominance of certain herbivores which revelled in the open conditions. The African savannas certainly have a rich fauna of this type, ranging from the giraffes and zebras of the open woodland areas to the gazelles of the desert margins. The time available would not have permitted the evolution of new species adapted to the conditions, but the change would have affected the relative dominance of the various groups.

Theories like this one concerning the origin of major grassland vegetation due to the effects of burning and grazing emphasise not only the effect which man can have on his environment, but also the interdependence of so many complex natural factors, as well as the lack of knowledge in many of the relevant fields. The simple answer is not likely to be the full picture of any natural situation.

Man and animals

Another effect of man in the world of nature has been his selection of certain animals and the eradication of others. Thus, certain grazing herbivores, like cattle, sheep, pigs and goats, were domesticated as a source of meat, milk, hides and wool, whilst others were used as beasts of burden or for sport (horses, oxen and camels). Large carnivores, important controls in the natural ecosystem balance, were restricted to the undeveloped parts of the world by the end of the nineteenth century.

The greatest effects in the extinction of species have occurred since the spread of western man through the world in the last 400 years. Species have been hunted for their fur (mink, silver fox), for their hides and skins (bison, alligator), for their oil (seal, whale), for food (caribou), or just for their shells or tusks (triton marine snail, elephant). Such hunting pursued beyond a certain point leads to the upsetting of the natural balance, and many species have become extinct, or almost so, unless protection has saved them. Other groups have been slaughtered as pests (rabbits), and others, it seems, in a purely wanton way. Thus the North American bison was almost exterminated by gangs of sharp-shooters, leading to the disruption of the Indian way of life. The passenger pigeon which once formed migrating flocks of thousands of millions became extinct when their northern forest habitat in North America was destroyed.

Removal of species affects the balance in nature. Just as the removal of forest must be paralleled by appropriate farming practice to prevent loss of the soil, so the removal of animal species must be compensated so that herbivores do not overgraze in the absence of carnivores, or so that too much inflammable underbrush does not become established due to the absence of large herbivores. Another situation where the balance has been upset is affecting many of the coral reefs at the present moment: the crown-of-thorns starfish, which was controlled by the triton snail, has become a plague threatening many of the coral reefs with extinction since the tritons were removed for the sale of their shells to tourists. If the coral reefs die, many of the islands they protect will be eroded soon by the sea. It is calculated that over a quarter of the Great Barrier Reef and up to 90 per cent of the corals around islands like Guam have been killed in this way, and since coral reefs are the homes of a rich variety of life the effect is passed on. As in the cases of soil destruction, it is difficult for man to restore the balance once it has been upset. But there are signs that he is beginning to assess each situation and its effect on the overall balance and economy. Man's conscience has been aroused, and the agitations of conservationists emphasise just how dependent he has become on the resources of his environment.

Man and the future of the biosphere

Man now uses about 41 per cent of the total land surface; up to 33 per cent is unusable since it is too dry, too cold or too mountainous; and there is a further 26 per cent, still largely forest, which could be used in a more productive way for man's benefit (Figure 18.5). Farmland will be used more intensively, but will probably not expand appreciably in area. Increasing demands for mineral resources will alter the distribution of many sources: iron, for instance, is being used up rapidly from the continental supplies, but the great use which is made of this material means that increasing quantities pass into natural waterways and thence to the ocean-floor deposits.

Increasing demand for food, and intensification of farming methods, involves the growing use of fertilizers and pesticides. The USA farms yield twice as much per unit area as African farms, but use eleven times as much pesticide; Japanese farms yield twice the USA quantities per unit area, and use a further ten times as much pesticide as the Americans. Whilst world food production increased by 34 per cent in the years 1951 to 1966, the use of phosphates increased by 75 per cent, nitrates by 146 per cent, and pesticides by 300 per cent (Figure 18.6). The effects of such pollution on the environment are becoming widespread: natural predators are particularly sensitive to the changes brought about and the inadvertent killing of wasps or other predatory insects may lead to plague outbreaks of the scale insects or the herbiverous mites on which they feed. DDT has been shown to accumulate in carnivores, and to affect the reproductive stages of many animals from oysters to ducks (i.e. especially those at the ends of food chains). The unintentional killing of pollinators on a large scale, and the overenrichment of lakes leading to the restriction of varieties to the less edible fishes — and even the extinction of these — are other effects of these additions to natural systems.

The situation in the middle 1970's is not serious, or beyond rectification, despite the many cries of conservationists, but the enormous projected growth rates, both in total human population, and in technology, will lead to parallel demands on the environment. It is thus important to understand more of what is happening, and to make the relevant investigations rapidly. Something is known about the effects of DDT, mercury, oil spills and the increasing eutrophication of lakes by the addition of phosphorus, but more must be done in the realm of monitoring and regulation by world

Use	Present	Potential
Cropland	11	24
Rangeland (pasture)	20	28
Managed forest	10	15
Reserves (80% forest)	26	0
Not usable	33	33
Total land	100	100

FIGURE 18:5 *Present and potential uses of the land surface of the Earth.*

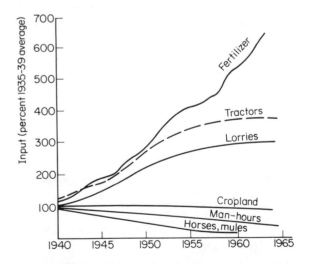

FIGURE 18:6 *Changing technology and US farming over 30 years. Notice how important the increased use of fertilizers has been.* (After Pratt, in Ehrlich, 1971.)

agencies. The increasing pressure on environmental resources is making such measures vital for the future of life on this planet. The living planet stands in danger of destruction unless the most intelligent and foresighted of the creatures to inhabit it can control his own expansion and utilisation of the wealth of resources provided.

The development of life on the Earth

Man is seen by biologists as the highest point reached by the process of evolving organisms through geological time. The concept of evolution is important in terms of understanding the present make-up of living forms in the biosphere, and is relevant in the context of a consideration of competition. Charles Darwin's greatest contribution to an understanding of this process was his theory of natural selection. In essence he said that within a particular species there is competition amongst a vast number of offspring for the available resources. Those which survive do so because of some advantage which enables them to outlive the others to reach maturity and so breed another generation. The natural environment (including both physical and biotic factors) can thus be said to select the most fitted, or best-adapted individuals. This can apply also to competition between species for the same environmental 'niche'.

An investigation of the Earth 3500 million years ago would find it largely barren of life, and even of the complex organic substances (carbohydrates, proteins) which compose living forms. An immense jump in time to 1000 million years ago would show little outward change, particularly on the land where no life existed. Even in the seas microscopes would be needed to find living forms — single-celled plants photosynthesising, releasing oxygen to the atmosphere and reproducing precise replicas.

By 500 million years ago the picture had changed dramatically. Life was still confined to the sea, but there had been a great development of the small non-chordate forms. Every group (i.e. phylum) known today had its representatives, although many of them looked very different and there were also other groups which have since become extinct. Thus the trilobites (arthropods) were probably the dominant carnivores, and the brachiopods the main shelled filter feeders fixed to the sea floor; but the molluscs, worms, coelenterates, echinoderms and sponges were also present. By approximately 300 million years ago further changes had occurred. The land areas were becoming the home of increasing numbers of living forms. Forests of pteridophytes and primitive seed-bearing plants (cycads and pteridosperms) were becoming widespread in the lowlying swampy areas of warmer lands: this was the period during which the main coal seams (Upper Carboniferous age) were formed. These forests were inhabited by numbers of insects and also by amphibious chordates: large, lumbering, low-slung creatures. Aquatic environments, both on land and in the sea, had seen the development of a wide variety of bony fishes and cartilaginous sharks, so that the chordate group had become well-established. At a point 100 million years ago the land was more completely covered with plants, including the early representatives of the most advanced land plants, the angiosperms (flowering plants), appearing on the scene parallel with the pollinating insects. Large dinosaur reptiles and numbers of birds also populated the landscape at this period, whilst the seas showed fewer changes. More advanced groups of fishes together with active carnivorous cephalopod molluscs (ammonites and belemnites) formed the diet of the great marine reptiles (ichthyosaurs and plesiosaurs). Mammals existed on the land, but as small, rat-sized insect-eating varieties, quite insignificant beside the reptiles. Coming nearer to the present time, to 50 million years ago, there had been further changes, and the living forms resemble modern groups to an increasing extent. The land and marine reptiles had been replaced by mammalian forms, the flowering plants had largely taken over from the earliest seed plants and other more primitive forms, and there were large numbers of flightless birds. Earliest forms of man emerged between 10 and 2 million years ago.

The fossil record thus provides evidence for a series of changes and developments affecting life on the Earth, but the process of natural selection has been put out of gear by man's actions. By carefully looking after his young and the poorer members of society man is changing the trend completely.

19
Ecosystems

Ecology is the science of studying the conditions in which living organisms exist 'at home' in the environment to which they are best adapted. It includes the examination of the inter-related factors studied in chapters 15-18 — physical, chemical and biological; plants, animals, soil, water, light, food and space. The study of ecology thus involves a knowledge of geology, meteorology, biology and the background physics and chemistry. This seems ambitious, but serves to emphasise the complexity of the factors, and demonstrates the importance of considering them together.

The individual organism is one of a **population** made up of its own interbreeding species. Several populations may exist in the **community** of plants and animals in one area. The biological communities are part of an **ecosystem**, which involves the complete relationships of living creatures with each other and the non-living environment of an area. Ecosystems are units within the biosphere: one can study an oak tree as an ecosystem, or a small area of a few hundred square metres, or divisions of the biosphere on a world scale. The ecosystem concept is particularly useful to the biogeographer, since it provides a basis for integrating the factors involved in the distribution of life on the Earth, and can also constitute a foundation for making zonal divisions.

Living systems on the Earth have never been immutable entities set in unchanging backgrounds: living things have evolved through geological time, adapted to changing environments, but also having a marked effect on the soils, rocks, atmosphere and oceans. Ecosystems are also dynamic entities, studied today at a stage of their development in which the environment is being continually modified so that it becomes more suitable for the next stage.

A particular sequence of ecosystem development, related to a particular set of physical and chemical conditions, is known as a **sere**, which is composed of a number of **successions**, replacing each other in the course of time. The last succession in a sere is known as the **climax**, or **climatic climax**, since the ecosystem is then linked closely to the local climatic conditions. There are, however, few, if any, parts of the world where a climatic climax succession can be found today. Many seres have not reached this stage, which may take over 1000 years to establish, and man has interfered with the natural vegetation to such an extent that many have been destroyed: little woodland was left in Britain by the early part of the nineteenth century after prolonged felling for fuel, shipbuilding and charcoal, and a large proportion of the present woodland has been planted since then. In addition it seems that the current enrichment of the atmosphere with carbon dioxide, nitrogen and sulphur gases is encouraging plant life to develop to higher climaxes than previously, and so every ecosystem is less than fully developed. The climax succession is often defined by the point reached when the annual production of living matter by the ecosystem or community is balanced by the re-cycling of decayed remains: it is a situation where the complex interacting forces are balanced, and checks can act to maintain that balance.

The concept of climax succession is restricted commonly to the plant aspects of the ecosystem, but it is impossible to divorce this from the bacteria and animals which play such an important part in the overall development. Animals may depend on the plants for their food and shelter on the land, but in turn keep the forest in check; insects are required for pollination; and the birds and rodents help in seed dispersal. Plant ecology, as a separate study, is a biased view imposed by the specialist approach of the botanist, as opposed to that of the biologist or biogeographer.

Primary and secondary successions
The successions which develop in a particular area may have different starting points.
1) **Primary successions** are changes experienced when living things become established on a previously sterile area such as a lava flow or newly exposed sea-floor.

2) **Secondary successions** are changes experienced in areas which have been occupied previously by living forms, but where some alteration in the conditions leads to a new emphasis. Thus farmland may be abandoned, or a pond habitat may become filled and change to a swamp.

It must be said that the distinction between these two is not always clear, and that it is often difficult to tell which was the original state. In addition, the case of the pond filling might also be represented as a primary succession, since ponds are always doomed to being filled eventually with sediment. Nevertheless it is helpful to consider the two successions from an ideal standpoint, since an attempt can then be made to identify successional stages in the vegetation and animal associations.

The **primary succession** may begin on a bare, recently cooled lava flow, and the following series of events could possibly be seen over a period of years. Compare this with the observations recorded in Figure 19.1.

Year	Total number of plant species*	Coast	Lower Slopes	Upper Slopes
		(Coastal woodland climax) ↑	(Lowland rain forest climax) ↑	(Submontane rain forest climax) ↑
1934 1932	271		Mixed woodland largely taken over from Savanna	Woodland with smaller trees, fewer species taking over
1928 1920	214			
1919			Scattered trees in grassland: single or in groups with shade species beneath. Thicker development in ravines	
1908 1906	115	Wider belt of woodland with more species, shrubs, coconut palms	Dense grasses up to 3m high. Woodland in ravines	
1897	64	Coastal woodland develops	Dense grasses	Dense grasses with shrubs interspersed
1886	26	9 species flowering plants (as on newly emerged coral reef)	Ferns and scattered flowering plants. Beneath on ash: blue-green algae.	
1884		No life		
1883	0	Volcanic explosion: all life killed; hills deeply gullied by rain		

FIGURE 19:1 An example of primary succession on the island of Krakatoa, which suffered a catastrophic volcanic eruption in 1883. This destroyed all the life on the island. It was visited subsequently by groups of botanists in 1884, 1886, 1897, 1906, 1919 and 1932. The number of species (*) include angiosperms and gymnosperms, and were taken from two nearby islands as well as from Krakatoa.
(After Richards, 1964.)

1) The rock surface is weathered, broken and pitted by the action of wind, rain and frost. Simple plants of the lichen form are blown in by the wind and become established: the surface is too exposed and the weathered layer too shallow for anything else.
2) The lichens add to the weathering process, helping to dissolve more rock, and filling small hollows with dust-sized rock fragments and organic debris. The first soil micro-organisms become established here, re-cycling the nutrients.
3) The pioneer community is enlarged as insects (mites, ants) and spiders move in and further enrich the scanty soil by helping to break down and distribute the organic debris.
4) The secondary community becomes established when mosses begin to anchor the soil more

firmly: it is now a physico-organic entity, and new soil organisms arrive as it develops, breaking down the organic debris at a more rapid pace.

5) Biennial, and, later, perennial grasses appear, and these provide food and cover for nematode worms. They also reduce the degree of exposure by creating shade and raising the humidity for the tiny organisms inhabiting the area.

6) Small shrubs and heaths are then established.

7) Tree seeds are then able to germinate successfully in the shade of these shrubs, and when they mature the environment is modified still further (cf. chapter 8).

8) Various woodland and forest successions lead to the climax succession after several hundred years. This completes the history of the sere, containing a wide variety of plants and animals very much dependent on each other and the general environment they have created.

The fact that the environment has been modified needs to be emphasised. The original state was too exposed and tree seedlings, for instance, could not grow successfully until some shade was available. Naturally there is a variety of possible detailed successions beginning with the same lava flow in the same area; when these are multiplied by the number of different climatic zones one can understand the variety found in nature. Whichever plants and animals are involved, each stage creates a new environment from which the next can grow (Figure 17.8).

A **secondary succession** follows a climatic change, forest fire, major flood, the abandoning of farmland (Figure 19.2), or some other means by which the earlier community is replaced. The early successions are distinctive, since they often contain a mixture of species, but later ones resemble the primary succession more and more closely. Thus when farmland reverts, the grasses grow longer and coarser; then bracken or broom may become established replacing the grasses; and tree seedlings then spring up, with silver birch or pines eventually giving way to hardwoods in temperate zones. Vegetation successions resulting from man's interference (e.g. grassland on the uplands of Britain) are known as **plagioseres** and may reach a condition of **plagioclimax**.

Special successions are identified with excess water (**hydroseres**), or insufficient water (**xeroseres**). In both, the emphasis on environmental modification is again evident. A pond, or shallow lake, soon silts up and is changed to a swamp, then a marshy thicket and eventually to a forest (Figure 19.3). Very dry regions may have few plant species, but these may be sufficient to build up the soil moisture so that other plant species can grow, and eventually a more normal community may be established.

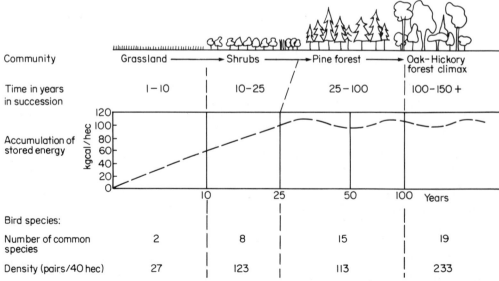

FIGURE 19:2 *Some features of secondary succession on abandoned farmland in eastern USA. The bird numbers give an indication of the animal life developing in association with the plant community, and reflect the increasing number of environmental niches available in the later stages of the succession.*

(After Woodwell, in Tivy, 1971.)

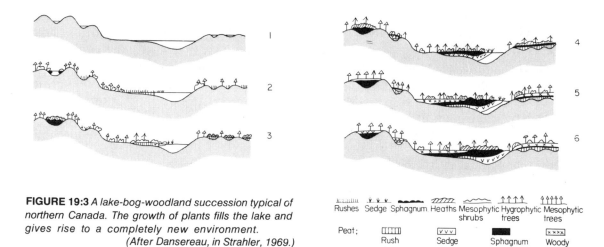

FIGURE 19:3 *A lake-bog-woodland succession typical of northern Canada. The growth of plants fills the lake and gives rise to a completely new environment.*
(After Dansereau, in Strahler, 1969.)

Dominants and stratification

The succession stage reached in the sere is often referred to by one or more of the larger species of plants. These are the dominant species, and they fulfill precisely that function in the community: they are the large trees which have a major influence on the local environment, or they may be the most numerous plant forms. The dominants are different for each successive stage in the sere, and are usually fewer in harsher conditions. Whilst two species account for 90 per cent of the trees in many temperate forests, tropical forests may contain over a dozen dominants.

The climax succession is the most complex and stable. For a particular area there are more species, a greater total biomass, and more organic matter in the soil. Delicate and subtle controls are established, ensuring a wider and more diverse range of foods and a variety of niches for animals. The forest communities are most clearly divided into layers at this stage (Figure 19.4). Aquatic communities may be layered in a similar way.

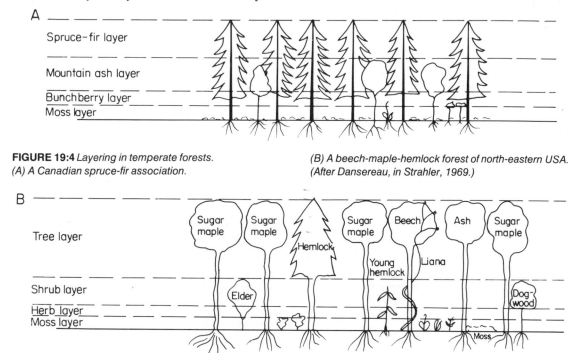

FIGURE 19:4 *Layering in temperate forests.*
(A) A Canadian spruce-fir association.
(B) A beech-maple-hemlock forest of north-eastern USA.
(After Dansereau, in Strahler, 1969.)

Energy flow during the succession

The increasing complexity of the community as successions advance towards the climax means that more and more energy is absorbed, but that more is also 'lost' to the system through respiration.

The autotrophs are the basic transformers of energy, which is then made available to the rest of the community. The total energy gain of the system is partly used up in metabolic activities and partly in the synthesis of organic compounds (Figure 19.5, equation 1), i.e. anabolic activities. When the ecosystem is considered as a whole the autotrophs use energy as just described, but the consumers use some of the net product they acquire from the autotrophs in metabolic activity and may use the rest to produce an increase of new matter which enables them to develop further (Figure 19.5, equation 2). If the consumers expend all the energy made available by the primary producers there is no energy left over for further development: climax has been reached (Figure 19.5, equation 3).

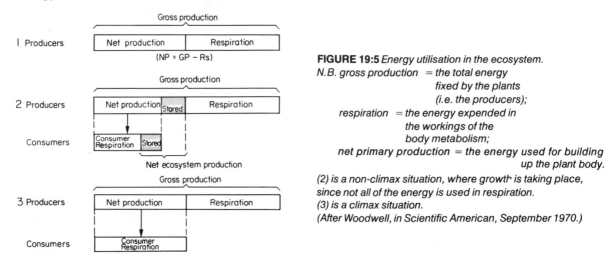

FIGURE 19:5 *Energy utilisation in the ecosystem.*
N.B. gross production = the total energy fixed by the plants (i.e. the producers);
respiration = the energy expended in the workings of the body metabolism;
net primary production = the energy used for building up the plant body.
(2) is a non-climax situation, where growth is taking place, since not all of the energy is used in respiration.
(3) is a climax situation.
(After Woodwell, in Scientific American, September 1970.)

Measurements on this basis have been made by ecologists. Figure 19.6 shows one such result. The gross product of this New England woodland is 2650 g living matter/m²year, of which 1450 is respired and 1200 is therefore the net primary product. The consumers respire 650 g of this (to give a total respiration of 2100 g and a net ecosystem product of 550 g). Thus this particular ecosystem respires only 80 per cent of the total energy gained: it is not climax, but late succession. In fact no ecosystem measured in this way has fulfilled the climax criterion.

The net product of the producing autotrophs (1200 g, or 0.9 per cent efficiency in taking energy

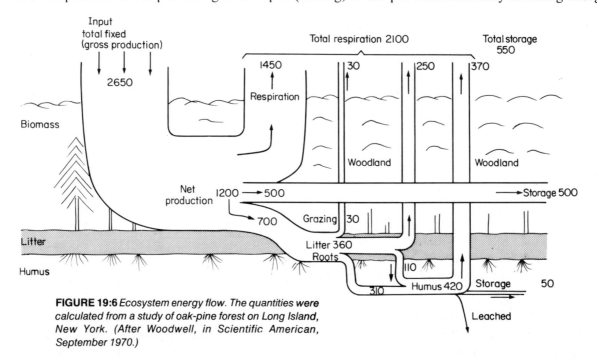

FIGURE 19:6 *Ecosystem energy flow. The quantities were calculated from a study of oak-pine forest on Long Island, New York. (After Woodwell, in Scientific American, September 1970.)*

from the amount supplied to the situation depicted in Figure 19.6) is typical for a small stature, midlatitude forest, and compares with the larger statured forests where nutrients are more abundant and growth continues for longer each year (3000-5000 g living matter/m²year, or 3 per cent of the energy supplied). This more normal forest production figure is approximately equal to the production from farmland, although sugar cane crops (growing over a full period of a year) may produce 9000 g/m²year and rice or maize 6000 to 10 000 (including canes, stalks, chaff, leaves and roots!). A variety of primary production figures can be compared (Figure 19.7).

Compared with the land yields, those of the oceans are relatively small. The highest levels reach 300 g of fixed carbon per m²year (approximately equal to 700-800 g dry matter/m²year in the land figures). From this it has been calculated that the maximum fish yield is 100 million tonnes, which will be reached by 1980, if pollution does not reduce the possible total before that date. On this view the oceans are very near the limits of exploitation for food, although others would suggest that they are a future source of increased food yield (chapter 24).

FIGURE 19:7 Ecosystems and net primary production: figures for some of the major ecosystem types, together with estimates of total biomass in each zone. Compare the figures, and attempt to explain the variations. (After Whitaker, 1970.)

Major ecosystem	Area (10^6 km²)	Net primary production/unit area (dry g/m² year)		World net primary production (10^9 dry tons/year)	Biomass/unit area (dry kg/m²)		World biomass (10^9 dry tons)
		Normal range	Mean		Normal range	Mean	
Lake, stream	2	100-1500	500	1.0	0-0.1	0.02	0.04
Swamp, marsh	2	800-4000	2000	4.0	3-50	12	24
Tropical forest	20	1000-5000	2000	40.0	6-80	45	900
Temperate forest	18	600-2500	1300	23.4	6-200	30	540
Boreal forest	12	400-2000	800	9.6	6-40	20	240
Woodland, shrubland	7	200-1200	600	4.2	2-20	6	42
Savanna	15	200-2000	700	10.5	0.2-15	4	60
Temperate grassland	9	150-1500	500	4.5	0.2-5	1.5	14
Tundra, alpine	8	10-400	140	1.1	0.1-3	0.6	5
Desert scrub	18	10-250	70	1.3	0.1-4	0.7	13
Extreme desert, rock, ice	24	0-10	3	0.07	0-0.2	0.02	0.5
Agricultural land	14	100-4000	650	9.1	0.4-12	1	14
Total Land	149		730	109.0		12.5	1852.0
Open ocean	332	2-400	125	41.5	0-0.005	0.003	1.0
Continental shelf	27	200-600	350	9.5	0.001-0.04	0.01	0.3
Attached algae, estuaries	2	500-4000	2000	4.0	0.04-4	1	2.0
Total Ocean	361		155	55.0		0.009	3.3
Total Earth	510		320	164.0		3.6	1855.0

The time factor

The early stages of a primary succession are generally very slow, due to the prevalence of severe conditions. Later stages succeed each other more rapidly, but there is then a further slowing towards the climax. It is thought that up to 1000 years are necessary to establish the climax ecosystem in a temperate region, but 250 years may suffice in the tropics where conditions for growth are better and it continues throughout the year. A secondary succession may take a shorter time (e.g. about 200 years for forest to recolonise farming land in temperate regions, but once again the time will be less in the tropics). So much damage is done to the soils by farming that there is often no prospect of re-establishing anything like the original ecosystem.

Successions are briefest where organisms can have little effect on the environment, as in the sea. In fact there are no real successions recognisable in the oceans — unless they can be reduced to the seasonal changes which recur again and again.

Some successions are so slow, and are subject to so many constant interruptions such as fires and destructive storms, that the climax is delayed indefinitely. This sort of interruption is most devastating in the early days of a succession. Thus the climatic climax association is not always bound to appear eventually: it should be seen more as a general tendency within a particular climatic zone, or in relation to a particular soil where the edaphic factors (i.e. soil, rock, drainage) are more important: in that case it would be an **edaphic**, rather than a climatic climax. Examples of this are found in the savanna grassland areas described in chapter 22.

Ecosystems and man

Knowledge of ecosystems and their development can be put to practical uses in a number of ways.
1) The climatic climax situation is the most stable and resistant to adverse changes, but production is stable. The early stages of a succession are more productive of new materials, and so, for example, reafforestation schemes aim to maintain that type of situation: much wood is cut from young stands.
2) It is important to think in terms of a complete ecosystem rather than of isolated parts of it. It makes sense to leave or plant forest on steep slopes, retaining the flatter land for arable farming. If the soils on the steep slopes are exposed by felling trees they are soon washed downhill by rain water.
3) A knowledge of such principles might have averted tragedies in the past. Rabbits were introduced to Australia, where there was no predator to touch them, and they multiplied without a check so that the food they shared with the sheep, but had eaten more rapidly and efficiently, soon disappeared. The red deer introduced to New Zealand gave rise to a similar situation of overgrazing and erosion. It is important to preserve as complex a situation, with all its checks, as is possible. Thus the New England woodland (Figure 19.6) has been deprived of its large herbivores (deer), and only insects and small mammals are left. This means that there are more low-growing shrubs and thus a far greater proportion of plant production is not consumed and is diverted directly into the 'detritus pathway' via leaf-fall. In addition one further check within the system has been removed: the net product is allowed to fluctuate so that the reduced numbers of herbivores may increase or decrease more intensively, paralleled by carnivore reactions. Thus, plagues of certain animals may be followed by wholesale deaths as the food source is terminated (Figure 19.8).
4) Similar principles are now being used increasingly in the fight against crop pests. Whereas these have been controlled by chemical means, it has been found that strains of pests resistant to chemicals are overcoming this threat to their existence, whilst the chemicals themselves are accumulating to **dangerous** levels in some parts of the biosphere. It is now seen that it may be better to control the pests by setting a natural predator loose on them.

Locusts: an ecological study of a pest

For centuries the desert locust has caused misery and starvation. As recently as 1957 Ethiopia lost 167 000 tons of grain in this way — enough to feed 1 million people for a year — and over 15 000 people died as a result of the plague. This pest can still affect nearly 20 per cent of the world's land surface and the lives of 1 in 10 of its people (Figure 19.9).

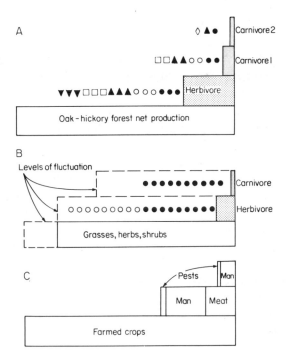

FIGURE 19:8 *Ecosystems and stability.*
(A) A natural ecosystem, with several stages of consumers in which 10-20 per cent of the energy received at one level will be passed to the next. There is a complex species structure (shown by the symbols).
(B) A degraded ecosystem containing few stages and a simple species structure. The annual production fluctuates, leading to fluctuations in the consumers with dangers of plagues and of complete consumption of the producers.
(C) Farmed land ecosystem, yielding a large production and maintained by cultivation, fertilizer and pesticide inputs. (After Woodwell, in Scientific American, September 1970.)

It was only in 1921 that the nature of the locust life-cycle was realised. Before that date people had been puzzled by the sudden swarming habit which alternated with periods in which the locusts seemed to vanish completely. This was found to be due to changes in numbers. In dry years the locusts look like green grasshoppers and are widely dispersed. When it rains, the eggs which lie buried in the sand hatch out and numbers climb rapidly. The locusts crowd together and change colour from green to black, yellow and red. At this point they swarm and destroy every bit of plant life. A knowledge of their life-cycle and the relationship of this to environmental conditions has enabled man to reduce the damage they cause. The Anti-Locust Research Centre in Kensington, London, keeps a careful watch on the general situation in northern Africa and the Middle East, and carries out research which has assisted in the prediction of locust swarms.

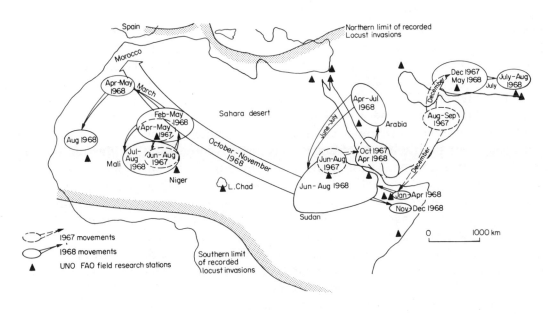

FIGURE 19:9 *Locusts: breeding areas and migrations, 1967-68, a bad plague year.*

It is now known that the female locust lives on average for four months (although this can be extended by lowering the body metabolism in time of drought). Towards the end of her life she lays three sets of eggs, with approximately 70 eggs on each occasion, burying them in the sand and protecting them with a soluble plug to prevent them drying out. After rain wets the ground, these eggs hatch into wingless hoppers, moulting five times before the adult, winged form is attained. If the wet weather is followed by dry the locusts remain in the green, solitary stage, but if there is continued rain vast hordes of locusts will hatch and give rise to swarms. The swarms take to the air when their body temperature rises to 20°C, flying with the wind at speeds up to 15 km/hour for up to 12 hours at a time. Locusts can range up to 4500 km in their lifetime, breeding as they migrate and spreading damage rapidly. They have some natural enemies — beetles, flies, wasps and birds — at various stages in their development, but none of these can cope with the swarms: birds fall to the ground gorged with locusts and too heavy to fly, but without making any appreciable difference to the size of the swarm.

In 1967 and 1968 bad locust plagues were experienced after a series of good years in the early 1960's. Wet weather encouraged breeding and migration (Figure 19.9), and much of northern Africa was threatened. Very little damage resulted, however, because of the organisation which is now available to combat the threat. This is a combination of the research and co-ordination facilities of the Anti-Locust Research Centre and the financial resources of the Food and Agriculture Organisation (FAO) of the United Nations, which has established a series of field research stations and provided aeroplanes and spraying equipment which can kill rapidly. The Anti-Locust Research Centre monitors a range of factors: everyday, weather conditions (particularly rainfall) in the likely breeding areas are recorded and compared with field reports of hatching locusts. Once a plague begins, swarms can be plotted and their movements forecast by comparison with weather maps and satellite photographs: the convergence of winds leads to congregation of swarms and intensified damage. All this provides the field teams with information which can keep them in touch with, and even one step ahead of, the locust swarms, enabling them to hit them from the air with pesticide sprays before they can reach cultivated land. This is why the damage in 1967-68 was so small.

There is, however, no likelihood that locusts can be eradicated entirely. This is because of their extreme degree of dispersal during the dry years. Drought will eventually kill the locust swarm, but it will also disperse the locusts to the extent that pesticide spraying is not practicable. Thus a continuous watch must be kept.

Ecosystems of the past

Changes took place in the past which led to different distributions and associations of living things. Life forms were often very different in the geological past (chapters 15 and 18), continents probably changed their positions relative to each other and the climatic belts and climatic changes within the last million years caused major advances and retreats of ice sheets (chapter 14).

The study of present and ancient populations of a valley in Montana (Figure 19.10) also points out the difficulties facing the ecologist studying and attempting to reconstruct past ecosystems — and these difficulties become more important as one goes further back in time. Many biologists would say that **palaeoecology** (the study of ancient ecology) is impossible for this reason. Present day ecosystems, however, can be used as analogies to discern something of the ancient patterns. This reliance on the studies of present ecosystems brings us to the next major problem: most of the ancient ecosystems which might be preserved are of sea-floor communities about which we still know comparatively little, and which are quite different in nature from the land ecosystems.

A palaeoecologist must first obtain his fossil record, and then has to sort out the forms which actually lived in the place where they became entombed in the sediment from those which were brought in after their death by marine currents, or 'dropped in' after a planktonic or swimming mode of life. It is often difficult to do this unless certain groups of shells have been broken or worn by transport, or even merely overturned, and thus stand out from the rest. When one examines a rock formation like the Oxford Clay (Upper Jurassic age, approximately 150 million years old) in the

ECOSYSTEMS 223

cliffs east of Weymouth, one finds several types of ammonite (extinct swimming forms), belemnite (also extinct swimmers), oysters (which live on top of the sea-floor sediment) including some in the living position and others which have been flipped over, and some sediment-burrowing bivalves. One can also remove the muddy sediment and examine the microscopic fossils, produced largely by former planktonic algae. One has thus discovered a record of marine life at that stage in time, including a similar general distribution of forms as today (i.e. plankton, nekton, benthos), but it is jumbled together. In spite of the mud-grade sediment at this spot, indicating fairly calm conditions of deposition, there are few forms which can be definitely said to have been preserved in their actual life position. Information is also lacking concerning the animals which have not been preserved (presumably these would have included worms, jellyfish, larger fishes and reptiles) due to their mobility or lack of hard parts, and concerning the physico-chemical environmental factors — temperature, depth, etc.

For much of geological history the information regarding land ecosystems is even poorer. Plants are seldom fossilised, and only a general outline of their evolution is available, reconstructed from occasionally preserved accumulations of leaf and stem material. One such occurs in the Old Red Sandstone (approximately 380 million years old) rocks near Rhynie in north-east Scotland, where the waters of a spring must have infiltrated one of the earliest peat bogs. This became permeated with silica and 'petrified', preserving the details of some of the oldest known land plants:

FIGURE 19:10 Ecological niches and population densities.
(A) Fossils obtained from Pliocene deposits (approximately 5 million years old) in a river valley in western Montana, USA.
(B) Statistics of animals living today in the same valley, which would be likely to give rise to fossils under similar conditions.
Compare the two situations. What changes have taken place in 5 million years, and what similarities are evident? What has caused the changes? (After Konizewski, in Laporte, 1968.)

			A. MIDDLE PLIOCENE											B. TODAY							
				Habitat zone											Habitat zone						
Animal	Food habits	Population density *	Terrace		Flood plain		River bank		Coniferous		Animal	Food habits	Population density *	Terrace		Flood plain		River bank		Coniferous	
			p	dp	p	dp	p	dp	p	dp				p	dp	p	dp	p	dp	p	dp
Pronghorn	Large grazers	2	9	18	7	14					Pronghorn	Large grazers	2	9	18	7	14				
Horse (2 species)		2	9	18	7	14					Bison		2	6	12	8	16			2	4
Camel (2 species)		2	9	18	7	14					Rocky mt. sheep		1	8	8	8	8				
Rabbit	Small grazer	3	8	24	8	24			2	6	Rabbits (2 species)	Small grazers	3	8	24	6	18			2	6
Peccary (2 species)	Large rooters	2			8	16	4	8	4	8	Mule deer		2	3	6	5	10			8	16
Proboscidean	Large aquatic plant eaters	1			6	6	10	10			White tailed deer	Large browsers	1	3	3	8	8			5	5
Rhinoceras		1			6	6	10	10			Elk		1	6	6	7	7			3	3
Beaver	Small bark-eater	1			1	1	14	14	1	1	Moose	Large aquatic plant eaters	1			7	7	6	6	3	3
Ground squirrel (2 species)	Very small granivore	3	8	24	2	6			6	18	Muskrat	Small aquatic omnivore	2					16	32		
											Beaver	Small bark-eater	1			2	2	12	12	2	2
											Pocket gopher	V. small granivore	3	5	15	8	24			3	9
											Ground squirrel (2 species)	V. small granivore	3	6	18	6	18			4	12
Total: 13 species 111 specimens	s:		8		13		5		6		Total: 14 species	s:		11		13		3		11	
	t (dp)		158		156		46		63			t (dp)		133		174		50		91	
	%t (dp)		37%		37%		11%		15%			%t (dp)		29%		39%		11%		20%	

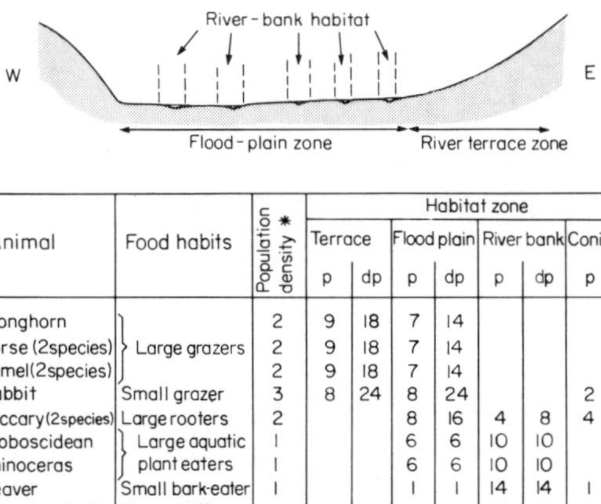

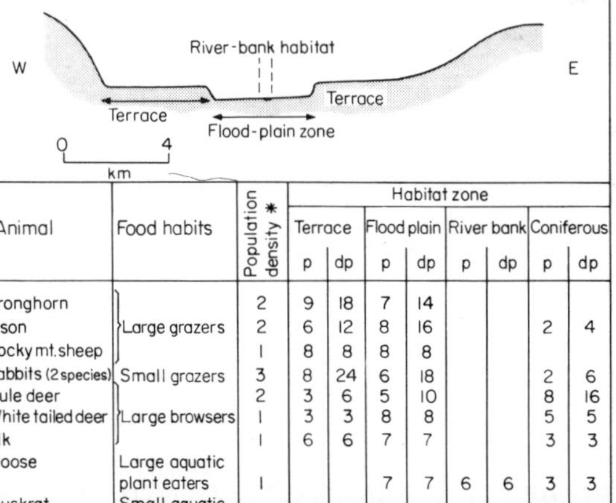

*   Total species population: 1 (low) – 3 (high)
p   Species habitat-zone preference: 0 (none) – 16 (total)
dp   Species population density within particular habitat zone
s   Species representation (ie. no. of species in habitat zone)
t   Habitat zone population density relative to total population density
%t (dp)   Percentage of t

low-growing, very primitive types with stems growing along the ground to take up moisture and with few 'leaves'. The Coal Measures (Upper Carboniferous age, approximately 300 million years old) present us with another such snapshot of past plant forms. By this time forests had developed with individual plants reaching over 40 metres high and forming the habitats for early forms of land animal life. The plants composing these forests were very different from those found today: giant ferns, horsetails and the first seed-bearing trees. Further rock sequences of deltaic origin in the Middle Jurassic of eastern Yorkshire (approximately 160 million years old) show evidence for the development of plants related to conifers, like the ginkgo. Yet another British scene can be reconstructed from evidence in the early Tertiary rocks (approximately 50 million years old) on the Isle of Sheppey in the Thames estuary. At this stage Britain was covered by forest similar to that now mantling the Malay Peninsula: so much can be suggested from the seeds, fruits and leaves, together with the microscopic spores and pollens preserved in the London Clay formation. Flowering plants had taken over completely; palm trees were the dominants; 73 per cent of all genera have living relatives in Malaysia; and one-third of the families are restricted to the tropics today. The occurrence of magnolias, however, suggests more of a warm temperate climate. The animals associated with these plant communities include relatives of the present crocodile and hippopotamus, together with many large flightless birds and *Eohippus* (a horse-like creature the size of a dog). It is unusual, however, to find the plant and animal evidence together, and most reconstructions of past ecosystems are pieced together using evidence from different sites.

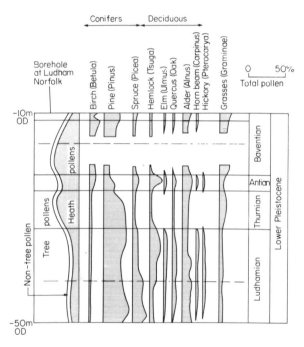

FIGURE 19:11 *Plants and the changing climates before the onset of the Ice Age in Britain (cf. Figure 14.5). The evidence comes from pollens in the shelly marine sands and silty clays extracted from a borehole at Ludham in Norfolk. Warmer periods have higher proportions of tree pollens; colder periods have lower proportions of tree pollens. Note also the occurrence of some trees (hemlock, hickory) not again found naturally in Britain: they did not return after being forced southwards by the advancing ice. (After West, 1970.)*

Quaternary ecosystem changes in the British Isles

The best reconstructions are based on studies of recent sediments, using evidence provided by analysing pollens contained in peats and offshore muds. From such studies a general impression is built up of the plant formations and their development through the fluctuating climatic conditions during and since the Quaternary Ice Ages.

At first the fluctuations involved changes from forest (milder climate) to heathland (worsening conditions) (Figure 19.11). Evidence for this comes from the study of pollens in the offshore muds preserved beneath the East Anglian ice deposits, whilst further support comes from associated foraminifera and molluscs. The forests at this time were evidently dominated by a mixture of coniferous and broad-leafed forms, including hemlocks and hickories (neither of which are found later occuring naturally in Britain), pines, alder, birch, oak, spruce and elm. The 'oceanic heath' was formed of increasing numbers of grasses and *Ericas* (crowberry, ling) which at times exceeded the contribution of the tree pollens. These early climatic fluctuations eliminated the hemlocks and hickories from Brtiain, and they have persisted only in North America where the north-south migration routes formed by the trend of mountain and lowland made re-establishment easier.

The ice then advanced and retreated across the British Isles, leading to a progressive impoverishment of the flora. All forest was removed by the ice, and tundra occupied the extreme south during the glacial periods. The intervening interglacial periods began with weather which was still very cold resulting in abundant herbaceous pollens, but gave way to plants adapted to warmer conditions in closed forest ecosystems, and then finally returned to vegetation reflecting increasing coldness and the re-establishment of icy conditions. Pollen diagrams (Figure 19.12) show the sequence of plants in the communities. The Hoxnian interglacial is particularly interesting for its record of man's first appearance in Britain: temporary local increases of herbaceous pollens suggest the effects of burning off a section of forest. The deposits of the last interglacial trapped the remains of a warm climate fauna (elephant, hippopotamus and rhinoceros with insects and molluscs) in areas like the terrace gravels beneath Trafalgar Square, London.

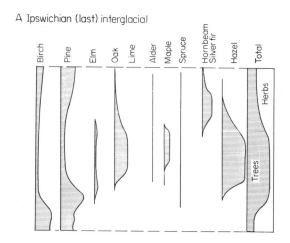
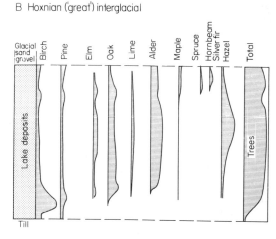

FIGURE 19:12 *Pollen diagrams for two interglacial periods. How did the plant communities change through each interglacial period? Note the similarities and differences. (After West, 1970.)*

The last glacial advance (Devensian, Figure 14.5) did not reach as far south as the previous two, and much more is known of the sequence of events due to the facts that the deposits lie on top of earlier formations, and that radiocarbon dating covers the period in question (i.e. 70 000 to 10 000 years ago). The sea level fell as the ice advanced, so that plants were able to retreat to Europe, leaving only the hardier groups. The deposits formed in lakes beyond the ice front contain fossil material redolent of insects and animals in today's northern boreal forests, tundra and open grassland; plants include many species found today in the tundra, and there is no record of trees apart from dwarf willow and dwarf birch (*Betula nana*).

The end of the Devensian glaciation is now dated 8300-8200 BC, but the temperatures had started to rise for at least 3000 years before this as the ice withdrew. Lakes associated with the melting ice front across Sweden deposited varved clays yearly, enabling a chronology to be established. This phase of cool temperate conditions is known as the **Allerød**, after a Danish lake deposit of the time. The climate fluctuated and the dominant vegetation has been described as 'park-tundra': this was more like true tundra during colder phases, but birch, willow, erect juniper and poplar copses intruded in the milder times. Some of the plants common at this time can be found still in areas sheltered from grazing like Upper Teesdale. At the very end of the Allerød phase the climate became colder again. Ice probably accumulated in the Lake District corries, and solifluction deposits covered some of the Allerød peats dated by radiocarbon methods.

The true post-glacial phase began with a fairly 'sudden' rise in temperature. Lake deposits, which are the source of pollens, show a marked increase in tree pollens within a few centimetres of the vertical profile, indicating the development of true soils on the glacial and solifluction deposits. Grass, sedge and herb pollens decline in favour of juniper, birch and eventually pine. Trees, of course, produce far more pollen than the smaller plants, which remained in association for some time.

The **post-glacial period** (or Holocene) is divided according to the stratigraphy of the bog deposits in Scandinavia, confirmed by wider studies of pollen and peat profiles, and radiocarbon dating, and extended across northern Europe.

Phase	Date	Climate
Subatlantic	c.500 BC	Cold and wet: oceanic
Subboreal	c.3000 BC	Warm and dry: continental
Atlantic	c.5500 BC	Warm and wet: oceanic
Boreal	c.7600 BC	Warming up, dry
Preboreal	c.8200 BC	Subarctic
Late glacial (Allerød phase)		

a) The **Preboreal** phase saw the first forest advancing, as the sea-level began to rise due to the final melting of the ice. England was still connected to Europe and much of the North Sea bed was dry land (postglacial peat has been dredged from its floor). Some relatively warmth-loving plants migrated in at this stage.

b) The **Boreal** phase is characterised by the increase of hazel, elm and oak pollens (Figure 19.13), though much of the woodland at this time may have been hazel scrub, replaced gradually by oak and elm later in the period as the soils developed and the climate warmed up. Eventually the most sensitive of all British trees, the lime, entered, though it did not reach Ireland which had already been cut off by the rising sea. The entry of the alder, together with increasing thicknesses of peat, suggest an increase of rainfall at the Boreal/Atlantic transition.

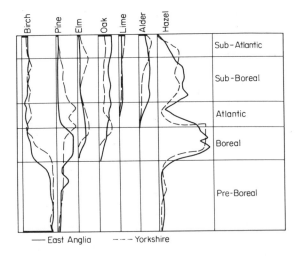

FIGURE 19:13 *Postglacial changes in climate, with evidence from pollen records in two areas of eastern England. What are the similarities and differences? (After West, 1970.)*

c) The **Atlantic** phase, beginning approximately 5500 BC, was notable for increasing wetness and for the final isolation of Britain from Europe. This was probably the best period of climate between the end of the glacial period and the present day; deciduous forest reached up to over 1100 metres above sea level, and few open habitats remained. Dominants included oak, elm, alder and lime ('climatic climax') with some birch left in the north and west, and the pine restricted to north-east Scotland. Lakes were filled and sphagnum peat bog succeeded normal hydroseres inland as well as coastal marshes. The rising sea level submerged coastal forests and led to the formation of the Straits of Dover: henceforward Britain was cut off from further natural migrations on a large scale from the continent.

d) The **Subboreal** phase was heralded in Britain by the decline of the elm (Figure 19.13), which gave way to ash. It has been suggested that this may have been due to climatic factors, but it could have been due to the activities of the early Neolithic farmers, who had to gather leaves and leafy branches of elm to feed their stock in the absence of grassy areas. Elm pollen would be reduced, and so would the base-rich litter necessary for regeneration, so that the less-demanding ash could take over. Extensive Neolithic settlement is known to have taken place in Britain by 3000 BC. Thus began man's overwhelming effect on the natural vegetation cover of Britain. Axes had also developed to the stage where forest trees could be felled.

The first activities of man began to differentiate regions within the once almost uniform forests: elm did not regenerate on the poorer soils, and some areas never returned to forest. Lime suffered a similar fate to the elm, and both were replaced by the ash. The Bronze Age began in the favoured south-east of Britain by 1700 BC and metal tools spread from there. Areas of light soils (e.g. the chalk ridges) were cleared most easily and the soils suffered irreversible changes.

e) The **Subatlantic** phase witnessed some deterioration of the climate: Bronze Age settlements (e.g. on Dartmoor) seem very spartan today. A period of colder and wetter conditions had arrived. Peat bogs which had largely dried out and been colonised by ling, pine and birch in the Subboreal period were replaced by sphagnum, covering up many Bronze Age trackways. The increasing effect of man's occupance led to fires in the forest and the extension of birch, especially on the poorer soils. Extensive clearances began just before the Romans arrived, but were still confined largely to the limestone ridges of the south-east. The Anglo-Saxons and Scandinavian invaders extended the settlements and clearings to the lowland clays, the midlands of England and the northern coastal areas. Forest clearances progressed at an increasing pace as Britain's navy, and the charcoal-smelting industry were enlarged in and after Tudor times. By the nineteenth century hardly any woodland was left, and the shortage of available wood in the 1914-18 War led to the establishment of the Forestry Commission. Many private woodlands of beech, pine, larch, fir and sycamore were planted in the latter half of the nineteenth century. Surrey, now regarded as a well-wooded county, was almost denuded of all woodland before 1830. The Forestry Commission have also introduced the Norway spruce, Sitka spruce and Lodgepole pine, which have a greater economic value. The successful planting of so many 'exotic' species in British woodlands demonstrates the fact that they were excluded not by the climatic conditions, but by a combination of barriers to dispersal and the effects of the fluctuating ice sheets which prevented many species from returning once they had been pushed southwards.

20

The distribution of life

Various aspects of the biosphere have been studied:
1) The **nature and complexity of living things** (chapter 15).
2) The **requirements** for the continued existence of living systems: solar energy and Earth supplies of mineral nutrients (chapter 15).
3) A range of **physical factors** affecting the occurrence and distribution of plants and animals in the oceans or on the land (chapter 16).
4) **The soil**, seen as a basic supporting medium for life on the land (chapter 17).
5) **Biotic factors** whereby plants and animals either compete with each other for the space and resources available, or help each other in some way (chapter 18).
6) A consideration of the inter-relationships in the biosphere between living and non-living factors, giving rise to developing **ecosystems**, a product of the local conditions and evolution during the past (chapter 19).

This leads to a consideration of the distributions of living creatures on the Earth's surface, often termed **biogeography**, — i.e. the study of patterns of the distribution of living things within the biosphere. There are two major approaches to such a study.
1) The first of these examines the worldwide distribution of particular species of plants and animals, and attempts to explain these distributions in terms of climatic differences, migrations and past movements of the continents. Such an approach is essentially taxonomic: i.e. each animal and plant is seen as a member of a particular group of similar features and habits, and not in terms of its relationships with other forms of life.
2) The second approach is more closely related to the discussion of ecosystems and behavioural associations (chapter 19). This sees the biosphere as a series of environments changing from place to place.

Both approaches have their value. In the case of the first it is useful to be able to apply considerations of migration and time to a particular species, and this is relevant to the spread and control of disease. In the second a complete environment-life association is considered as an entity: particular forms become more important than species. In the first case the student may discuss and account for the distributions of placental and marsupial mammals, or of the different varieties of deerlike creatures or of the species of trees; the world regions are divided by major barriers to movement and dispersal. In the second he would look at different types of, say, tropical forest developing around the world as basic forms of ecosystem without considering too deeply the actual species composition of the trees, or the plants and animals living in that ecosystem; life-form related to the total environment is most important, and world divisions are based on similar associations of life forms.

The world distribution of land plants

Few plant species have a worldwide distribution, and plant geographers have divided the world into six major floral kingdoms (Figure 20.1). Even within these there are major variations, so that the larger kingdoms are divided into subkingdoms and provinces, reflecting differences between forest, grassland, desert and tundra groupings within the major division.
1) The **Australian Kingdom** is particularly distinctive with its plant varieties dominated by different species of *Eucalyptus* (75 per cent of all Australian trees) and *Acacia*. The *Eucalyptus* varieties include 600 species and range from large shade-giving trees to dwarf desert forms. The Eucalyptus is a relation of *Mimosa*, some of which also occur in South America.
2) The **Cape Kingdom**, a tiny area at the southern tip of Africa, has many plants with bulbs or tubers.

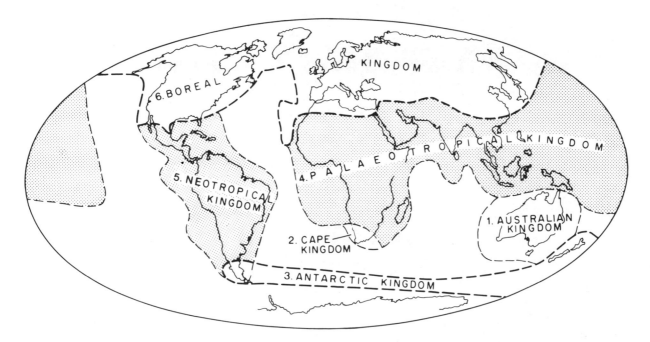

FIGURE 20:1 *Floral kingdoms. Relate these to the faunal realms (Figure 20.2), and to changes in continental distributions (Figure 20.5). The floral kingdoms are based on the distribution of the higher plants (i.e. the seed-bearing varieties), and on species distribution rather than on life-form distribution. (After Neill, 1969.)*

3) The **Antarctic Kingdom** occurs in a ring round the north of that landmass, from Patagonia and southern Chile, through a line of oceanic islands to New Zealand. The most notable plant found in all these is the *Nothofagus* (southern beech).

4) The **Palaeotropical Kingdom** includes three major subkingdoms — African, Indo-Malaysian and Polynesian, each of which is subdivided into several provinces. Few plants are common to all these regions.

5) The **Neotropical Kingdom** includes most of South America.

6) The **Boreal Kingdom** is the largest of all, including North America, Europe and Northern Asia.

Each of these kingdoms is bounded by a well-marked barrier — ocean, mountain range or desert.

The world distribution of land animals

Whilst plants are immobile during life, and rely on seed or spore dispersal for migration into new areas, the larger animals are more mobile, and even the smaller forms can often migrate for long distances over a series of generations. Like the plants, however, animals became adapted to particular environmental niches, and their migration is prevented by climatic and physiographical barriers, such as deserts, mountain ranges and oceans. Islands thus have rather special faunas, isolated from other areas, but the introductions of outside animals to places like the Galapagos and to New Zealand has had disastrous results for the native populations (chapter 19).

The main zoogeographical regions were established by Wallace in 1876 (Figure 20.2), and are artificial in some senses, but do reflect real contrasts. The barriers which separate the main regions also vary. A region may contain **characteristic groups** of animals: thus Australasia has a wide range of marsupials; the Neotropic region has armadillos and sloths; antelopes are common in the Ethiopian region; sheep and goats abound in the Palaearctic; and ducks and geese occur in both Nearctic and Palaearctic. There may also be distinctive **relict groups**: the flightless birds form an interesting comparison, with the rhea in Neotropic, ostrich in Ethiopia and emu, cassowary and kiwi in Australasia. **Absences** may also be striking: no bears or deer are found in the Ethiopian region; no placental mammals in Australasia; and no primates (apart from man) in the Nearctic or Palaearctic. Sometimes a region may contain a **very few species** of a group which is well-represented in the adjacent region: tigers occur in Manchuria; opossum and racoons in the Nearctic, and the speckled

Ecosystems

Plate 166 Ecosystem succession. Dry oakwood (*Quercus petraea*) near Elgin in Scotland is being invaded by beech (*Fagus sylvatica*). The beech was derived from four mature trees at one corner of the wood: it has greater stature, tolerates shade better when small, and it is likely that it will replace the oaks.

Plate 167 Ancient ecosystems. Old pine roots exposed in a peat cutting at Lettoch in Glen Livet, Scotland. The layers of peat have covered the remains of a forest, which probably grew during better climatic conditions in the past. (166 & 167 Forestry Commission)

Plate 168 Ecosystem boundaries. The ridge at 2000m above Francis Canyon in Utah shows a contrast between slopes facing into and away from the Sun. The slope which receives less direct sunlight on the right is able to retain more soil moisture, and supports shrubs and trees, in contrast to the sparse sage bush on the drier and sunnier slopes. (USDA Soil Conservation Service)

THE DISTRIBUTION OF LIFE 231

169

170

171

172

173

Equatorial forests. Some examples of evergreen rain forest.

Plate 169 Forests east of Lake Kivu in Zaire, with the tallest trees reaching above the general canopy level.

Plate 170 Forest margin in Brazil. Notice the tall, slender trunks, the epiphytes and lianes.

Plate 171 Rio Pasteza in eastern Ecuador, on the margin of the Amazon Basin.

Plate 172 Mangrove swamps in the Niger delta, West Africa.

Plate 173 Detail of mangroves. (171 Ewing Galloway, New York; others Aerofilms)

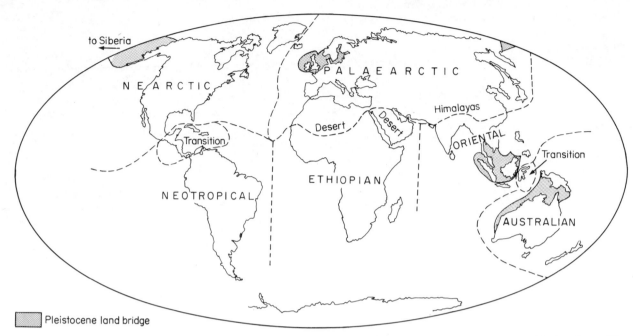

FIGURE 20:2 Wallace's zoogeographical regions of the world. Where are the major boundaries? (After Neill, 1969.)

bear and ten species of deer in the Neotropic. In other cases the similar groups are more widely separated: the marsupials occur only in the Australasian and Neotropical regions; tapirs in Ethiopia and Neotropic. Figure 20.3 summarises much of this information.

The Palaearctic and Nearctic show particularly close relations, having common circumpolar species and close species in regions farther south. European and American bison interbreed; and they are the only two regions with salmon and trout. Some biogeographers would combine the two as the Holarctic region, but the Nearctic originally had no horses, pigs, goats or sheep. One can also see similarities between other regions. Thus the Oriental and Ethiopian have rhinos, antelopes, monkeys and hornbills in common.

The regions can also be divided into subregions on the basis of climate: some, like the Manchurian are virtually transitional; others, like Madagascar and New Zealand are islands.

1) The **Palaearctic region** includes Europe and northern Asia. It has only 28 chordate families represented, and of these 9 are widely distributed elsewhere. In particular there are few reptiles, none of which are exclusive.

2) The **Nearctic region** (North America, Greenland) shares many similarities with the Palaearctic, the two having been connected in the Tertiary and Pleistocene. Interchange with the Neotropical region was not possible until the Pleistocene uplift of the connecting land link. It has some distinctive forms (pocket gophers, pocket mice, pronghorns, wild turkeys), and is rich in reptiles.

3) The **Oriental region** is characterised by tropical forms existing in an area of peninsulas and islands. Its tropical position gives it links with the Ethiopian region, and the Himalayas form a sharp northern boundary over much of the area. There is a transition towards the Australasian region through the East Indies. The fauna includes an elephant, two rhinos, several species of deer and antelope, pheasants, tigers and a rich variety of lizards and snakes. Tree shrews, gibbons, orangutans and tapirs are unique.

4) The **Ethiopian region** is also largely tropical, Africa south of the Sahara and South Arabia, but is a more compact unit and has the most varied fauna of all the realms (though it has no moles, beavers, bears or camels), sharing many forms with the Oriental region. The hippopotamus, aardvark, ostrich and groups of rodents and insectivores are found exclusively here.

5) **Australasia** may or may not have been connected by land to the Oriental region, but it has few

Vegetation type	Nearctic	Palaearctic	Oriental
Tundra	Caribou, musk-ox, lemming, Arctic hare, wolf, Arctic fox, polar bear	Similar genera, different species	
Coniferous forest	Moose, mule, deer, wolverine, lynx.	Similar genera, different species	
Temperate grassland	Bison, pronghorn, jack rabbit, prairie dog, gopher, fox, coyote	Saiga, wild ass, horse, camel, jerboa, hamster, jackal	
Deciduous forest	Racoon, opossum, red fox, black bear	Similar genera, different species	
Desert	Lizards, snakes, kangaroo, jerboa, hamster, hedgehog rat, cottontail		Gibbon, oranutang, monkey, Indian elephant, sunbear, porcupine, tiger, snakes and lizards
Tropical forest			
All vertebrates	Total families No. 122 / Unique 12 / % 10	136 / 3 / 2	164 / 12 / 7
Mammals	Total genera No. 74 / Unique 24 / % 32	100 / 35 / 35	118 / 55 / 46
Birds	Total genera No. 169 / Unique 52 / % 31	174 / 57 / 33	340 / 165 / 48

	Neotropical	Australian	Ethiopian
Temperate grassland	Guanaco, rhea, viscacha, cavy, fox, skunk		
Desert	Guanaco, rhea, armadillo, vulture	Marsupial mole, jerboa, parakeet, lizard	Springbok, porcupine, jerboa, rock hyrax
Savanna		Emu, red kangaroo, bandicoot, wombat, cockatoo, parrot	Zebra, eland, gemsbok hartebeest, gnu, giraffe, elephant, ostrich, lion, cheetah
Tropical forest	Monkey, kinkajou, pygmy anteater, sloth, tree snakes, parrot, humming bird	Tree and musk kangaroos, wallaby, koala, opossum, cassowary	Okapi, gorilla, chimpanzee, monkey, forest elephant
All vertebrates	Total families No. 168 / Unique 44 / % 26	141 / 30 / 21	174 / 22 / 13
Mammals	Total genera No. 130 / Unique 103 / % 79	72 / 44 / 61	140 / 90 / 64
Birds	Total genera No. 683 / Unique 576 / % 86	298 / 189 / 64	294 / 179 / 60

FIGURE 20:3 *Some of the data forming the basis of Wallace's zoogeographical regions: animal distributions are related to major vegetational belts, and unique animals in each region are shown. (After Selby, 1971.)*

placental mammals and is distinctive because of the ways in which the marsupials have filled similar niches in similar ways (e.g. marsupial mole and wolf). Of the 9 families of mammals present, 8 are unique! New Zealand has been even more isolated and has a very small fauna: bats are the only mammals, and flightless birds dominated the scene before white man's arrival. The reptiles are unusual survivals, like the gecko and the *Sphenodon*, which disappeared elsewhere in the Cretaceous.

6) The **Neotropical region** (South America) is another mainly tropical region and has the highest number of exclusive animal mammal families. Half of the 32 families of marsupials (different from the Australasian forms), together with many monkeys, birds and rodents are quite distinct.

7) The **Antarctic** is now regarded as a further region by many biogeographers. It has an extremely impoverished fauna.

Animals and plants in the oceans

Distributions in the oceans form a great contrast with those on the land. There is little plant variety in the sea, where the plant kingdom is formed mostly of the microscopic phytoplankton and bottom-living algae in shallow waters. The oceans form a continuous medium, in which movements of water encourage mixing and equalisation of temperature. Perhaps the greatest distinctions are between the bottom-living forms, and those which swim or float. A general division between the cold water and tropical forms can be made, but only the shallow water forms are markedly zoned or restricted to particular areas of the world. Thus the tropics are much richer in living forms than the colder regions and many genera and families are confined to the warm water surface zone. This region can be divided into Atlantic, Indian, West Pacific and East Pacific zones, since colder waters could prevent migration from one to the other, but despite this many forms occur in all the zones. The greatest variety of shelf-living forms occurs in the **Indo-West Pacific zone**, and it is the fish and planktonic forms which give the homogeneity to this region. Islands like the Hawaiian group are isolated from the rest and have some unique forms, but many of the Indo-West Pacific corals are missing there. The **East Pacific** and **Atlantic zones** are very similar, since they have only been separated by the Panama land bridge in recent times (Figure 20.4).

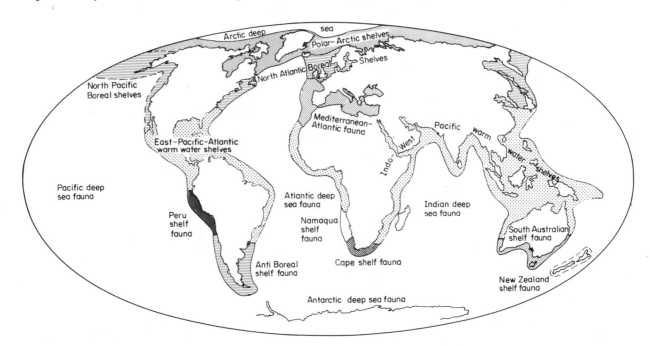

FIGURE 20:4 *The zoogeographic zones of the oceans. Continental shelf and deep ocean areas are distinguished by bottom-living (benthonic) forms. Planktonic and swimming forms involve somewhat different distributions. (After Ekman, 1953.)*

In similar ways the faunas of the shelf seas of the **northern Atlantic** and **northern Pacific** can be related, and are different from the margins of **Antarctica**. **Transitional groups** between cold and warm waters are found in the Mediterranean Sea and around the margins of the southern continents.

Apart from the shallow water forms there are the swimming forms and the deep water benthonic forms. These are often related closely to the temperature conditions which determine their food supply, and may be confined to certain water masses. This is important in the search for fisheries.

Zoogeography and continental movements

Many aspects of the patterns of distribution outlined in this section have puzzled biogeographers. Why should the animals and plants living in Australia be so unique? Why should there be more similarities between the animals and plants of Australia and South America than between those of Australia and south-east Asia? Many other questions of this nature suggest themselves.

The theory of continental drift has been one of the suggested solutions to the problems posed by these present-day distributions, and the heightened status which theories of continental movement across the Earth's surface now enjoy (*The Earth's Changing Surface*) has prompted a renewed interest in its relationship to biogeography (Figure 20.5).

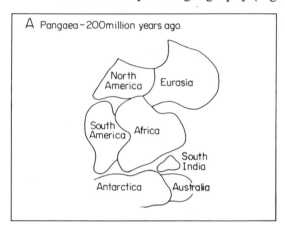
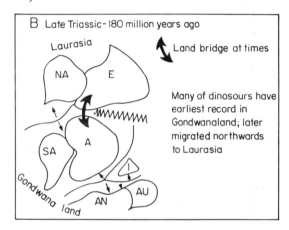
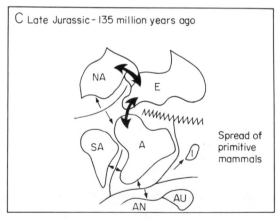
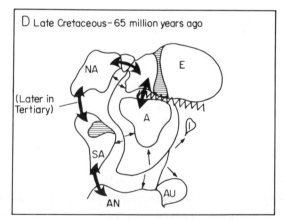

FIGURE 20:5 Animal migration and continental movements. Whilst the continents were close, or joined, migration could take place and common faunas could occur. After (D), the continents parted, leading to isolation and diversification of mammalian faunas in the early Tertiary. Later in the Tertiary some continents came closer again (e.g. North and South America were joined; Eurasia and North America had contact across the Bering Straits; and Eurasia and Africa came close). Mingling occurred once again. (Data from Kurten, in Wilson, 1972.)

It would seem, for instance, that the great development of reptile life on the Earth took place at a time when the continents were much closer together. The main phase of their diversification began in the Triassic (cf. Figure 20.5(A)) and lasted until the continents had parted in the Cretaceous. The number of reptilian orders over this period (i.e. over 100 million years) is somewhere between 7 and

13 (different authorities disagree), but this compares with up to 30 orders of mammals which have developed in the succeeding period of time (the last 65 million years). The first mammals left fossils in Triassic rocks, but were insignificant until the Tertiary, diversifying when the continents had parted (Figure 20.5(D)). Different groups of mammals occur in each continent, but there are many cases of these different groups becoming adapted to similar conditions in similar ways: the classic case is that of four distinct ant-eating mammals in four continents belonging to four different orders but having very similar shapes with low bodies and long snouts. Such results are often referred to as due to 'evolutionary convergence'. At a late stage in the Tertiary, North and South America were joined, as were Africa-Arabia and Eurasia, leading to mingling, new migrations and competition-with-extinction. It is thought that man's ancestors spread from Africa to Eurasia and then via the Siberia-Alaska land bridge to the Americas at this time. The evolution of continents and living forms can be traced further back in time (Figure 20.6).

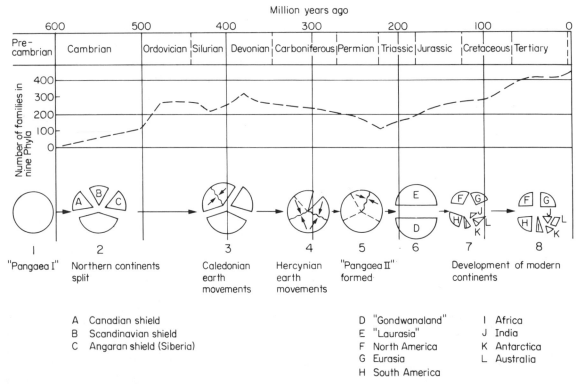

FIGURE 20:6 *Continental evolution and the development of life during the last 600 million years of Earth history. The continental shapes are generalised to emphasise the meetings and partings which have taken place. The nine phyla, used to show the relative diversity of life, are the most common fossil groups. How did the combination and separation of continents affect the diversification of life? (After Valentine and Moores, in Hallam, 1973.)*

The geography of disease

One very practical aspect of zoogeography is the study of diseases affecting man and the animals he domesticates. There have been cases where the actual cause of a disease is not clear, but where detailed mapping of its occurrence has led to some understanding of its causes and then to its restraint. The best-known example of this is **cholera**, which still remains largely a mysterious disease. Before 1816 it was confined mainly to India (Figure 20.7), but then spread throughout the world as trade developed, so that within 50 years it became a scourge in most areas. The root of the trouble was not spotted until 1854, when Dr. John Snow mapped the occurrences of 500 deaths in the Soho district of London: they clustered around a stand pump in Broad Street, from which drinking water was obtained. When this stand pump was stopped up the deaths ceased. Drinking water had been identified as the source, and within 20 years the disease began to retreat. It is now confined within its original area.

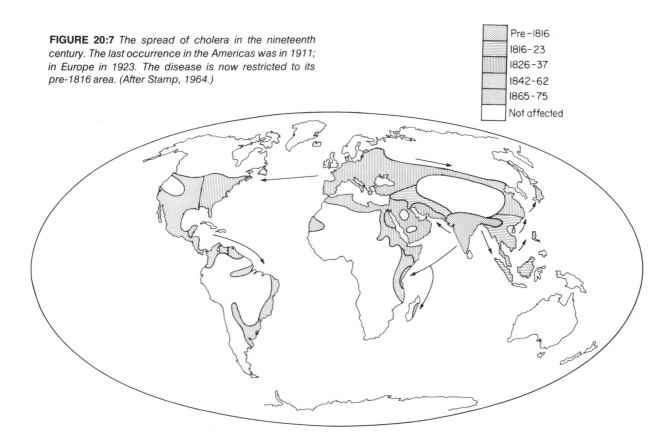

FIGURE 20:7 The spread of cholera in the nineteenth century. The last occurrence in the Americas was in 1911; in Europe in 1923. The disease is now restricted to its pre-1816 area. (After Stamp, 1964.)

Malaria has been the greatest killer of all diseases with world-wide distribution; where it does not kill, man is considerably debilitated. Areas which have been prosperous, have been abandoned when malaria takes over: thus the northern, once-irrigated parts of Ceylon and many of the coastal areas of southern Italy, were abandoned to malarial swamp after periods of prosperity before the Middle Ages. The malaria parasite depends on being taken into the stomach of the blood-sucking *Anopheles* mosquito for fertilisation. Offspring are transferred to mouth saliva and thence to the human victim of the next bite. The *Anopheles* thus has to bite twice: once to obtain the parasites from a carrier, and secondly to transfer the young to the victim, who suffers periods of alternate high temperature and shivering fits. Personal preventative measures have ranged from the use of mosquito nets to taking quinine and covering the body with repellent oils like estronella. Effective drugs, developed since 1930, now give some immunity to the disease. The most important developments in malaria control have concerned the mosquito breeding grounds, which are nearly always in standing water. The eastern part of the Ganges delta which has faster-flowing waters and regular annual floods has been freer of malaria than the western, more stagnant area. Drainage of Italian coastal swamps in the 1930's removed many malarial areas from that country, and elsewhere the films of oil on stagnant water surfaces, and the use of insecticides have proved efficient. The World Health Organisation has done much to help remove this threat to life and working efficiency. Many of the inhabitants of the tropical developing countries have learnt to live with the disease in the past, but this has taken its toll of the individual's stamina (Figure 20.8).

Medical geography is at an early stage, but the study of world and local distributions has practical applications, and the increasing emphasis on the ecology of man — his relationship to his environment — will surely be productive of further advances. One of the most important factors in the history of mankind over the last 100 years has been the decline in mortality rates: this has been largely responsible for the rapid population increases leading to extreme pressure on the land available. Malnutrition is perhaps a greater threat than many of the diseases as strictly understood.

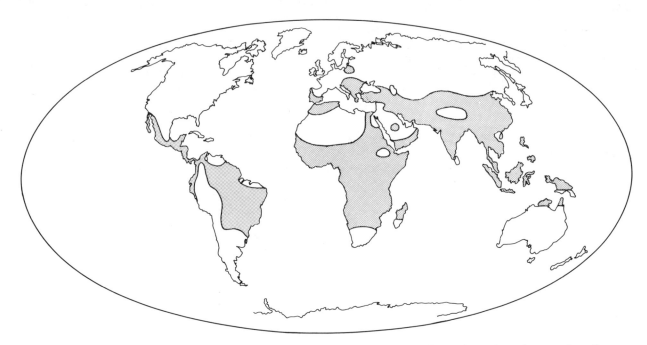

FIGURE 20:8 World malaria: areas still liable to the disease in 1960. Few areas of the world apart from the deserts and very cold regions have always been immune from the disease. (After Stamp, 1964.)

World ecosystems

There are two main approaches to the differentiation of ecosystems into regional types having boundaries between. It is common to find, in the first, that different ecosystems are characterised in terms of distinctive environmental conditions, life-forms or species composition. On the other hand it is also valuable to examine the changes as one passes from one ecosystem to another: this method not only emphasises the fact that all such boundaries are zones of transition, but also makes it possible to measure the nature and rate of the changes taking place. This concept of changing conditions across an ecosystem boundary is often known as '**gradient analysis**', since the changes can be plotted with the degree of variation in a particular direction.

Gradients measured in this way can involve populations of plants and/or animals; or they can be related to environmental factors such as altitude or moisture supply; or they can include both types of approach. The combined approach is most relevant to the study of ecoclines. Some ecoclines on a world scale are shown in Figure 20.9, and these can be related to the division of world ecosystems. On a more local scale ecoclines between small-scale ecosystems like a woodland and river can be distinguished; or in the intertidal zone on the seashore where there is an increasing exposure gradient from low tide, up to high tide, where there is almost continuous exposure (Figure 20.10). Suggestions for studying local ecoclines are to be found in the Hultons Field Biologies, recommended in the Bibliography.

Each species of plant or animal has an individual distribution (Figure 20.11). When two species compete for the same 'niche' one will be eliminated from the area and will live in what is perhaps a less favourable position, but one where it will be able to maintain itself more effectively. There are thus few sharp boundaries between natural ecosystems, which are recognised by a few dominant plant species. Even these may overlap slightly with those of other ecosystems, but smaller species may overlap more completely. Any sharp boundaries which occur between ecosystems are due to relief features like a coast or inland cliff, or to man's activities such as logging or burning.

Thus ecoclines can provide a most useful approach to the study of the distribution of ecosystems. The study of the major world-scale ecosystems introduces more problems in terms of the availability of information, but it is valuable to be able to form a world picture of the overall structure of the biosphere in its natural state, and the relative productivity of the different world areas.

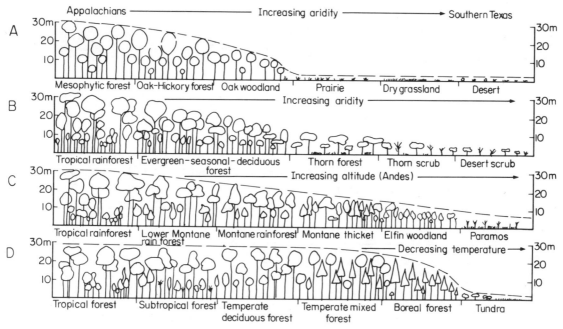

FIGURE 20:9 *Profile diagrams of four ecoclines on a major world scale. Compare the effects of increasing aridity, increasing altitude and decreasing temperature. (After Whitaker, 1970.)*

Major ecosystems within a continent are known as **formations** (relating to plants only) or **biomes** (plants and animals). Thus a biome is a grouping of (normally terrestrial) ecosystems on a given continent that are similar in vegetation structure, in the major features of the environment to which this structure is a response, and in some characteristics of their animal communities. Similar biomes are found to occur in several continents: e.g. tropical rain forests, hot deserts, temperate deciduous forests. These can be related as **biome-types**. Biomes and biome-types are thus related basically to vegetational distributions — and plant life is after all the foundation for life and the most bulky aspect of life on the land — but the vegetation structure provides varying numbers and types of niches for animals, and is interdependent with them. The application of the concept to the oceans is more difficult, but has been attempted (chapter 24).

FIGURE 20:10 *Some factors which change along an ecocline: the figures given here take the extremes of a tropical forest and a desert.*

	Favourable environment	Ecocline — Increasing aridity →	Extreme environment
Productivity	High (2000 dry g/m² year)	→	Low (5 dry g/m² year)
Community size	Large (Trees over 40m high)	→	Small (Plants under 1m high)
Biomass	Large (> 40 kg/m²)	→	Small (< 1 kg/m²)
Coverage	Complete (> 100%)	→	Partial (< 10%)
Community structure	Complex (Many species, stratified)	→	Simple (Few, separate plants)

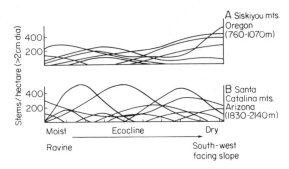

FIGURE 20:11 *Species distribution along environmental gradients. These two records show how species intermingle, and how it is difficult to draw ecosystem boundaries on this basis alone. (After Whitaker, 1970.)*

The major biome-types are studied in chapters 21 to 24 (Figure 20.12). They can be related to climatic humidity and temperature (Figure 20.13).

I Land Biome-Types

Optimum land biome-types (Growth throughout year)	1 Tropical evergreen rain forest		Chapter 21
Land biome-types seasonally deficient in heat and/or water (growth ceases for part of year)	Tropical Seasonal forests	2 Semi-evergreen rain forest	Chapter 22
		3 Deciduous tropical forest	
		4 Thorn forest	
		5 Thorn scrub	
	6 Savanna		
	'Mediterranean' lands	7 Mixed evergreen woodland	
		8 Sclerophyllous scrub	
	Temperate forests	9 Northern coniferous forest	
		10 Mixed forest	
		11 Broad-leaved forest	
	12 Temperate grassland		
Land biome-types permanently deficient in heat and/or water	Arid regions	13 Subtropical	Chapter 23
		14 Temperate	
	15 Tundra		
	16 Alpine, mountain biome-types		

II Oceanic Biome-Types

Warm waters	17 Continental shelves	
	18 Open ocean	
		Chapter 24
Cold waters	19 Upwelling waters	
	20 Continental shelves	
	21 Open ocean	

FIGURE 20:12 *Major world biome-types.*

THE DISTRIBUTION OF LIFE 241

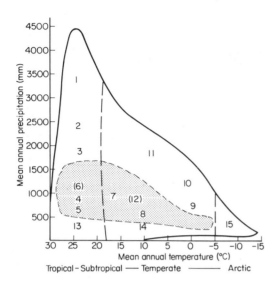

FIGURE 20:13 *A pattern of biome-types related to rainfall and temperature: numbers used refer to Figure 20.12. The shaded area is that in which forest may give way to grassland if burning, grazing or soil depletion have an effect. (Related to a diagram by Dansereau, in Whitaker, 1970.)*

The biome-types to be studied are different from the older concept of 'natural regions', where climate, vegetation and soil-types all fitted the same boundaries. The activities of man and the geological evolution of different parts of the world have led to too many exceptions to a general rule of this nature, and that concept should be abandoned. The ecosystem — and biome-type — concept emphasises the interactions taking place in each area shown on the map (Figure 20.14) but allows for the intervention of the other factors discussed in chapters 15 to 19.

FIGURE 20:14 *The world distribution of biome-types.*

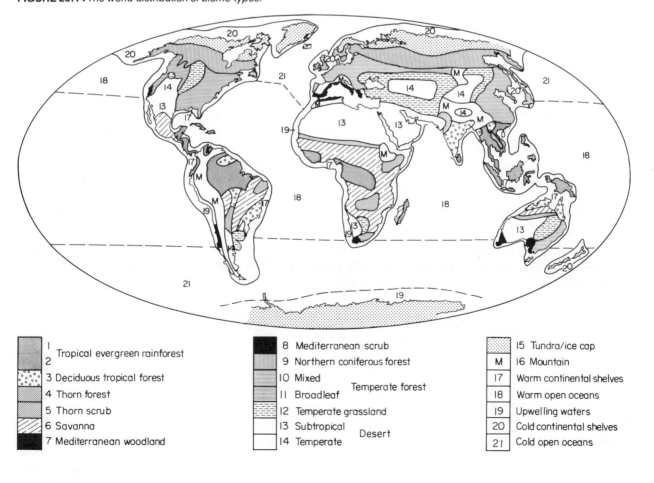

21
Optimum land biome-types

A relatively small proportion of the world's land surface, in the Tropics, has temperature conditions which are suitable for the continuous growth of plants throughout the year. It is only in the equatorial areas of uninterrupted rains that water supply is also sufficient at all seasons (chapter 10). Three main regions (Figure 21.1) possess these characteristics.

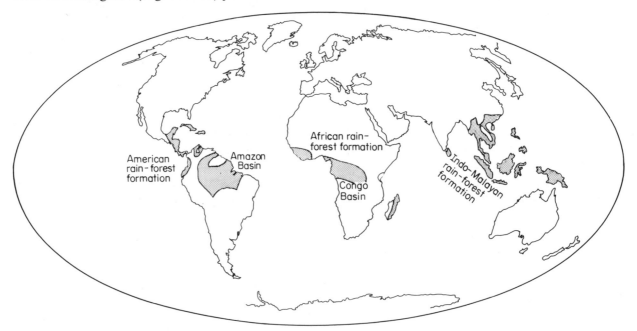

FIGURE 21:1 *The distribution of optimum land biomes, shown in terms of the plant formation types. Compare this distribution with that of the equatorial rainy climates (Figure 10.1) and attempt to explain the differences.*

1) The lowlying centre of the **Amazon Basin** extends into tributary valleys, along the Andean foothills and around the Brazilian coastal lowlands. The vegetation of this area is classed as the American formation of the tropical evergreen rain forest, and includes an outlier on the north-west coast of Colombia and adjacent Central America.
2) In Africa a similar formation is found in the **Congo Basin**, whilst large areas have been cleared in the last 50 years from the coasts of West Africa and those of Tanzania and Malagasy.
3) The **Indo-Malaysian area** is the most extensive, ranging from Assam in the north-west to northern Queensland. The forest is most typical in Indonesia and the Malayan peninsula, although, as in Africa, large areas have been cleared, or greatly modified by man.

Whilst these forests are associated dominantly with the equatorial rainy climatic zones, they occur beyond these limits when local conditions of relief bring rain throughout the year.

The plant formation-type of all these regions is evergreen rain forest: deciduous species cannot compete where soil moisture is available for growth at all seasons. Biological activity is at a high rate, growth is rapid and continuous, and competition between species is at a high level. The three main regions display differences of species composition due to effective isolation from each other, but the life-forms and overall structure of the forest formations are similar (Figure 21.2).

The evergreen rain forests contain a bewildering **variety of species**: 6000 species of flowering plants are known in west Africa, 20 000 (different from the west African) in Malaysia, and 40 000 in

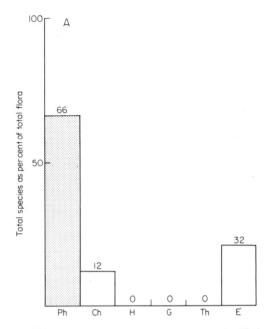
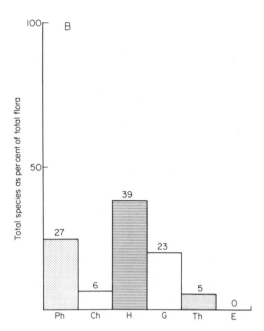

FIGURE 21:2 *Raunkiaer spectra. Compare the life-form features of each type.*
(A) A tropical rainforest in Guyana.
(B) A deciduous woodland in Germany (average of size associations.)
(N.B. the key is explained in Box, p.246-9.
(After Richards, 1964.)

Brazil. Whilst a natural deciduous forest in north-west Europe may contain up to 12 species of tall trees, a Malayan forest may have 2500. Over 300 species of trees have been plotted in 2 square kilometres in Brazil. One outcome of this variety is that there are few stands where trees of economic importance like the mahogany occur in great numbers. Apart from a few trees used in the making of furniture (which are often named after the shipping port or district from which they were first exported), the names sound strange to European readers, and few are mentioned here. In addition the plants are packed closely together, utilising the resources of light, water and nutrients to the full.

The **plant forms** in this forest are similar in all the three main areas where it grows. The leaves of the trees are particularly uniform, mostly being dark green, leathery and oval in shape — rather like those of the laurel. The leathery nature is due to a heavy cutin covering which gives the leaf rigidity in conditions of intense insolation and transpiration, since it might otherwise wilt and collapse. So-called 'drip-tips' are found on the leaves of shorter trees: this characteristic may assist them to shed rain water rapidly, thus preventing other plants from occupying the leaf surface as epiphytes. The bases of the trunks are often distinctive. Roots are short and confined to the uppermost soil layers; many trees have buttresses, beginning 3-9 m above the ground and spreading out to the base (Figure 21.3), and these may bring the advantage of strengthening the whole tree structure, or of assisting the movement of water and nutrients to the upper part of the tree. Another characteristic of the trunks is that some bear flowers and fruit (e.g. cocoa). The bark of trees in these regions of warmth and high humidity is often less than 2mm thick.

The **forest structure** is characterised by the closest possible packing of the trees in several layers beneath the tallest which form the canopy (Figure 21.4). Species occupying the lower levels are adapted to live in lower light intensities. Little light reaches the forest floor, and the shrub and ground layers are very restricted. This type of forest is thus not 'jungle'. Thousands of seedlings grow, but remain virtually dormant until an old tree dies and crashes down, letting in the light. When this happens there is a sudden burst of growth, dominated at first by a group of plants adapted for rapid growth and the production of seeds before the trees emerge and once again assert their dominance. The stratification (Figure 21.4) is not obvious to the traveller through these forests, who can observe only a tangle of vegetation. It becomes evident after careful measurement of a small

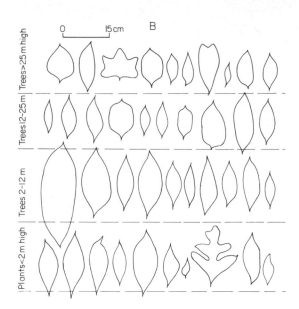
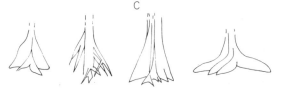

FIGURE 21:3 *Floristic features of tropical rainforest.*
(A) Cauliflory: cacao seed pods growing from the trunk.
(B) Leaves of forest plants of different heights. Between 70 and 80 per cent of these leaves have areas of 20.25-182.25 cm² and are known as mesophyllous.
(C) Buttresses of forest trees.
(D) Rates of growth: compare these with temperate trees. What is the difference between forest and open conditions?
(After Richards, 1964, and Selby, 1971.)

D. Species	Diameter class (cm)	Number of individuals measured	Average annual diameter increase (cm)	Average number of years in diameter class
Parashonea malaanonan	0-5	4	0.07	71
(in forest)	5-10	10	0.27	19
	10-20	21	0.38	26
	20-30	19	0.49	20
	30-40	12	0.74	13.5
	40-50	28	0.82	12
	50-60	7	0.94	11
	60-70	10	0.75	13
	70-80	1	0.84	12
(in open)	0-5	13	0.42	12
	5-10	27	0.55	9.1
	10-15	7	0.73	6.8

strip of forest. The tallest trees reach up to 60 m high, but these are not the world's tallest, since the Californian sequoias reach well over 100 m high.

A group of unusual plants, adding further diversity to the structure of these forests, possess adaptations enabling them to reach the light, or to do with less light in their growth. The **lianas** are woody, climbing plants which begin life in the shade and reach up to the brightest equatorial sunlight. Their stems may be up to 100 m long, draped from tree to tree, and may extend to the uppermost part of the canopy layer. Large trees may still stand in place, although the trees

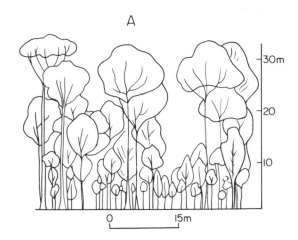
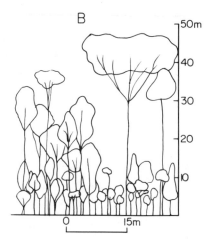

FIGURE 21:4 The layered structure and species composition in tropical rainforests. Trees over 4.6 m tall are included. Compare these diagrams with Figure 19.4.
(A) Guyana, compiled from a strip 7.6 m wide.
(B) Southern Nigeria. Suggest a division of this diagram into strata. This was also compiled from the survey of a strip 7.6 m wide.

(C) Guyana. The letters stand for tree species: how many are represented? Compare the numbers of species in each layer. E = Eperua falcata ('soft wallaba'); Eg = Eperua grandiflora ('ituri wallaba'): these are the dominants in what is known locally as wallaba forest. (After Richards, 1964.)

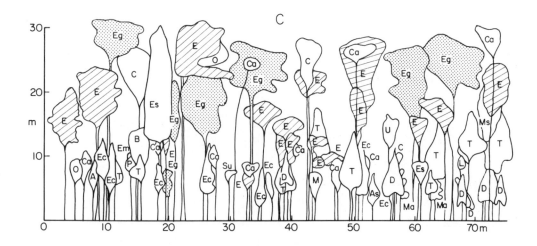

themselves are dead and rotting, by virtue of the enveloping lianas. **Epiphytes** are small shrubs or herbs which live in the boughs of the upper storey trees. They suspend a tangled mass of fine roots and catch falling leaves and moisture to nourish the plants. Scores of species may be found in one tree: many orchids belong to this group. **Semi-parasitic plants** may also lodge on branches and penetrate the host tree with suckers. **Stranglers** begin life like the epiphytes, germinating high up in the tree, but then extend roots down into the soil, forming a dense network which eventually destroys the host tree. **Saprophytes** have adapted to a heterotrophic way of life, obtaining their organic materials and store of energy from the forest floor humus by means of symbiotic fungi in their roots.

Competition in such optimum conditions is thus intense. Only those trees and plants which possess the common, successful features can survive. Hence the apparent paradox of so many species occurring in the forests, but having so few differences of form.

In the **natural state** the forests would remain undisturbed for a long time. The high production rate results in a high rate of litter accumulation on the forest floor, but this is soon decayed by the active bacteria and taken up once again by the trees in the re-cycling process. For a long time human occupation was limited to shifting agriculture: a patch of forest was burnt or cut down and planted

Vegetation life-form analysis

Vegetation life-form may be used as a basis for mapping the different types of plant, rather than their botanical names. This is a useful basis for ecological studies, and three different systems are summarised (Figure B-1, Figure B-2 and Figure B-3). Each has its virtues, and may form the basis of a local study, or another may be devised to suit the individual, who will then construct a chart for the field recording of informations.

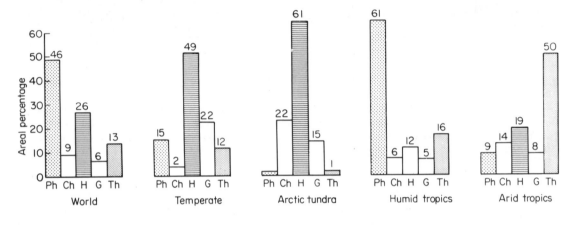

Category	Symbol	Description
Phanerophyte	Ph	Exposed plants: bud high above ground. All trees, shrubs.
Nano phanerophyte	N	Dwarf shrubs: 25cm - 2m high
Mega phanerophyte		Trees over 20m tall
Meso phanerophyte		Trees 10 - 20m tall — Subdivided on height
Micro phanerophyte		Trees, shrubs 2 - 10m tall
Chamaephyte	Ch	*Ground plants*: herbaceous, woody plants. Buds near soil. Common in dry, cold climates.
Hemicryptophyte	H	*Half-hidden plants*: grow in favourable season, then die back to bud at soil level. Grasses, herbs, rushes, sedges.
Cryptophyte — Geophyte	G	*Earth plant*: die back below soil level due to cold or drought. Bulbs, corms, tubers, etc.
Cryptophyte — Hydrophyte	HH	*Water plant*: bud immersed in water.
Therophyte	Th	*Summer plant*: annuals, producing seed and inactive in unfavourable seasons. Common in deserts.
Epiphyte	E	Grow on trunks, branches of Ph forms above ground level.

FIGURE B-1 *The Raunkaier classification of life forms together with selected spectra for major climatic zones.*

Vegetation life form analysis

The Raunkiaer method (Figure B-1) is simple and accurate, based on the height above ground of the bud: this reflects the degree to which it is protected, and thus whether the plant is subject to seasonal or prolonged inhibiting factors. On the other hand this method is highly selective in the features it includes, and is difficult to apply where continuous growth is possible. In addition no distinction is made between types of inhibiting factor (i.e. cold or drought).

Dansereau's scheme (Figure B-2) divides plants into five categories based on general form, and these are subdivided according to height, vegetative function (i.e. length of year in leaf), leaf form, size and texture. These are all given letters and symbols. Kûchler's more recent refinement of this system (Figure B-3) provides a greater range of possibilities and allows more detailed records to be made.

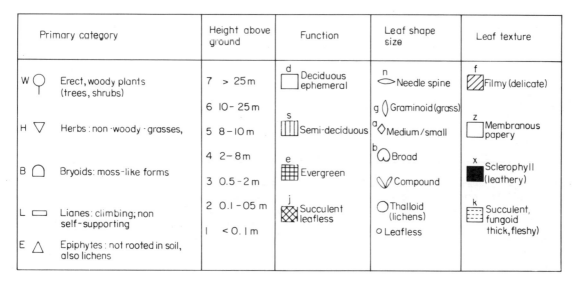

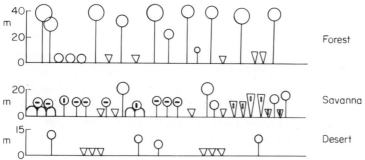

FIGURE B-2 *The Dansereau scheme of life-form classification, with three typical profiles.*

Vegetation life form analysis

1 Life form categories

Woody plants	Herbaceous plants	Special categories
B Broadleaf evergreen	G Graminoids (grasses, sedges)	X Epiphytes
D Broadleaf deciduous	H Forbs (herbs, ferns)	C Climbus
E Needle leaf evergreen	L Lichens, mosses	K Stem succulents (Cacti)
N Needle leaf deciduous		P Palms
O Aphyllons		V Bamboos (woody grass)
M Mixed (E + D)		T Tuft plants (tree ferns)
S Semideciduous (B + D)		

2 Leaf characteristic	3 Height (at maximum development)	4 Coverage (each life form)
h Hard (solerophyll)	8 > 35m	c Continuous (> 75%)
w Soft	7 20-35m	i Interrupted (50-75%)
k Succulent	6 10-20m	p Parklike, patchy (25-50%)
l Large (> 400 cm^2)	5 5-10m	r Rare (5-25%)
s Small (< 4 cm^2)	4 2-5m	b Barely present (1-5%)
	3 0.5-2m	a Almost absent (< 1%)
	2 0.1-0.5m	
	1 < 0.1m	

FIGURE B-3 *The Kuchler life-form classification. He gives an example, related to forest in western New York State:*

$$D7c4r \quad H2p \quad L1r$$
$$(a) \quad (b) \quad (c)$$

(a) D7c4r: two layers of broadleaf deciduous trees. One, 20-35 m high, covers over 75 per cent of the area; the other 2-5 m high covers 5-25 per cent.
(b) H2p: ferns beneath this are on average 10-50 cm high and cover 25-50 per cent of the ground.
(c) L1r: Mosses and lichens on the ground (less than 0.1 m high) cover 5-25 per cent of the area.

with root crops, but after a few years the lack of leaf-fall led to the exhaustion of the soil. Wild plants also began to spring up and stifle the crops, so that it was easier to clear another patch. Natural regeneration to climax forest would take at least 250 years, even in these optimum conditions, so that forest which is cut every 5-20 years is not going to have the same character. Recent years have seen increasing pressure on the land, so that the cutting of a particular area takes place more often and the introduction of tree crops and plantations has completely modified the forest scene. It is important to remember that if the soil is deprived of the leaf-fall and protecting canopy for even a year or two, the soil structure is disturbed. The surface soil particles are removed by run-off in a very short time, leaving a hard crust of dried accumulated iron salts which prevents the re-establishment of the forest. In some areas a **secondary forest** will be established after human cultivation, having lower dominant trees, more undergrowth and less stratification than the primary type. It is thus more like 'jungle'. If the soil structure has been destroyed, however, xerophytic shrubs and herbs may dominate the area.

The **limits** of tropical evergreen rain forest occur where it gives way to seasonal forest or mountain forest. The essential criterion seems to be sufficient rain falling throughout the year, and the length of the dry season is therefore particularly critical. The change is rarely abrupt and a broad ecotone (transitional ecosystem) intervenes in areas where there are three months each with only 50-100 mm of rain in the Amazon Basin — though typical rain forest grows in Nigerian areas with over 5 months under 100 mm rain. In the latter case the humidity remains at a high level, as does the soil water table, so that the forest is never deprived of the requisite moisture. There is often a relationship between the soil permeability and the forest development in these marginal areas: very permeable soil will support only seasonal forest, even where the dry season is short. Rainfall as such is not the fundamental factor, but the rate at which water drains away or is lost to the atmosphere. The increase of humidity, cloudiness and windiness on mountain slopes leads to a further transition (chapter 23).

The **primary productivity** of tropical evergreen rain forest is the highest of all the natural biome-types: up to 5000 dry g/m^{2}year, which is nearly 40 per cent of the world net primary land production on only 13 per cent of its area. Much of the production is wood, and much is diverted through the detritus pathway for re-cycling, but there is still plenty of food to support a wide variety of **animal life**, and there is a multitude of environmental niches. The fact that growth continues throughout the year means that animals do not have to migrate in order to obtain a continuous supply of food. Most of the animals live in the canopy trees where forage is plentiful, and there are small, mobile forms like monkeys, lemurs and snakes together with a host of birds and insects, each with their own adaptations and particular habitat. Studies in the Central American forests have shown that nearly 200 bird species live wholly in the trees, whilst less than 20 forage on the ground. Of the 50 species of mammals in Guyana, 30 are arboreal. Ground-dwellers, on the other hand, are still small, but shy and inconspicuous. Of the few larger varieties the African elephant, buffalo, okapi, leopard, tapir and jaguar are the best known, but the pig family are the most characteristic along with rodents and large insects.

Coasts and rivers

The margins of tropical evergreen rain forest along the edges of water courses, and along the coasts, are marked by an increase in the sunlight reaching to lower levels, and by the raised height of the water table. These communities are therefore **hydroseres**. The availability of sunlight at lower levels means that shrubs and smaller trees can grow, giving rise to impenetrable thickets along the river banks. The early stages of a water's edge succession of this type are dominated by reeds; mangroves then invade, including a range of shrubs and trees producing long stilt-like roots extending down through the water and often allowing their seeds to germinate on the branches before falling off at the seedling stage. Both reeds and mangroves lead to increased rates of silting around their roots and palm trees seem to be the ultimate successional forms of these hydroseres (Figure 21.5).

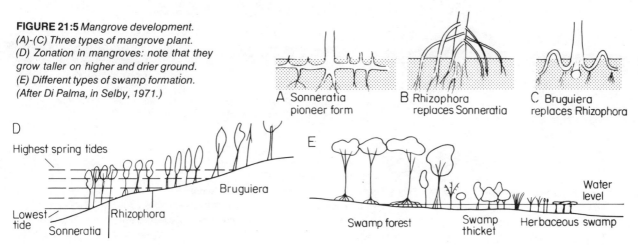

FIGURE 21:5 Mangrove development.
(A)-(C) Three types of mangrove plant.
(D) Zonation in mangroves: note that they grow taller on higher and drier ground.
(E) Different types of swamp formation.
(After Di Palma, in Selby, 1971.)

22

Land biome-types with seasonal climates

As soon as the tropical dry season becomes marked, the optimum conditions for plant growth give way to a situation where there is a season of limited, or no growth. Plants adapted to these conditions replace those occurring in the evergreen forests, with a gradation from humid to extremely dry environments. Many parts of the tropics with seasonal drought are now occupied by savanna grasses, but these have probably been greatly extended by the activities of man: ecologists see forest as the truly natural vegetation.

The warm temperate west coasts of continents (i.e. the 'Mediterranean' regions) present another case of seasonal drought. In this case it is the summer season which is arid, and the climatic regime is so unusual that a distinctive type of vegetation is associated with such regions.

Cooler temperate regions, which are so extensive in the Northern Hemisphere, and which have been associated closely with the development of western civilisation and the richer industrialised nations, have a seasonal deficiency in solar radiation. This acts as a limiting factor in a way very similar to drought: a seasonal fall in temperature inhibits plant growth, since there is no tree growth below 6°C in these regions. Many adaptations to cold seasons can be related to those in regions with dry seasons. Thus deciduous trees drop their leaves, preventing transpiration in time of drought, and the cold season often creates a physiological drought by freezing the water in the soil.

A large proportion, perhaps 40 per cent, of the Earth's land surface is occupied by biome-types with such seasonal regimes of climate and plant growth. They include the following major groupings (Figure 22.1).

FIGURE 22.1 *The distribution of biome-types having a seasonal regime of plant growth due to a shortage of heat or water at one season. Is there a noticeable pattern in the distribution of the group as a whole?*

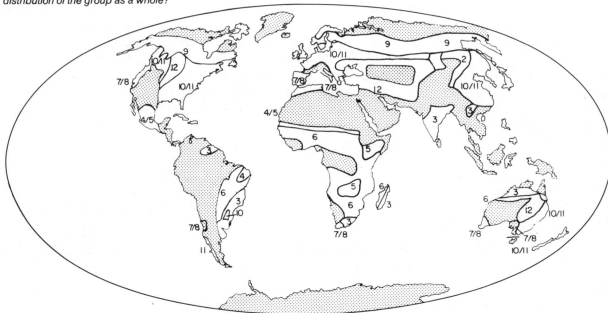

- 3 Tropical deciduous forest
- 4 Thorn forest
- 5 Thorn scrub
- 6 Savanna
- 7 Mediterranean woodland
- 8 Sclerophyllous scrub
- 9 Northern coniferous forest
- 10 Temperate mixed forest
- 11 Temperate broad leaf forest
- 12 Temperate grassland
- Other biome-types

1) Tropical seasonal forests and savannas.
2) Mediterranean evergreen forests and sclerophyllous (i.e. 'hard-leaved' and thus drought-resistant) scrub.
3) Temperate forests and grasslands.

In each of these groupings there are what may be regarded as the climatic climax association (the forest), and the man-affected association (grassland or scrub). Man has been particularly active in these regions, and the effects of his modifications have been felt mostly on the drier margins, where fire can have its greatest sway, and where grazing animals have been important or ploughing disastrous.

The tropical seasonal forests and savannas

Forest and woodland continues from the edge of the tropical evergreen rain forest to the borders of the deserts and temperate regions. This is a broad transition, known as an **ecocline** (i.e. a gradient along which communities and environments change) (Figure 22.2). During the course of this transition the number of forest layers are reduced, deciduous varieties increase, and eventually drought-resistant thorn trees and cactus become the dominants. This transition is found in all parts of the tropical world, but differences between the American, African and Indo-Malaysian formations are greater than in the case of the tropical evergreen rain forests, due partly to the greater variation in the range of species, but also to the range of effects caused by man.

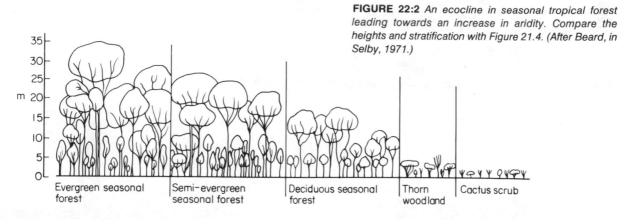

FIGURE 22:2 *An ecocline in seasonal tropical forest leading towards an increase in aridity. Compare the heights and stratification with Figure 21.4. (After Beard, in Selby, 1971.)*

Semi-evergreen seasonal rain forest

This vegetation formation-type occurs in areas where there is a drier season (i.e. 25-100 mm rain per month for at least 5 months), and the total annual rainfall is under 1500 mm. In Central and South America there are two storeys of trees rather than the three of the evergreen forests, and the open upper storey contains up to 30 per cent of deciduous trees. Even the evergreens here lose their leaves in unusually dry seasons. The lower storey, which forms a more compact and complete canopy, is wholly evergreen, including some microphyllous (i.e. 'small-leaved' — possibly reducing transpiration) trees (Figure 22.3). Africa possesses few large areas of this type of forest: those which

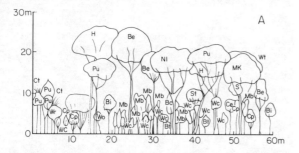
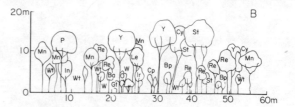

FIGURE 22:3 *Tropical seasonal forests in Trinidad.
(A) Semi-evergreen seasonal forest.
(B) Deciduous seasonal forest: compare the layering and species composition with (A). (After Beard, in Eyre, 1971.)*

174

Tropical seasonal biome-types

Plate 174 Monsoon forest in the foothills of upper Thailand. Why is there an area of lower vegetation along the river banks?

Plate 175 Sudan: savanna with *Anogersuss herocarpus* trees.

Plate 176 Thick acacia-dom scrub at Lodwar, Kenya.

175

176

LAND BIOME-TYPES WITH SEASONAL CLIMATES 253

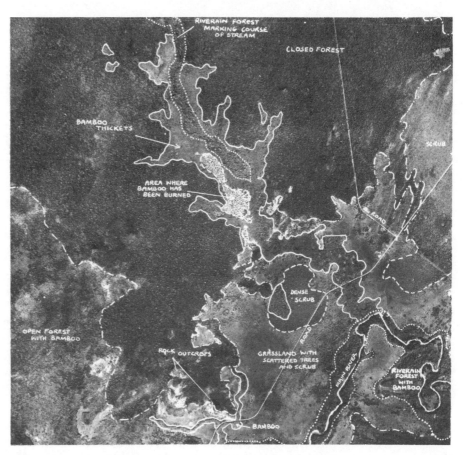

Plate 177 Vertical aerial photograph of part of Tanzania, showing the distribution of vegetation types in relation to relief.

Plate 178 A giraffe in Nairobi National Park; Baobab tree in East Africa.

Plate 179 Banyan tree, covering over half a hectare, in India. (178 Ewing Galloway, New York; others Aerofilms)

C	Species (percentage)		Individuals (percentage)	
	'Upper story'	'Lower story(s)'	'Upper story'	'Lower story(s)'
Evergreen seasonal forest	24	Negligible	6	Negligible
Mora forest	12	0	Negligible	0
Semi-evergreen seasonal forest	33	10	16.5	10
Deciduous seasonal forest	50	10	66	25

(C) The proportion of deciduous and semi-deciduous trees in Trinidad forests.

do exist are scattered along the northern margins of the evergreen forest. A discontinuous upper layer is all deciduous, and the true canopy is composed of two lower layers of mostly evergreen trees. In both these continents there are few buttressed trees, and epiphytes become uncommon, but lianas still occur regularly. The more open nature of this forest allows greater light penetration, and an undergrowth of flowering shrubs develops. The dry season is characterised by the accumulation of leaf litter, which decays in the wetter part of the year.

Wetter varieties of seasonal 'monsoon forest' are widespread in south-eastern Asia. Interference with the natural vegetation by man seems to have favoured fire-resistant trees like the teak, but these forests resemble those in the other continents in several points. Wet teak forest occurs in areas of Burma with 1500-2000 mm rain each year. The broken canopy layer is mostly evergreen, including up to 10 per cent teak, and the lower tree layer is all evergreen, being underlain by discontinuous bamboo thicket. Lianas occur, but few epiphytes. In areas with under 1500 mm rainfall, or on more permeable soils, teak increases in importance, since it can stand up to 5-6 months drought: this is the dry teak forest of Burma.

Deciduous seasonal tropical forest

Areas with longer dry seasons have increasingly deciduous forest. Total rainfall may be as low as 750 mm, or as high as 2000 mm if the soils and underlying rocks are extremely permeable. The essential feature is the nature of the dry season, which lasts 4-7 months, including 2-5 months in which under 25 mm rain falls per month. Forest still provides a continuous cover of trees, but the deciduous element increases, and the dominant trees are not so tall. In the Americas the upper discontinuous layer is entirely deciduous, and the trees often have a crooked, gnarled appearance. The lower tree layer is still largely evergreen and many trees in both layers are microphyllous. There are few lianas and herbaceous plants (e.g. grasses) and epiphytes are rare. Once again there are few areas of this type in Africa, where more open woodland, passing into savanna, is common: trees are never dense enough to exclude grasses. Fires affect these regions almost annually. In south-eastern Asia the teak, in and sal trees form a discontinuous upper layer beneath which shorter evergreen trees and bamboo thickets — but few grasses — give a variable cover.

Thorn forest

Increasing drought leads to the next stage in the ecocline sequence. Small deciduous trees are now the dominants, having a gnarled, spreading habit. Many species found in wetter areas have disappeared, but a closed canopy forest is still possible despite an annual rainfall total of as little as 700 mm and a dry season of over 6 months. In South America the *caatingas* of north-eastern Brazil and Venezuela have trees up to 10 m tall, including both evergreen and deciduous varieties: the former are adapted to dry conditions (microphyllous), and have the advantage of being able to make the greatest use of the short wet season. In this way the evergreens of dry margins are similar to the conifers of high latitudes. *Mimosa* and cacti are also found, especially on the drier soils. Epiphytes like *Spanish Moss* are common again, having developed hard leaves to reduce transpiration. India, Burma and Thailand have areas of similar thorn forest with the sha, dahat and thorn trees growing to 10 m, infested with climbers and having a grassy undergrowth, degenerating to sha thorn scrub in the driest parts (only 2 m high with much bare ground between the trees). Africa and Australia have small areas where acacias are common in open, grass-dominated communities; denser thorn scrub occurs along the river valleys. The Australian bottle tree is a bizarre inhabitant of these areas, water being stored in the tissues of its thick trunk.

Thorn scrub

Semi-desert conditions give rise to thorn scrub, in which succulent plants form a significant proportion of the flora, reaching only 2 m high and widely spaced owing to the roots taking up so much of the ground near the surface. 'Cactus scrub' is a feature of the south-western USA desert margins, and of the fringes of the Atacama and Mexican deserts. It includes a range of communities from an almost complete cover of succulents and shrubs to sparsely vegetated areas, and grasses are rare or absent throughout. Few areas in south-eastern Asia are dry enough, though euphorbias grow on gypsum-rich clay soils in Burma. In Australia there is a wide belt with 250-400 mm rain per year, and a distinctive group of eucalyptid and acaciaid plants with thick bark, like the wattle and mallee, together with spinifex grass, have made this their home. The largest area of all stretches around the southern margin of the Sahara-Arabian-Thar deserts: the sparse, impoverished communities include thick-trunked euphorbias with tiny leafless branches, thorny bushes intertwined with sprawling vines and creepers together with large tuberous plants. Similar thorny grassless scrub occurs in south-western Africa.

There is thus a forest-woodland-scrub series of plant formations related to decreasing rainfall in tropical areas. Evergreen trees give way to deciduous and increasingly drought-resistant varieties; dominant trees decrease from over 50 m high to under 2 m; and individual plants become more widely spaced. Grasses are important in few of these formations.

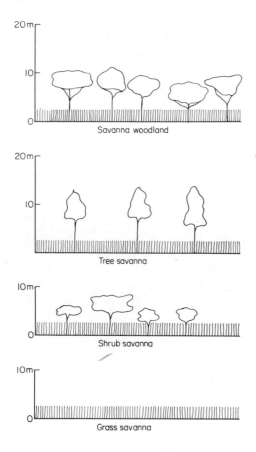

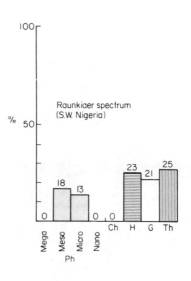

FIGURE 22:4 Types of African savanna. The key to the Raunkiaer spectrum is in Chapter 21, Box p.246. (After Hopkins, 1965.)

It is nevertheless a fact that **tropical grassland, or savanna**, is extremely widespread, particularly in Africa (Figure 22.4). It is clear, however, that these grass-dominated communities are not always the simple result of climatic conditions, since they cover such a wide range of conditions, and forest formations occur throughout this range.

Savannas vary from well-wooded, high grass (elephant and other tussock grasses to 2-4 m) formations, through a range of tussock-grass-with-acacia-tree varieties, to the acacia-desert grass savanna in which the grasses are discontinuous and the trees very scattered. Trees, like the baobab, acacia and palm, are significantly fire-resistant. Regular fires can modify the climatic climax vegetation in these areas (Figure 22.5).

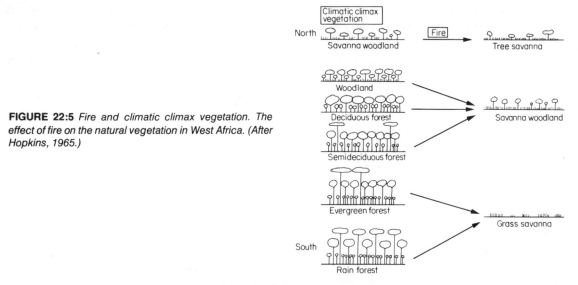

FIGURE 22:5 *Fire and climatic climax vegetation. The effect of fire on the natural vegetation in West Africa. (After Hopkins, 1965.)*

This picture is most true of Africa. In South America, where man has not been an agent in vegetation control for such a long time, there is more evidence of a relationship between landform, soil and plant community (Figure 22.6). The greater prevalence of savanna-type communities in Africa suggests that fire may be the over-riding factor there, though local soil (edaphic) factors may

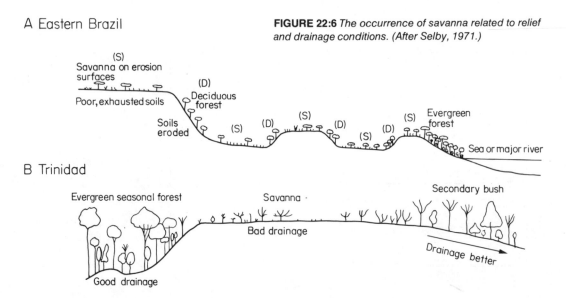

FIGURE 22:6 *The occurrence of savanna related to relief and drainage conditions. (After Selby, 1971.)*

be important: forest continues to grow along the water courses. Elsewhere the edaphic factors are often more pronounced, although some fire modification may also be evident. It is a matter of emphasis between the regions.

Animal life in the seasonal tropics

The range of vegetation types, from moderately tall and dense forest to open grassland and semi-desert scrub, provides a wide variety of ecological niches. Animal varieties range from the tropical forest types which have increased their ranges to only slightly drier conditions, to those adapted for living on the desert margins. The most distinctive group are associated with the savanna grasslands, and just as the greatest area of savanna is found in Africa, so is the widest variety of large grazing herbivores. The rich spectrum of antelope, zebra, giraffe, wildebeest, rhinoceros, hippopotamus and elephant contrasts with the small deer, guanaco and rodents of South America and the wallabies of Australia. The African forms are adapted to various niches — browsing on the higher leaves (giraffes), trampling between the forest and open savannas (elephant), living in the more open grassland areas (wildebeest), or inhabiting water sites (hippopotamus). These African forms are accompanied by a range of carnivores hunting in families (lions) or in packs (hyenas, jackals). Their task is not made easy by the fleetness, size and herding instincts of the prey. The range of savanna animals illustrates a clear energy pathway from the primary producers (grass, forest) to the herbivore and carnivore consumers.

As man has expanded his occupation of these areas, and has introduced settled and mechanised farming methods, the natural savanna and seasonal forest has been reduced, together with the habitats of the animals they contain. At the same time the animals have been the targets for hunters, so that populations are much reduced. The designation of special reserved areas has been necessary to ensure the survival of savanna faunas. The reserves in east Africa are several thousand km^2 in extent: some are as big as England.

The Mediterranean lands

The climatic regime of the hot summer drought and mild wet winters, which is so unusual in the overall pattern of Earth climates, is associated with distinctive biome-types. It is even more difficult to find examples of true natural vegetation in these areas than it is in the African savannas. Various types of secondary succession have developed in the Mediterranean-type environments, shorn of their original woodland and subsequently eroded of their soils. In short, these areas are amongst the most greatly modified by man. The summer drought, the heavy shower type of rainfall, and the coincidental fact that slopes are often steep in these regions, make them particularly vulnerable to soil erosion once the natural forest cover is removed. The shores of the Mediterranean Sea have the most extensive development of these conditions, and this is where the Greek and Roman civilisations had their bases. The lesson spelt out here was not heeded, and areas occupied more recently (e.g. central California, central Chile, southernmost Africa and south-western Australia) have suffered similar fates to a greater or lesser extent.

Some clue concerning the original vegetation can be found, but one cannot be definite concerning its constitution since man has had so much influence. The European and African areas bordering the Mediterranean Sea still have some trees growing amidst areas largely cleared for farming. Evergreen oaks (e.g. cork oak, holm oak) and pines on the hills (parasol pine, maritime pine, Aleppo pine) form a mixed woodland with drought-resisting characteristics (small, hardened leaves, deep roots), but it cannot be known if this was once more extensive since the soils would have been richer in humus and the whole scene different. The cedars of Lebanon were once fabled for their size and must have covered extensive areas. A pine-fir association is still found commonly on higher slopes, and this was probably the original cover between 1000 m and 2000 m.

The suggestion that woodland could have been the original vegetation in these areas is supported by observations in other parts of the world. The tree varieties are different, but have similar life-forms. Thus in central Chile evergreen angiosperms such as the evergreen beech, Chile gumbox, sumac and Chile soaptree form mixed forests with Araucarian pines and southern 'cypresses'. A remnant of similar structure occurs in South Africa, but the conifers and evergreen angiosperms are mixed with shrubs, lianas and epiphytes. This is more reminiscent of the outstandingly luxuriant forests which originally covered the North Island of New Zealand in a somewhat different climatic regime. The kauri pine has been removed completely, but there is a mixture of other distinctive native conifers (tatara, kahikatea, miro), together with broad-leaved evergreens, tree ferns, palms and epiphytes.

Few 'Mediterranean' areas still bear this mixed evergreen woodland: by far the greater area is now occupied by **sclerophyllous scrub vegetation** — where it is not cultivated or built over. Such vegetation has developed in widely separated parts of the world, involving very different groups of plants, but it always has a striking similarity. The processes of evolution have given broadly comparable results in comparable environments (i.e. 'convergent evolution'). Some communities of this type may have originally been a local climax in the driest parts of southern Italy, south-eastern Spain or north Africa, but it also presents many of the features of vegetation affected by man's presence in the area.

Sclerophyllous scrub is typified by the possession of small, thick, leathery leaves which are thought to reduce transpiration. Most of the plants are evergreen, low-growing (up to 3 m high), and woody like the wild olive, myrtle, carob and arbutus. This type of vegetation, combined with heathers and broom, is known sometimes as **maquis** in southern Europe. Permeable rocks like limestone may have a distinctive variety, **garrigue**, which is dominated by aromatic herbaceous plants like thyme and small, more widely-spaced maquis varieties like juniper and yew.

The **chaparral** of central California occurs naturally on the lower slopes of the coastal range and mountain flanks farther south, but the characteristically fire-resistant plants have advanced to take over from the woodland and grassland in areas with up to 1500 mm rainfall. The species involved occur elsewhere, but have adopted a sclerophyllous habit. One, the dwarf oak, is deciduous, reaching just over 3 m in height and combining with taller evergreen oaks and piñon pines in wetter areas, but is often under 2 m in drier parts; bush oaks, roses, buckthorns, heaths and the wild cherry are also common in the richly varied flora. The Australian **mallee scrub** is most extensive along the southern margins of the desert area around the Bight. Again the plants are sclerophyllous, 2-3 m high, and there is a rich variety, but they are composed of local varieties of eucalyptus occurring with spinifex grasses. South Africa and Chile exhibit similar features.

The long summer drought and the thick cutinisation of the leaves suggests that even in the original state the Mediterranean areas would not have been very productive: animal life was probably restricted. No evidence remains in the European areas, but the Californian areas were occupied by some herbivores (ground squirrel, deer, elk) and carnivores (grizzly bear, mountain lion) in the early nineteenth century, whilst marsupials occupy these niches in Australia, and the Chilean and South African areas also possessed a native fauna.

Temperate forests and grasslands

Some of the most complex distributions within the land-based biosphere occur in the cool temperate parts of the world. These are extremely extensive in the northern hemisphere, but restricted in the southern due to the shapes of the continents, so that similarities between the two hemispheres are few. The European, Chinese and North American areas have also seen a great deal of modification and removal of natural vegetation, leaving very few patches from which the original vegetation and climatic climax can be reconstructed. These areas were also affected most by the Pleistocene ice sheet fluctuations (chapter 19), which have been responsible for some of the differences. At least four varied plant formation-types, together with their related animal life, can be distinguished.

The northern coniferous forests

Conifers are gymnosperms, which have a history dating back over 300 million years (cf. angiosperms just over 100 million years), but they still form an important feature of the Earth's vegetation and compete successfully in certain environments.

The **boreal forest** of Eurasia and North America is found in regions with cold winters and a short growing season. The needle-shaped leaves reduce transpiration when the soil is frozen, whilst the evergreen habit allows the trees to begin growing as soon as the spring sunshine supplies sufficient light (for photosynthesis) and warmth. The forest provides a dense cover of vegetation and reaches 12-21 m high, but there is little undergrowth. The floor is characterised by a few shrubs, together with a litter of leaves and decaying wood, but the humus is low in nutrients, biotic activity is low, and the soils are highly leached, acidic podzols. At first sight the appearance of this forest is monotonous over thousands of square kilometres, but there are variations. Bogs occur in lowlying areas; forest gives way to heath on exposed hilltops; sandy soils may be dominated by one particular variety of tree (e.g. Scots pine in north-western Europe); loams and clays will have other varieties (e.g. spruce); and rainfall varies from 750 to 3500 mm per year, favouring a range of different varieties. Thus species tend to be limited to their own distinctive environmental demands and are segregated. Stands of one species are common, and this has made the forests the chief source of wood pulp for over 70 years. Regeneration has become a problem, and where the cutting has been wholesale the conifers are replaced first by deciduous birch and aspen, which regenerate more rapidly and have become much more common in these areas since the coming of large-scale lumbering. It has been discovered that the best practice is to cut conifer stands in strips so that seeds are available for the regeneration of the forest in the cutover areas.

The species-composition of conifer forests varies. The European sector west of the Urals is much poorer than the Asian. Monotonous stands of Scots pine and spruce in the west are replaced by a variety of fir, spruce, pine and larch species, increasing towards Japan. The birch and smaller larches (both deciduous) are the only trees to occur in the coldest heart of forested Siberia: this is an area where conifers cannot exist. North America also has a richer flora than Europe, and the eastern areas of Canada have more species than the west.

British Columbia	Ontario-Quebec	
	white spruce	better drainage
lodgepole pine	balsa fir	
alpine fir	black spruce	
	tamarack	impeded drainage
	jack pine	

The northern margin of the boreal forests is quite sharply defined. The place at which the winters become too long for trees to grow is known as the 'tree line'. It may be a zone of thinning trees becoming more stunted and gnarled by the winds which damage buds on the exposed side, thus allowing most growth on the sheltered side. The southern margin of the belt is less clear, especially where, as in western North America, the mountain ranges run north-south, continuing the conditions southwards at increasing height, and in Eurasia the southern margin is associated with the limits of continuous cultivation.

The long cold winters with temperatures as low as $-40°C$ limit productivity in the forests. In fact it is almost surprising to find such luxuriant forests under these conditions. But detritus feeders are few, and animal numbers and varieties are limited by the severity of the climate and the small range of habitats provided. Many animals live underground, and can just exist in their burrows when temperatures fall to a minimum of $-70°C$. Few cold-blooded amphibians and reptiles can stand such an experience, and well-furred and feathered rodents and birds are the most numerous forms. Deer are the largest herbivores, including a range of species, and there are also the accompanying carnivores — lynxes, wolves, weasels, stoats, minks and sables. Migratory birds and insects swarm into these regions in summer.

Coniferous forest also occupies extensive areas of North America outside the northern, boreal, belt. The **lake forest** once stretched around the Great Lakes and included quite different species of tree (white pine, red pine and eastern hemlock growing up to 60 m high), but the whole formation was removed by lumbering at the end of the nineteenth century. On the west coast the conifers

180

181

Mediterranean vegetation. It is difficult to tell what is original in an area which has been so well settled by man.

Plate 180 *Pinus brutia* forest in Cyprus. A typical stand of old trees with wide spacing and some regeneration in the open patches.

Plate 181 *Cupressus macrocarpa* on the Monterey Peninsula in California, USA. This is a native stand showing the mature form. How is this vegetation association adapted to the summer drought?

Plate 182 The river Golo valley, looking towards Monte Cinto, in Corsica. Few trees remain in an area which was once forested. What has happened to the soils?

Plate 183 A plantation of *Pinus pinea*, reaching 20m in height, with an understorey of *Cupressus sempervirens* on a terraced hillside near Palermo, Sicily. This healthy association suggests that forest is the natural vegetation of the region. (All Forestry Commission)

182

183

LAND BIOME-TYPES WITH SEASONAL CLIMATES 261

184

185

186

187

188

189

North American temperate forests. North American forests are characterised by a great variety of types and species.

Plate 184 Near the limit of forest in Labrador: well-spaced coniferous trees of short stature. (Ewing Galloway, New York)

Plate 185 A felled-over area around Lake Opongo in the Algonquin National Park, Ontario. The best white pine (*Pinus strobus*) and white spruce (*Picea glauca*) have been removed, leaving regeneration dominated by Balsam fir (*Abies balsomea*) and some white pine. White birch was also a common feature of the regeneration, but is now dying from disease. Black spruce (*Picea mariana*) is common in the damper hollows.

Plate 186 Virgin Douglas fir (*Pseudotsuga taxifolia*) with some western hemlock (*Tsuga heterophylla*) in a stand with trees reaching 80m high and with trunk diameters of 1-2m. The sword ferns of the ground vegetation indicate a good site on the drier eastern side of Vancouver Island, British Columbia.

Plate 187 A damper western part of Vancouver Island, with Sitka spruce reaching 60m high and having trunk diameters of 3m. The moist atmosphere encourages mosses hanging on the branches. (185-187 Forestry Commission)

Plate 188 The Great Smoky Mountains National Park in North Carolina. Mixed deciduous-coniferous woodland covers the landscape. (Ewing Galloway, New York)

Plate 189 Cypress and hardwood trees in coastal Georgia. These swampy conditions extend into Florida and Louisiana. (USDA Soil Conservation Service)

include the world's tallest trees — the giant redwoods (*Sequoia*) reaching over 100 m. These **western forests** also show changes in composition as the climate varies: the Sitka spruce is dominant in the cooler north, giving way southwards to the western cedar and western hemlock (over 60 m tall), and then the redwoods; to the drier east and in burned-over areas, the Douglas fir becomes a dominant. The humid coastal conditions encourage the taller trees, but their heights are less on mountain slopes farther inland.

A further isolated and distinctive development of conifers takes place in **south-eastern USA**, particularly in the special conditions of the sandy coastal plain with its immature or marshy soils. Almost pure stands of pines (loblolly, shortleaf, pitch, longleaf, slash) are found, with cypress increasing in the wetter regions of Florida and the Mississippi delta. These may be non-climax communities within the dominant broad-leaved forests of the region and associated with Indian burnings in the past. They formed an important source of pitch and resinous timber for the navies of the eighteenth and early nineteenth century, and are now used for wood pulp.

Conifers occur in such distinctive stands only in the northern hemisphere, but are found in a wide range of climatic conditions: rainfall totals vary from under 500 mm to over 2000 mm; the spruce and boreal forests withstand temperatures of $-40°C$, whilst those on the flanks of the Sierra Nevada in California have summers with temperatures over 25°C; and soils range from poor sands to good loams. The conifer life form is thus extremely adaptable, and has the particular advantage of demanding less in the way of nutrients from the soil, but germination is slow compared with the broad-leaved angiosperms, and so the conifers compete best in regions of seasonal adversity, where the growing season is less than half the year.

Mixed forests: ecotones

An **ecotone**, or series of transitional ecosystems, exists to the south of the boreal forests, where mixed conifers and broadleaved forests are dominant formation-types. Rainfall totals are higher and competition between the two main tree forms is in a state of balance. The European varieties of these forests contained a mosaic of either coniferous or broad-leaved stands. The pines often grew on sandy soils, whilst the oaks and beeches grew on better, nutrient-rich loams. Similar belts existed in eastern Asia (i.e. north-eastern Manchuria, northern Korea and north-central Japan), but once again included a greater variety of native species compared with Europe, and there was another development in North America to the south of the lake forest formation-type. These areas also show a close relationship between the varieties of tree and the soils (e.g. white or red pine on sandy soils, and oak with hickory on heavier soils in New England). Nearly all these forests have been removed to provide fuel or timber and to free land for cultivation.

Broad-leaved forests

These occupy more humid and warmer zones to the south of the mixed forests in the northern hemisphere and include both deciduous and evergreen stands. Conifers are largely ousted from such communities.

The deciduous broad-leaved forests are common in the northern hemisphere, occupying areas between the boreal forests and steppes in the west of continents, and giving way southwards to more tropical forests in the east of Eurasia. In Europe, forests of this type characterise the lowlands west and south of the Baltic-Danube line. Dominants vary. The pedunculate oak is common in Britain and northern France, with the beech and ash on calcareous and well-drained soils; the sessile oak and birch occur on shallower, siliceous soils; and the elm and lime become more important farther south. The composition of the forests in southern Europe is quite different from those in the north. Soils are related closely to the types of tree: podzols develop beneath oak-birch forest; brown forest soils are common on heavier soils with oak and beech; and grey forest soils develop with species growing in the drier conditions towards the steppe margins.

The structure of temperate broad-leaved forests includes typically an open upper storey, which allows light to penetrate to a rich lower shrub layer. North American deciduous forests, covering much of the central and southern Appalachians, have a similar structure to the European forests, but are richer in species. Whereas 12 species of dominant trees are common and widespread in Europe, there are 50-60 in the USA. In the USA the oak and hickory are common throughout, together with maples, beeches, chestnuts, the sweetgum and tulip in the north-east, giving way southwards to increasing chestnut and hickory. In eastern Asia, too, there is a similar belt of deciduous forest, but without hickories.

The northern hemisphere deciduous summer forests have always proved more attractive to civilisation than the boreal coniferous forests. This fact is probably connected with the farming qualities of the brown forest soils as compared with the podzols, but the association has been carried through into more recent features of human geography, such as the siting of the administrative centres of the Prairie Provinces in Canada, and the course of the Trans-Siberian railway: both are in the former deciduous forest belts.

Broad-leaved evergreen forest is more characteristic of the humid temperate regions of the southern hemisphere. The southern beech (*Nothofagus*) dominates the western flanks of South Island, New Zealand, the southern tip of Chile, and once also occurred in south-western Australia. Tall trees, rivalling the California redwoods, provide little shade, and have a dense ground layer, including tree ferns. Their distribution on the tips of the southern continents provides the biogeographer with an interesting problem, which P.J. Darlington examines in '*Biogeography of the Southern End of the World*'. He cites a wide range of evidence, such as has been included in this book — geological, climatic and biological — and his interpretations will lead to discussion.

In the northern hemisphere, such forests, with evergreen oaks and magnolias, occur in Florida between the conifer stands, and may be the climatic climax formation of the region. There are also small stands of evergreen oaks, laurels and magnolias in southern China and southern Japan, remnants of a once luxuriant vegetation.

The productivity of broad-leaved woodlands is much higher than the conifers: an average 1300 dry g/m² year as compared with 800. But a large proportion of the production is in the form of wood. Decomposers work slowly in the colder regions, where the forest floor retains a litter together with the mosses and ferns. This provides little forage for animals. Thus the introduction of deer into New Zealand 200 years ago eliminated the previous occupiers (grazing, flightless birds) but soon led to overgrazing of the limited food supply in the absence of a carnivore check.

The evolution of extratropical forests

The differences between the temperate forest formations in North America and Eurasia, and between the eastern and western margins of these continents, are due to a combination of differences in climatic and soil conditions, man's intervention, and the varying effects of the Pleistocene Ice Age fluctuations. This emphasises the idea that present divisions within the biosphere are not related simply to present climatic regions.

The Pleistocene changes affected British vegetation (chapter 19), and their effect extended to the entire temperate zone — but in different ways. The land connections persisting between Eurasia and North America from Siberia to Alaska allowed a considerable degree of mixing and uniformity in the floras throughout the temperate regions of these continents in the Tertiary. Subtropical forests of considerable variety stretched across both continents, indicating maturity and stability. Cooling in the Pleistocene brought higher latitude deciduous and coniferous trees southwards. The advancing ice caused even further migrations, and in western Europe many more delicate species were not re-established, impoverishing the species variety. The east-west trend of the mountain ranges in Europe (contrasting with the more open relief patterns of Asia and the north-south lineations in North America) prevented seed dispersal in a northerly direction and thus many species could not return. The fact that the pedunculate oak became dominant at the time of 'Optimum' climate in Britain during the Atlantic phase is due to the presence of few competitors. Trees like the sycamore grow equally well, but have had to be re-introduced by man.

Thus the forests of western Europe were most impoverished by these changes; those of North

America less so because the mountain ranges run from north to south. The only temperate broad-leaved evergreen forests are found in southern Japan, China and New Zealand, which largely escaped the effects of these glaciations. These are probably the best type of forest that could grow in temperate latitudes if given the chance.

A simple concept like the 'climatic climax' must therefore be used with caution. If man has not altered the original vegetation and prevented it from regenerating, natural processes themselves may have reduced the forest development below what is possible in a region by excluding more productive, but more sensitive, species. Similarly the concept of 'natural' vegetation might well be replaced by 'original' or some other term which discerns the differences met with in practice.

Temperate grasslands

Grassland formations are extensive in the drier interiors of the continents, particularly North America (the prairies) and Eurasia (the steppes), but also in the more variable conditions in South America (the pampas), South Africa (the veld), Australia (Murray-Darling basin) and New Zealand (Canterbury Plains). Most of these areas have a summer rainfall maximum, which suits the growth cycle of grasses: new shoots and plants grow up in spring and produce seeds before the cold and drought of winter. As with the savannas, however, many temperate grasslands are thought to have been either created, or extended, by the activity of man.

The North American **prairies** occur in a wide, north-south belt in the centre of the continent, reaching westwards to the Rockies, and including many local distinctions. Near the Rockies there is a drier belt due to the rain-shadow effect caused by air descending in the lee of the mountain ranges (Figure 22.7). Here, on the Great Plains, the zone of 'mixed prairie' includes grasses which reach as high as 1 m, together with dwarf varieties; they give way to shorter xerophytic forms in the warmer and drier south-west. Overgrazing in the late nineteenth century killed off many of the taller annual grasses and allowed scrub and even cactus to invade. Farther east the 'true prairie' has been almost completely destroyed by farming. The dominant grasses, of the tussock variety, grew to nearly 2 m, and the turf-forming types maintained a continuous sward. In addition to this main area there are also grassland communities in the Pacific states (e.g. central California and the Palouse prairies in Washington), but these were dominated by tussock grasses and had no sward. Overgrazing in these western areas has led to the invasion of sagebrush.

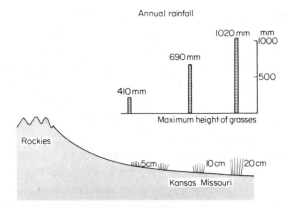

FIGURE 22:7 *Grass height in the prairies related to distance from the Rockies and annual rainfall. (After Selby, 1971.)*

When European settlement began to work westwards across North America the forest/prairie boundary was discovered to be well-defined. The combination of Indian fires and the grazing habits of the bison herds prevented the re-establishment of tree saplings and the extension of forest. When the fires were stopped, and herds reduced, trees began to grow in marginal zones. It seems that the effects of fire would have been greatest in the intermediate areas between wet (forested) and dry (no forest). Grasses could spread in from drier zones and would revive after the fires.

Similar features are found in the Eurasian **steppes**. The grasses have identical life-forms to those in North America, and there is also a transition towards drier regions. The 'meadow steppe' near the forests is formed of turf grasses and occasional tussocks, growing to 1.5m and associated with broad-leaved herbaceous plants; the 'Stipa steppe' is dominated by tussock grasses tapping deeper water sources; and the 'short-grass steppe' is formed of fewer, shorter tussocks together with dwarfed forms and semi-desert shrubs. Little evidence of these orginal divisions remains today, and the forest/steppe boundary is a complex feature related to both natural and human factors (Figure 22.8).

FIGURE 22:8 *The steppe-forest boundary, where the relationship between topography, soils and vegetation is clear. The area shown is typical of the region between the Dnieper and Volga rivers. (After Keller, in Eyre, 1971.)*

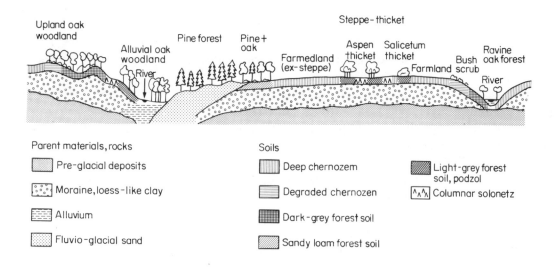

In both the prairies and the steppes the soils beneath the dominant grasses are distinctive black chernozems, due to the high nutrient return from the turf grasses and the consequent high base status of the soil. Towards the drier margins the humus supply is less and the soils become less dark. The soil-grass association is due largely to the local microclimate created by the grasses, and is not, as was formerly thought, directly related to the larger-scale climatic factors. Evapotranspiration rates are always high, even in the wet season, since wind and Sun have more effect on the soil than in the forests. Water movement in the soil is therefore mainly upwards, and the dark humus accumulates near the surface instead of being leached.

The **southern hemisphere grasslands** have also been much altered in the last 100 years. They occur in less extreme climates, and are more difficult to account for than their northern counterparts. The pampas formed an area dominated by tall feather grasses without any woodland except along the valley of the river Uruguay. Areas of bare soil occurred between the bunch grasses. Towards the drier south-west, shorter grasses and xerophytic shrubs took over. Rainfall is over 750 mm throughout, and it is possible that Indian fires may have been effective in maintaining the grasses.

The veld of South Africa seldom occurs without scattered trees. The original dominant red grasses, which were highly nutritious, have been reduced by overgrazing and competition with less useful, xerophytic forms. A similar pattern exists in Australia, where there are no extra-tropical treeless grasslands: scattered gum trees dominate the underlying sward to the west of the Great Dividing Range.

The Canterbury Plains of New Zealand were covered by remarkably uniform tussock grasslands with sparse shrubs and sedges when first occupied by Europeans, but have been transformed by cultivation and the rabbit plague. They extended into areas with over 1500 mm rainfall, and once again it has been suggested that their extent is due to fires lit by prehistoric man as he hunted the flightless moa birds: carbon-14 dating indicates that original forests were destroyed during the thirteenth and fourteenth centuries A.D. (Figure 22.9).

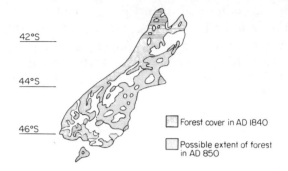

FIGURE 22:9 *The distribution of forest cover, estimated for 1840, together with the area burned off by immigrants since AD 850 in South Island, New Zealand. (After Cumberland, in Eyre, 1971.)*

The soils of the southern hemisphere areas are less subject to the drying effects which produce the chernozems, and include a wider variety. Veld soils are often dark, but none are true chernozems: black soils are restricted to areas of basic igneous rocks. Yellowish brown soils are common in the humid pampas.

All the temperate grasslands are very productive, though less so than the forests. The soil produces a rich humus layer, but is liable to dry out. The dominant animals in these conditions are large, migrating herbivores: the bison and pronghorn antelope of North America; the sage antelope, wild horses and asses of Eurasia; the guanaco of South America; the kangaroos of Australia; and the large flightless birds. All of these have been largely replaced as farming has spread across the regions, and many have become extinct or nearly so. The grasses themselves are often difficult to digest, so that the animals feeding on them are adapted by either possessing specially enamelled teeth or gut compartments in which symbiotic bacteria break down the tough cell walls, or feeding on the associated fruits, seeds, tubers and even nectar instead of on the grasses. North American carnivores include bears, wolves and a variety of predatory birds — hawks, eagles and vultures.

23

Land biome-types with permanent low temperatures or water shortages

Plants have become adapted to cold or dry seasonal interruptions to growth by shedding their leaves for the adverse part of the year, or by adopting a variety of other drought-resisting habits. Animals are able to migrate seasonally to better areas, and many hibernate during the difficult season. Both cold and drought have the effect of stopping the essential supply of water to plants, and thus prevent the passage of energy through the whole food chain.

Many parts of the world have low temperatures and water shortages for nearly the whole year (chapters 11 and 13). It is difficult to identify an absolute desert where nothing will grow — outside of the ice sheet areas — but large areas (approximately one-third of the world's land area) experience conditions where the temperatures are so low, or water is so scarce, that distinctive biome-types develop. High mountains and the coldest tundra regions are included as well as the arid parts of the world. Many would include the boreal forests in the 'very cold' environments, but they have a marked seasonal regime, whereas extreme conditions are experienced virtually throughout the year in the tundra and desert areas.

There are three main biome-types in this category (Figure 23.1).

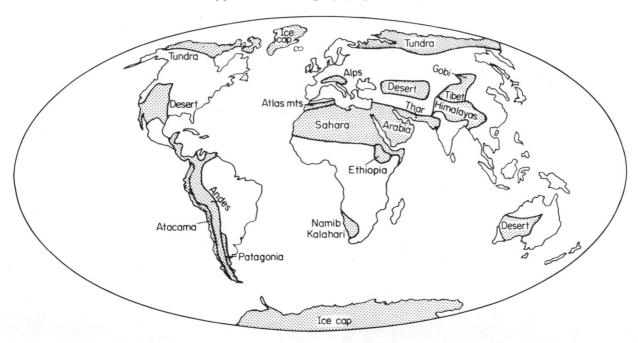

FIGURE 23:1 *The distribution of biome-types which are constantly short of heat and/or water.*

1) The arid regions of the subtropics (e.g. Sahara-Arabia; Atacama desert; west-central Australia) and the extremely dry regions of middle latitudes, mainly in central Asia.
2) The cold deserts of the extreme north — the tundra.
3) The high level alpine deserts of the highest mountain ranges of the world. For convenience the altitudinal zonation of these mountain ranges will be studied at the same time; these could be discussed in connection with the adjacent lowlands, but they experience extreme conditions throughout the year in terms of relief, immature soils, high wind velocities and high light intensities.

Grasslands. A variety of USA rangelands.

Plate 190 A good stand of Tanglehead grass on sandy loam soil in Texas.

Plate 191 A sandy area in New Mexico supporting sand sage (main type), with some grama (small bunch grass), yucca (pointed leaves) and sideoats grasses.

Plate 192 An upland area at 1600m in Arizona with 500mm precipitation each year on gravelly loam soil. Sparse oak trees occur with a rich variety of tall grasses.

Plate 193 A range area in Montana, with an average of 300mm precipitation per year: the grasses are mostly Bluebunch wheatgrass, with some Rabbitbrush plants — a more restricted species composition in cooler and drier conditions. (All USDA Soil Conservation Service)

LAND BIOME-TYPES WITH PERMANENTLY LOW TEMPERATURES

194

Tundra conditions involve difficult soil characteristics.

Plate 194 A pit near Fairbanks in central Alaska, caused by the shrinking of ice lenses in the soil following clearing and cultivation. The grasses are being grown to prevent such summer thaw. (USDA Soil Conservation Service)

Plate 195 Problems of food supply for a reindeer herd in an area of snow and frozen soils in northern Norway. (Wideroes Flyveselskap, Oslo)

195

The arid biome-types

No area of the world can be classed as an absolute desert where no rain falls. There is in fact a range of conditions of increasing rainfall from arid to semi-arid (chapter 11). All regions included here are short of sufficient water for continuous — or even regular seasonal — plant growth, and the living organisms adapt to this shortage in a number of ways.

One group of plants is dependent on the occasional shower of rain — once a month or once in five years. These plants may be annuals or perennials, and are dormant for most of the time. Roots are short, penetrating only the surface, and the plants complete their life-cycles of growth and reproduction rapidly. Many such plants are minute in size, thus requiring small quantities of water, and many lack all the so-called drought-resisting features except for the special manner in which the seed is protected and preserved. Some store water in the remains of leaves, stems or underground bulbs.

Another group of plants lives in lower depressions, wadis, oases and along lines of water seepage, extending deep roots down towards the local water table which may fluctuate violently in level between rain showers. Such plants are always highly xerophytic, having short, spiny stems above the ground and massive root systems below. They are known as succulents and include the cacti of the New World and the euphorbias of the Old. A range of the most unusual plants found on the Earth occurs under such conditions. The naras and *Welwitschia* of the Namib desert in South-west Africa are amongst the strangest: the naras can grow so rapidly that it may extend up through drifting sand dunes to establish its thorny bush at the top; *Welwitschia* has a hollow, woody base only 15 cm high but up to 1-2 m in diameter, and this produces flowers and just two long fleshy leaves.

The plants of the driest regions can scarcely be regarded as belonging to communities. Large patches of bare ground exist between, and it is seldom that there is competition even between their root systems. Areas in which the most extreme conditions hold are rare, particularly in temperate regions where evaporation rates are lower: even predominantly bare regions like Death Valley, California, and parts of central Mongolia, have plants growing where the wind has spared soil material (Figure 23.2). Semi-desert scrub is much more common, involving a variety of low-growing, thorny shrubs around the margins of the most arid sectors, where there will also be extensive zones of bare rock, sand and pebbles.

In the tropical zones the scrub of the semi-arid environment grades into **thorn forest** (chapter 22). Scrub vegetation includes a large proportion of succulents, which are often shallow-rooting forms, whilst the areas between are underlain by the wide-spreading root networks of other plants. The greatest expanse of this type of vegetation formation occurs around the margins of the Sahara-Arabian-Thar deserts. Sparse, impoverished 'communities' are dominated by the thorny, succulent euphorbias in an open structure growing to heights of 1-2 m. Plants with thick trunks and tiny, leafless branches, together with thorny bushes, are intertwined with sprawling creepers and vines to form impenetrable thickets in regions where moisture is supplied more regularly. Similar life forms occur in South-west Africa on the eastern borders of the Namib desert, though the species are different. In Australia there are large areas with less than 500 mm of annual rainfall, and droughts lasting 2-3 years are common. Varieties of eucalyptid, acaciaid and other local forms have developed a thick bark and spiny leaves, whilst the spinifex grasses invade drifting sands, providing a distinctive element to the vegetation of this continent. In the USA, and on the margins of the Mexican and Atacama deserts, the cactus scrub has a largely grassless aspect, similar to the African formations, but it contains a richer variety of species. Semi-desert regions may be regions of difficulty and special adaptations, but any degree of floral impoverishment is in terms of numbers of plants rather than of the variety of plant species.

In the **cooler temperate regions** plants may have to cope with interruptions in growth due to low temperatures as well as drought. Once again, individual plants grow up to 1-2 m high and are widely spaced with spreading roots to make the greatest use of water as soon as it enters the soil. The cactus,

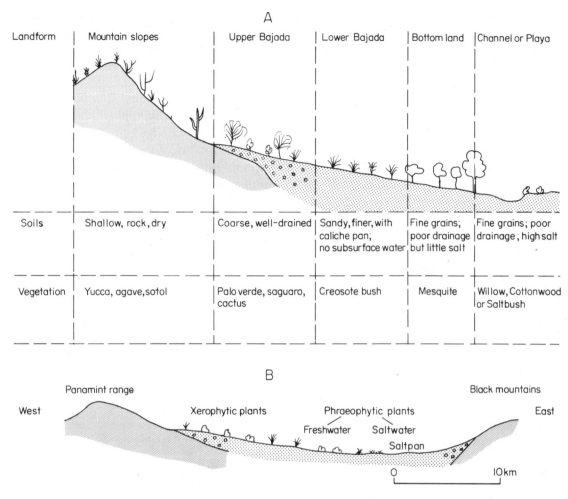

FIGURE 23:2 *Desert vegetation.*
(A) A generalised profile across a basin in Arizona.
(B) A transect across part of Death Valley, California.
Xerophytes are adapted to limited supplies of water;
phraeophytes obtain water from beneath the surface and are not dependent on rainfall.
(After Benson, Darrow and Hunt, in Selby, 1971.)

creosote bush and sage brush are common forms in western USA, whilst more grasses occur in the Eurasian area stretching from the Caspian Sea eastwards into the Gobi desert, together with the wormwood shrub which resembles the sage brush in form. Patagonia in southern Argentina has scrub vegetation with shrubs up to 1-3 m tall even in the driest parts. Most of these shrubs are deciduous, but the leaves are also tiny. Once again ground creepers render this formation impenetrable in places where it is fully developed.

All dry regions have an extremely low **primary productivity**, averaging 3-70 dry g/m²year (i.e. less than 10 per cent of that of grassland and 1 per cent of that of some tropical forests). Much of this production enters into woody tissues and moisture-retaining devices. Yet there is often a considerable diversity of animal life supported in deserts. The animals are commonly small and hide from the heat: insects and reptiles needing little water and protecting themselves from loss; rodents burrowing underground to eke out the water supply in the cooler conditions there (e.g. 17°C in a burrow 48 cm deep compared with 65°C at the surface). Towards the desert margins, in the semi-desert scrub, herbivores include varieties of deer, rodents, insects and flightless birds, whilst there are foxes, small cats, badgers and some carnivorous birds.

Desert and semi-desert biome-types are more extensive today due to man's activities. Overgrazing removes the natural vegetation and it is replaced by more xerophytic forms on the semi-arid margins. On the other hand the low desert primary productivity is due largely to the lack of water in the warmer parts of the world, and irrigation may introduce high levels of farming productivity if the basic soil materials are available.

Tundra biome-types

Land areas with extremely cold conditions throughout the year are found mainly in the northern hemisphere (chapter 13). A broad belt of tundra (lit: Russian 'marshplain') occurs in northern North America and Eurasia between the boreal forest and the ice sheet or Arctic Ocean. Winter conditions, when the frozen soils and strong winds impose a physiological drought, may last up to nine months. At this season the exposed plants can survive only if they are very low-growing; many die down completely to rhizomes, corms or bulbs in the ground. The short summer is used for rapid growth. Some plants flower early after storing food resources from the previous summer, whilst others flower late in the season having built up their supplies during the current summer.

Trees are thus generally absent, due to such a short growing season, the high wind force, and the shallowness of the soil which thaws in summer. Frozen tree stumps farther north testify to growth in the past, but changes in climate together with man's activities have forced the treeline southwards. Flowering herbaceous plants compete with dwarf shrubs, mosses and lichens, and the intensity of plant cover decreases northwards. Grasses and sedges are dominant in the better areas, but give way to mosses and lichens on poorer sites. The best-developed vegetation may be 10-25 cm high if left ungrazed. Variation in relief gives rise to a mosaic of communities. In the more sheltered areas with deeper soils a wide variety of sedges, cotton grass, flowering plants (anemones, marsh marigolds, buttercups, saxifrage, gentians, primroses), grasses, lichens and mosses form a complete cover; boggy areas will have cotton grass, sedges and sphagnum moss in greater abundance, giving way to a tussocky cotton grass surface under conditions of slightly less moisture; arctic heath on exposed sites includes the arctic bell heather; and arctic scrub often occupies areas near the forest margin, incorporating more bushy growth in terms of alder, scrub birch and dwarf willow (Figure 23.3).

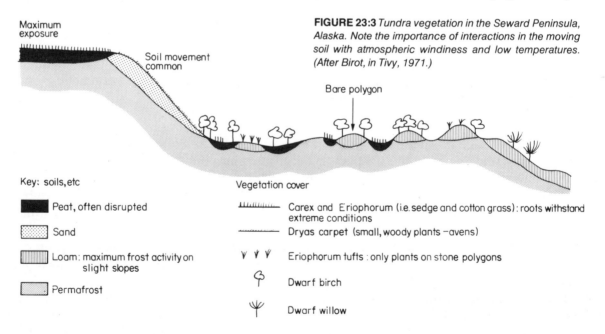

FIGURE 23:3 Tundra vegetation in the Seward Peninsula, Alaska. Note the importance of interactions in the moving soil with atmospheric windiness and low temperatures. (After Birot, in Tivy, 1971.)

Primary productivity is extremely low, as in the arid regions (as low as 10, but averaging 140 dry g/m²year) and so little food is available for herbivore consumers: the herds of reindeer, caribou and musk-ox roam over vast expanses of tundra to find sufficient food, and migrate southwards in winter. Peoples like the Eskimo hunters and the Lapps have based their way of life on these herds, but increasing numbers and European hunters depleted the caribou herds of North America by the late nineteenth century, and the reindeer introduced to replace them have overgrazed the poor tundra vegetation. Smaller herbivores are mostly rodents living on the underground rhizomes and bulbs (e.g. lemming, snowshoe hare). Hosts of insects and birds migrate into the regions in summer, the rock ptarmigan being the only permanent bird. Owls, foxes, ermines and lynxes are the main carnivores. The delicacy of the ecological balance in such regions is illustrated by the cycles of plant productivity and animal numbers. A good year for plant growth will encourage increased numbers of rodents, leading to overgrazing and increased carnivore activity; the animals then face a lack of food and die off, allowing the vegetation to recover.

In the southern hemisphere, the climate tends to be colder in the cool temperate regions, and there is no real summer south of 70 degrees South. Flowering plants are scarce, only two species of grasses exist on Grahamland, and permafrost is common to 60 degrees South. There is, however, little land in this zone, and the few islands like Kerguelen and Macquarie Island have a tundra-like vegetation of heath and moor with a very restricted number of species. The winds are too strong to allow trees to grow.

Alpine tundra and mountain vegetation

Vegetation immediately below the snowline on mountains extending from the tundra regions into warmer and even tropical areas bears many similarities to the Arctic tundra. These similarities extend to the species of plants, and it may be the case that many of the Arctic tundra plants have developed from species which became adapted to the new high mountain environments created in the middle Tertiary upheavals affecting the Alps, Andes and Himalayas. These plants were then able to colonise the cold areas as high latitude climates deteriorated in the Pleistocene Ice Age. In addition, the fluctuation of climatic zones during the late Pleistocene allowed considerable contact between the tundra biome-types and the midlatitude high altitude biome-types.

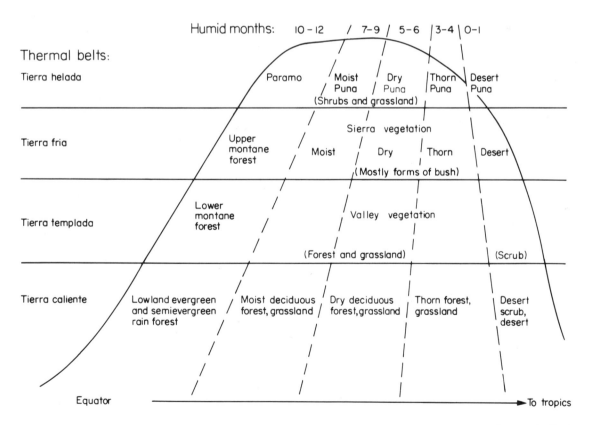

FIGURE 23:4 *Vegetation types in the various tropical belts through which the Andes pass.*

There are also differences between the two tundra biome-types, since the mountain areas experience even higher wind velocities than the tundra-covered plains around the Arctic, together with higher light intensities and daytime warming throughout the year in tropical latitudes. Steep rocky slopes also impose their own conditions and a general lack of stability; south-facing slopes in the northern hemisphere may be almost semi-desert in nature due to intense daytime heating, whilst the opposite north-facing slopes may remain in the shade for long periods. Permafrost is rare except in the higher latitudes. Such areas are, however, just as unproductive as the the Arctic tundra, and few animals or people can be supported there. The greatest extent of this biome-type is found in Tibet, where the ibex, yak and sheep exist with the carniverous wolf.

The alpine tundra is merely a single zone amongst a series occurring on mountains. In temperate regions there may also be one or two other biome-types between the alpine tundra and the lowlands (e.g. lowland forest giving way to mountain forest, then to tundra and the snowline), but in tropical regions there may be a wider range of biome-types ranging from tropical evergreen forest through to icefields. Tropical mountains have been regarded by some as ideal microcosms of world vegetation, allowing the ecologist to study transitions and contrasting types within a small area. The situation is not so simple, however, since seasonal changes are marked in temperate regions but do not impinge on the climates of tropical mountains, and temperate species of plants and animals are seldom found in tropical areas. Factors such as high light intensity and wind speeds, steep slopes and heavy cloud at

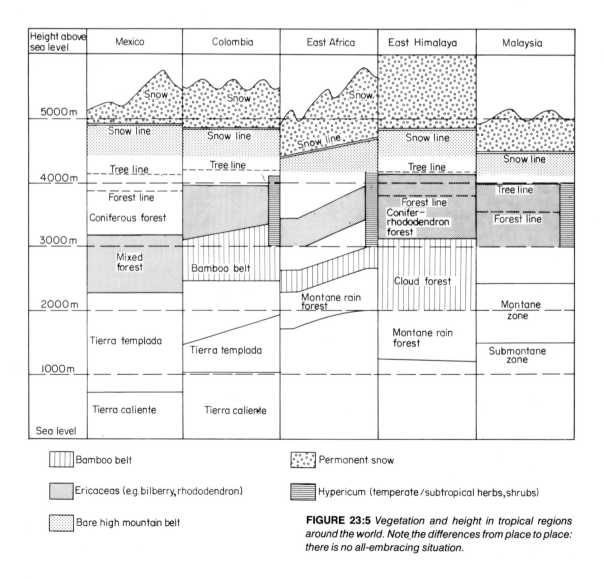

FIGURE 23:5 *Vegetation and height in tropical regions around the world. Note the differences from place to place: there is no all-embracing situation.*

certain levels impose their own modifications on the pattern which might be expected from the effect of decreasing temperature with height alone. Mountain vegetation zones are thus interesting for their own sake, rather than as a cross-section of world ecosystems.

A series of changes is experienced in a climb up tropical mountains in a variety of conditions in South America (Figure 23.4, cf. Figure 20.9). Beginning in the tropical evergreen rain forest with its tall trees and well-layered structure (cf. chapter 21), there is a transition into submontane forest with trees reaching 20 m and including some conifer species. The increasing cloudiness and humidity above this supports montane, or 'mossy' forest, where the single canopy layer of trees reaches only 10-15 m high and is commonly covered with mosses, epiphytes and lichens: these are densely wooded, wet areas enveloped in cloud. Above this a further transition includes lower-growing and gnarled species related to cooler conditions (conifers and evergreen oaks), with a dwarf woodland growing to only 1 m high and including trees with largely creeping habits. Finally the alpine zone is reached, where the dominant plants are low shrubs and grasses (see also Figure 23.5).

Drier areas within the tropics may begin with savanna or even with desert, but this gives way to forest (or grassland and scrub) as the humidity increases with height. The Ahaggar and Tibesti mountain blocks in the central Sahara have bare slopes beneath 1500 m, but the summits are clothed with evergreen sclerophyllous woodland related to Mediterranean forms.

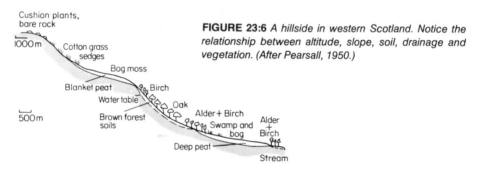

FIGURE 23:6 *A hillside in western Scotland. Notice the relationship between altitude, slope, soil, drainage and vegetation. (After Pearsall, 1950.)*

In temperate regions, only the subalpine forests intervene between the lowland forests and the alpine tundra. These include some of the coniferous forest species growing at lower levels, but in addition there are distinctive species related to special conditions. Winter days are longer and light intensities higher than in the boreal forests, whilst the summer days are shorter but hotter. Mountain ranges like the Pyrenees and Alps have pine, spruce and larch forests at heights between 1000 and 2000 m. More species variety is found in the mountains of Asia: the Caucasus, Tien Shan and Altai. In North America the north-south arrangement of the mountain ranges allows closer relationships with the species found in the boreal forests. Figure 23.6 illustrates some of the arrangements found on a temperate hillside.

Mountain vegetation

Plate 196 Arosa in the eastern Swiss Alps. The treeline on the far slope, facing northwest, is between 1850 and 1900m. Above this is pasture and bare rock.

Plate 197 An elevation experiment with Sitka spruce in Scotland, showing dwarfing in the marginal plots due to exposure to high winds. (Both Forestry Commission)

Underwater environments

Plate 198 The Tektite Habitat Project. An artist's impression of the underwater living quarters, together with an annotated aerial photograph of the site off St John in the Virgin Islands. (US Geological Survey)

Plate 199 The Florida Aquanaut Research Expedition (FLARE), which took place in early 1972. A small undersea laboratory, EDALHAB II, with working space of 3 × 2.5m, housed two workers, and was connected to the surface vessel with up to 15 scientists. The FLARE program covered the study of ocean conditions, bottom geology, pollution, reef communities and fishing methods. (NOAA)

24

Oceanic biome-types

The teeming and varied life in the oceans is different in many ways from the environments of the land-based communities (Figure 16.1). The composition of the ocean waters, their vertical stratification into water masses, and the movements to which they are subjected were studied in chapters 2 and 5.

The differences between the living forms in the oceans and on the land are marked (Figure 24.1): no fishes, sponges or echinoderms are found out of water in terrestrial environements, and even where groups like the mammals, arthropods (crustaceans and insects), or molluscs (clams, oysters, snails, squids) have representatives in both environments the forms are often distinct. The oceanic environment presents few sharp boundaries and changes in conditions, and the animals living in it are often adapted to narrow temperature bands. It is a buoyant medium, compared to the atmosphere, so that basic swimming and floating shapes are common in many different groups of marine animals. Animals on the land must support their bodies and be able to withstand sudden, drastic changes of temperature and moisture supply. In addition, the seas support many varieties of bottom-living (benthonic) animals, which live fixed to rocky outcrops, or crawl over or burrow into the soft sediment. Many of these creatures rely on the rain of organic matter from the surface waters for their food supply.

	Land life	Ocean life
Animals	Most mammals, reptiles, amphibians. Most insects. Some molluscs — snails, slugs, etc. Many worms, able to breathe air.	Most fishes, with whales, sealions. Most crustacea — crabs, shrimps, etc. Most molluscs — clams, oysters, squids, whelks, etc. Distinctive marine worms. Echinoderms — sea urchins, starfish, etc. Coelenterates — corals, jelly fish. Sponges.
Plants	Dominated by 'higher', vascular plants: trees, shrubs; also mosses, lichens. Ninety per cent species are flowering plants (angiosperms).	Seaweeds: multicellular algae. Phytoplankton: single-celled, microscopic forms. (i.e. all non-vascular).

FIGURE 24:1 *The contrast between life on the land and in the sea. Compare, for instance, the variety of plant and animal life on land and in the oceans.*

Perhaps the greatest difference between life in the two realms is that between the forms taken by **plants** on land and in the water. Most water-living plants, apart from the seaweeds living in shallow coastal locations, are tiny, single-celled organisms carrying out the photosynthetic conversion near the water surface. Whilst plant forms dominate the land biome-types and provide the niches and food supply for the animals, the water plants can often be seen only under the microscope, and the animals dominate the scene — though they are still ultimately dependent on the primary producing plants.

Ecosystem succession on the land depends normally on a community of plants and animals modifying their immediate environment so that it becomes more suitable for the next stage in the succession and gradually works towards the climax state. No successions, apart from annually or seasonally repeated cycles, exist in the oceans, where life can do little to modify the environment over the longer term. The sheer volume of water, together with rapid mixing, prevent local differences from becoming established.

Man has so far had less effect on the oceanic environment than on the terrestrial. He has been interested in the end products of the food chains as sources of food and raw materials, and has regarded the oceans essentially as an unlimited dumping ground. The oceans are far less productive than the land areas, with an average of 125 dry g/m²year, rising to 350 dry g/m²year on the continental shelves. This means that the oceans give only 34 per cent of the world's net **primary production**, although they occupy 71 per cent of the surface area. Whilst certain coastal areas have relied on oceanic food resources, the total contribution to man's needs is small, still providing less than 10 per cent of the total animal protein foods despite the fact that the world fish catch trebled from 21 million tonnes in 1938 to 64 million tonnes in 1968.

Certain characteristics of the marine realm must be borne in mind as a division into marine biome-types is attempted.

1) The environment shows **few sudden changes**, and has **few areas of extreme conditions**. A large proportion is near the optimum for the continuance of a varied range of life forms, and the shelf seas are the richest of any Earth zone in both numbers of individual creatures and the diversity of species. Ocean temperatures are nearly all within the range 0°C to 30°C (i.e. suitable for metabolic processes to take place), and life exists even in the deepest trenches. Seawater is rich in dissolved nutrient salts, especially along zones of upwelling deep water (Figure 24.2), and there are few areas which could be termed 'marine deserts'.

2) The **food chains** and **food webs** in the oceans depend on the availability of light, water, carbon dioxide and oxygen. These are present together only at the surface, and the light in particular

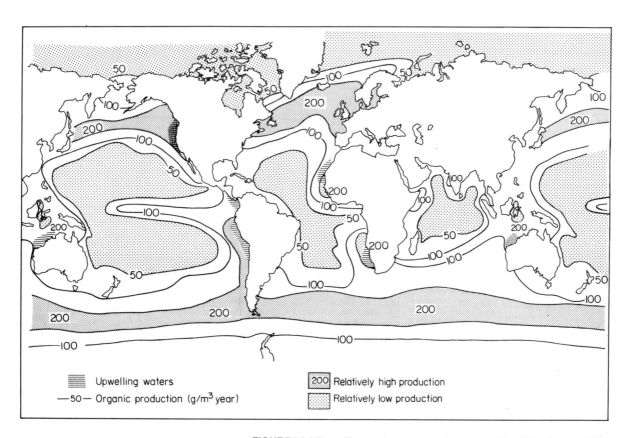

FIGURE 24:2 *Upwelling nutrients and marine productivity. (After Gross, 1972.)*

decreases to almost nil at depths of more than 200 m. This uppermost, **euphotic** zone is where the primary producers, the green plants (or phytoplankton), incorporate nutrients in the process of photosynthesis. This is where the food web begins: the whole system, from the upper layer downwards has been likened to the situation in a tropical forest, where the sources of all the trees' needs, apart from the nutrients, come from above. Figure 24.3 shows some of the connections within the marine food web, including the re-cycling process of decaying debris and wastes. This is a simplified version of the actual patterns. The primary producers are mostly microscopic diatoms and green flagellates occurring in vast numbers and reproducing rapidly (producing 'blooms' which colour the ocean waters) in time of optimum conditions, but the consuming zooplankton does not allow them to accumulate for long, and the bulk of phytoplankton at any one moment is always small. The animals living at depth, below the euphotic zone, are divided between the detritus feeders relying on the rain of organic matter from the upper levels of the ocean, and the predators and scavengers which are linked to these primary consumers.

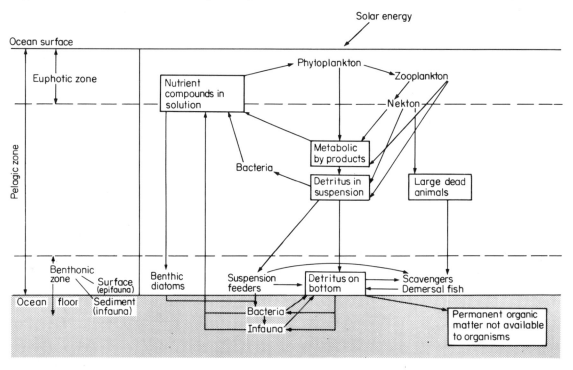

FIGURE 24:3 *The marine food web (generalised), showing the flow of energy and matter. Note the important vertical connections, and compare these with the situation on the continental biosphere. (After Gross, 1972.)*

3) There is an important distinction to be made, therefore, between the **main environmental niches**: the largely floating and surface-living phytoplankton and zooplankton; the more manoeverable pelagic forms — fishes, whales, squids; the bottom-living or benthonic forms in all their variety living on the surface or burrowing into the sediment; and the bacteria acting on the waste products and playing a vital role in the re-cycling process. Figure 24.4 summarises these vertically differentiated, but interdependent, marine environments.

Oceanic biome-types

This review of the major features of the oceanic environment, with particular reference to their effects on the distribution of living creatures, has shown that, although a certain vertical differentiation can be made in terms of plankton, pelagic forms and benthos, there are so many interactions between these levels that it is more realistic to regard them as part of a related ecosystem. Each aspect is dependent on another in a vertical rather than a horizontal dimension. A biome-type division, however, must be based on factors giving rise to the horizontal differentiation.

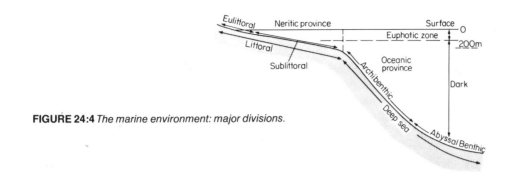

FIGURE 24:4 *The marine environment: major divisions.*

There is an important difference in species' forms and in numbers of individuals between cold and warm water areas. The shelf sea environment is also distinctive when compared with the open oceans. A third factor included in the present division is the supply of nutrients associated with the zones of upwelling deep water and the high productivity of such areas. The major division of biome-types in the oceans will therefore be as follows:
1) warm water continental shelves, largely within the tropics: surface temperatures over 20°C;
2) cold water continental shelves: surface temperatures less than 20°C;
3) oceanic areas with upwelling nutrients, in which the surface temperatures are colder than in the surrounding oceans;
4) cold open oceans;
5) warm open oceans.
This is the most general and wide division possible, and many variations are included within each biome-type, but this is inevitable in terms of the scope involved in this particular study and also of the state of present knowledge of life in the oceans. Figure 20.14 shows the world distribution of these oceanic biome-types.

Warm water continental shelves

These areas are the richest biome-types in the world in terms of the variety of living forms contained. The environment lies almost completely within the euphotic zone, and includes a wide variety of niches, from the shore zones exposed at low tide to increasing depths where the light intensity and colour spectrum changes (Figure 15.7). Local communities can be distinguished on mud, sand or rock at various levels, and depending on the nearness to freshwater flowing into the sea at river mouths. The richest communities of all occur where erosion of a nearby landmass provides a high content of lime, and where corals and other reef-building organisms become established. The complexity of the reef environment encourages a particularly varied fauna to exist, ranging from the greatest variety of non-chordates (corals, echinoderms, molluscs, crustaceans, sponges, worms) to vast shoals of myriad varieties of fishes and their predators such as the sharks.

Cold water continental shelves

Cooler waters mean fewer species (Figure 16.5), but the niches are no less numerous and a further variety of communities exists on the sea floors, related to the pelagic and planktonic conditions above (Figure 24.5). Although these waters contain fewer species than the warmer shelf seas, the coldness does not inhibit the activity of plankton varieties which provide the source of food for many types of commercial fish, and this results in greater numbers of individuals within a particular species. Temperate continental shelves are also the scenes of the greatest seasonal migrations due to changes in climate. In the summers of these regions the fish from warmer regions come to spawn, and in their winters colder water fish arrive. An area like the English Channel thus has a seasonal pattern of fish catches.

The continental shelf areas grade into freshwater ecosystems at river mouths. The types of animal and the number of species can vary as a river estuary is entered (Figure 24.6).

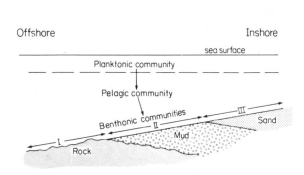

FIGURE 24:5 *Communities on the continental shelf area: all are closely connected in a food web.*

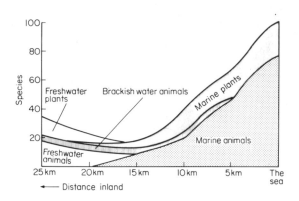

FIGURE 24:6 *The Tees estuary: distribution of plants and animals related to tolerance of increasing salinity. (After Alexander, in Gross, 1972.)*

Ocean areas with upwelling nutrients

Another biome-type with both a rich diversity of species and great numbers of individuals is that found in areas where offshore winds bring up colder waters from depth and in the process draw up mineral nutrients (cf. Figure 11.5). These upwelling waters become extremely turbulent in areas of uneven bottom topography, and this helps to maintain vast numbers of diatoms near the surface. The conditions are ideal for these microscopic primary producers, and the consumers feeding on them become very numerous: thus off the coast of Peru vast shoals of anchoveta feed on the phytoplankton and provide food for millions of sea birds, fish and even whales. Some of the regions have narrow shelf seas, and thus varied benthonic communities, but generally the floor descends to depth close offshore, and these are thus the most fertile of the open sea biome-types (Figure 24.2).

Open oceans in cold regions

Plankton numbers are more restricted in the open oceans. There is less phytoplankton since the mineral nutrient supply is not as good as it is near the continents, and the zooplankton consists of smaller forms than in the shelf seas, where it also includes many fish larvae. In addition, the benthonic communities exist at great depths (on average between 4000 and 6000 m, but down to 10 000 m), and are very restricted in their range of species. Seasonal changes regulate the numbers of living forms at the surface (Figure 24.7). In winter the surface waters have lower temperatures:

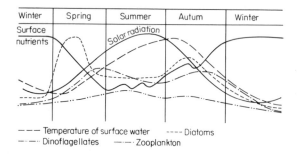

FIGURE 24:7 *The relationship of zooplankton numbers to production of phytoplankton (i.e. diatoms, dinoflagellates), and this in turn to surface nutrient and energy supplies. This diagram refers to conditions in the middle latitudes. What would be the differences in low and high latitudes? (After Tait, 1968.)*

for instance in the North Atlantic, between Ireland and Newfoundland (52 degrees north) the summer surface temperature will be approximately 15°C, and the winter 10°C. Convectional mixing results in winter, bringing up nutrients from depth, but the lower light intensities restrict the development of plankton. As the surface temperatures rise a few degrees in spring, and illumination increases, the rate of primary production is at its highest, using the nutrients concentrated during the winter. Phytoplankton increases, followed more gradually by the zooplankton — which consumes and reduces the phytoplankton — and at the height of its development may colour the ocean red or brown. During the summer the surface nutrients are low, and the phytoplankton is reduced in mass (though the dinoflagellate element is at a maximum). Autumn cooling leads to increased convectional activity again as the steep temperature gradient of summer breaks down: nutrients begin to be replenished at the surface. This seasonal effect is greatest in the middle latitudes, since the surface temperatures fluctuate little in the high or low latitude regions. The long summer days in high latitudes give rise to high primary production and rapid development of zooplankton in a very short season, but during the winter there is virtually no phytoplankton available in these seas.

The fertility of the ocean waters is at its greatest in the colder areas because of the greater extent of the vertical mixing of the waters. The effect of this mixing is most pronounced near the edges of continental shelves, since the mineral nutrients are most plentiful there.

Open oceans in warm areas

The seasonal changes in production found in the cooler parts of the oceans do not occur in the tropics, where the summer regime remains almost uninterrupted throughout the year. Surface waters remain consistently warm and well illuminated, but vertical mixing is restricted and nutrients are in short supply at all times. Production extends to greater depths because of the greater light intensities, and continues throughout the year, but at lower levels. Seasonal wind changes, rather than variations in temperature and lighting conditions, cause any slight differences which can be recognised. In spite of the lower productivity of these areas, the tropical open oceans contain a wide range of varieties: thus there are ten times as many species of zooplanktonic animals in warm waters as in the colder waters. This is a reflection partly of the general tendency of warmer waters to support more species, and partly of the greater extent of tropical oceans.

It is in the tropics that the few areas of oceanic semi-desert occur. The Sargasso Sea area, for instance, has a very low rate of primary production. The Mediterranean Sea is another area of low fertility, due to the outflow of nutrients in the surface waters moving across the sill at the Straits of Gibraltar.

The life in **deep ocean waters** below the euphotic zone is similar throughout the world: this is a dark, cold (0-4°C) realm with no seasonal changes and only slow movements of water. All animals living here are carnivores or detritus feeders. Zooplankton may be found down to 3000 m in the North Atlantic, but not below 1500 m in the Indian Ocean. The numbers of species and of individuals are both low: few varieties are adapted to the monotonous environment which provides small numbers of distinctive niches, and it has been estimated that the ocean floor biomass is only 1 per cent of that in the shelf seas per unit area. Animals such as the deposit feeding sea cucumbers and many worms, together with filter feeding sea lilies, sponges, coelenterates and some molluscs predominate. A large fish or shark falling dead to the ocean floor can attract large numbers of scavengers from the seemingly empty deeps within a short time.

Fisheries

There are 14 major world marine fishery areas (Figure 24.8). The vast increase in fishing since the nineteen forties has been due to the realisation that there were many fisheries ripe for development, and to the increasing size and efficiency of the fishing fleets owned by such nations as Japan, USSR and Peru. Figure 24.9 charts the main types of fish caught.

The fish are caught by a variety of methods. **Trawling** is used for bottom-living (demersal) fish. The most important modern development involves the building of ships with a stern trawl, by which the nets are drawn into the trawler up a slipway, rather than over the side. A greater degree of mechanisation and thus larger nets can be used to take larger catches. **Drift nets** and **ring nets** are used to draw in the surface, or pelagic fishes. In both cases the mesh of the nets is chosen according to the size of fish to be caught, and usually the younger fish are allowed to escape to grow for future years. Other modern developments to increase fishery yields include better methods of searching, such as the use of echo sounders and weather forecasts, better methods of catching, such as the use of

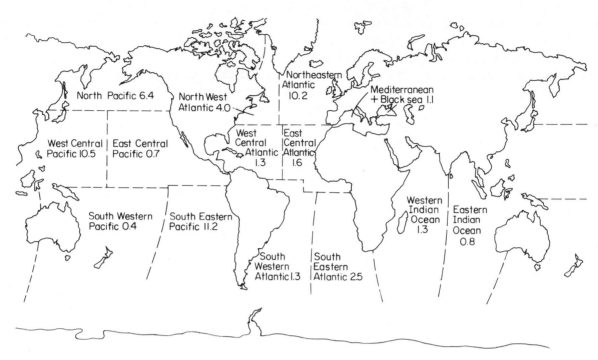

FIGURE 24:8 The world's major fishery areas. Numbers refer to millions of metric tons landed in 1967 (UN-FAO). Note the contribution of the northern temperate oceans, the Peruvian fisheries (virtually the whole catch from the south-eastern Pacific), and the tropical regions. (After Holt, in Moore, 1971.)

%		Total world catch, 1967: 60 million metric tons	
100	15%	Other fishes, invertebrates (oysters, squids, shrimps and prawns, clams and cockles).	
90			
80	14%	Flatfish, perch, mullet, jack, tuna, etc.	
70	3%	Mackerel	
	3%	Alaska walleye pollack	
60	10%	Others	Gadoid fish
	5%	Atlantic cod	
50			
40	26%	Others	Clupeoid fish
30	2%	South African pilchard	
20	6%	Atlantic herring	
10	16%	Peruvian anchoveta	
0			

FIGURE 24:9 The major types of fish caught in 1967. Clupeoid fishes live near the surface and are often small; gadoid fishes live on or near the bottom of the sea and are bigger, but can be fished only on the continental shelf areas. (After Holt, in Moore, 1971.)

nylon nets which are stronger and less perishable, and better methods of preserving the catch in refrigerated holds so that the ships do not have to return to port so often. These better methods and larger, highly capitalised ships have led to a decrease in the numbers of smaller vessels. The accompanying increase in the knowledge of fish habits has led to worries over future supplies in some of the more heavily fished areas of the world.

Areas which are not fished contain a far greater range of individuals, and more older fishes. As soon as fishing begins, initial catches are high, but they tend to fall as the intensity of fishing increases. If the smaller, younger fish are left they will begin to form a much larger proportion of the total population. Fisheries are at their most productive over a period of years when the highest sustained yield is obtained by either leaving the young and fishing the old heavily, or by fishing the whole range moderately. Fish lay vast numbers of eggs, whose survival depends partly on the supply of food: removal of large numbers of fish makes more food accessible to those which might not survive otherwise. Only the most intense forms of over-fishing will remove all the fish or reduce yields drastically. Individual fishermen, however, do not see the overall picture, and control is difficult. The discovery of new sources, and the degree to which management of the existing grounds can be successful will therefore determine the future expansion of the fishing industry.

At the moment too little is known about the possibilities in terms of oceanic production. Some scientists predict that this is reaching its limit, whilst others say that present production is only one-third of the ultimate. Better methods, including the fish farming (mariculture) which is being carried out in Japan and Scotland, may also increase the fish production. Approaches to this include stimulating the food web at various points, such as the addition of more mineral nutrients at the best season, and increasing the numbers of edible fish varieties. At present there are many species of fish which are of no use to man.

This short survey of fishing in the oceans will be completed by examining two of the main kinds of fish caught, their life histories and ecological associations. They are amongst the best known of all marine forms, since their importance to the fisheries of the advanced countries of Europe and North America has meant that they have been studied closely.

The herring (*Clupea harangus*)

This fish is found in a wide area of the North Atlantic, from Newfoundland to the North Sea and northwards into the Arctic Ocean. A close relation (*Clupea pallasi*) lives in the North Pacific Ocean from Japan to British Columbia.

Herring congregate in shoals for spawning, when densities may be as high as 100 million per km^2. The females lay up to 60 000 eggs in shallow, gravelly sea bed areas and the sticky masses adhere to stones or weed. Oceanic forms live in deeper waters and spawn around the northern and western coasts of the British Isles in spring in water of 5-8°C. The shelf sea forms of the North Sea spawn in summer, autumn or winter, the shoals gathering in water of 8-12°C: the position of water with such temperatures moves southwards during the latter part of the year, and so do the main herring shoals. The two stocks — oceanic and shelf sea — have slight morphological differences, and are probably separate species.

The eggs hatch out in a few days (22 in colder water, but only 8-10 in warmer), and the larvae swim to the surface to feed on diatoms, copepod eggs and other larvae. As the herring larva grows, it feeds on the later stages of the zooplanktonic copepod larvae. At first the herring larvae are carried around in water near the shore, and do not move into deeper waters until they are a year old. They fatten up in the third year, retaining a taste for the surface plankton. Maturity is reached in years 5-8, when they leave the feeding shoals for the spawning shoals. During the day they remain near the sea bed, coming up to the surface to feed at night. There is also a seasonal migration from spawning ground to deeper water. The position of the shoals is difficult to forecast, since the migrations depend on a supply of zooplankton, and this in turn on the water temperature.

The rich herring fisheries which once extended down the east coast of Britain and into the English

Channel have been impoverished since 1950 by the adoption of intensive trawling, rather than selective drifting methods of fishing. This brings up great numbers of immature fish from the nursery grounds, and they can be used only for fish meal and cattle food. No herring has been obtained from the western areas of the English Channel for some years, but a few returned in 1973. This may be related to the return of cooler waters.

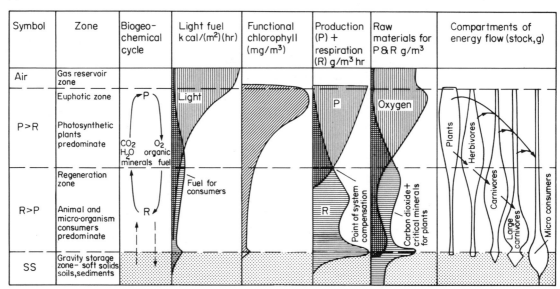

FIGURE 24:10 *Ecological systems.*
(A) The main vertical zones of ecological systems on land in in water.
(B) The principal vertical zones in ten contrasting ecosystems.
(After H.T. Odum, 1971.)

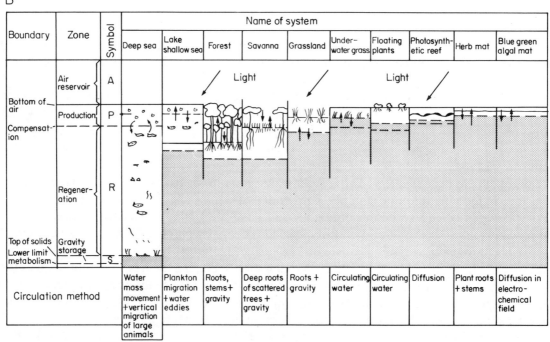

The cod (*Gadus callanus*)

This is another North Atlantic and Arctic form, ranging from Greenland to the English Channel, and from Newfoundland to Virginia: it is most abundant in seas with a temperature range within 0-10°C. It is the most important of all fish species in terms of a human food resource. Landings valued at £25 million reached the United Kingdom in 1965 — over half the total of all demersal fish.

In the North Sea, spawning occurs in the first four months of the year, when the fish gather in shoals near the bottom at depths of 60-100 m, so that most of the spawning grounds are in the northern parts of the Sea. The female sheds between 3 and 7 million eggs, whilst the male sheds milt (sperm). Fertilisation is external in the water. Fertilised eggs are buoyant and rise to the surface. Most eggs hatch out after 9-20 days, forming a larva 4 mm long, which grows to 2 cm in a planktonic phase lasting 10 weeks, feeding on copepods. Then the cod fry descend to the sea floor in rocky nursery areas and feed on small crustaceans. Further growth involves changes of diet to larger and larger prey: herring, sand eels, haddock and squid. The North Sea forms grow to 8 cm by 6 months; 14-18 cm by 1 year; and 28-35 cm in 2 years. Farther north, growth is slower (e.g. 30 cm after 3 years). Maturity is reached after 4-5 years (approximately 70 cm long, some fish growing to 1.5 m).

The study of life in the oceans is thus complex, and still based on little information apart from areas where fisheries have made research economic, or where other projects have been established. At present, more studies are being carried out than ever before, and diving gear is more widely available for work in shallow depths of water. The Sealab experimental stations off the Californian coast are also resulting in an increased understanding of the oceanic situation. The oceans represent a vast area within the biosphere: one which includes an outstanding variety of living species; one which is valuable as a producer of protein; and one which has biome-types quite different in character from those on the land, necessitating different methods of investigation (Figure 24.10).

Bibliography

Bibliography

Many books have been consulted in the course of writing this volume. They will be found of value to those following up a particular point or reading more deeply into the subject as a whole. In order to assist a degree of selection from a long list, an asterisk (*) has been placed against those which are particularly useful in the context of each list. The bibliography is divided into three sections:

Section A. Source books, and work books which provide a simple introduction to techniques for which there was not sufficient space in this text; collections of original review papers; general interest background reading; and some reference manuals.

Section B. General texts of approximately the same level as this volume, but giving a different or more restricted approach in terms of subject matter.

Section C. Advanced texts, mostly published since 1970, which give access to more detailed developments.

Section A

*Atkinson, B W (1968). *The weather business*. Aldus.
Bowen, D (1969). *Britain's weather*. David and Charles.
Brazell, J H (1968). *London weather*. HMSO.
Chandler, T J (1965). *The climate of London*. Hutchinson.
Costello, D F (1971). *The prairie world*. David and Charles.
Deacon, M (1971). *Scientists and the sea 1650-1900*. Academic Press.
*Detwyler, T R (1971). *Man's impact on environment*. McGraw-Hill.
Dorst, J (1970). *Before nature dies*. Collins.
*Ehrlich, P R (1971), Editor. *Man and the ecosphere: readings from Scientific American*. Freeman.
Ehrlich, P R & Ehrlich, A H (1972), Editors. *Population, resources, environment: issues in human ecology*. Freeman.
*Eyre, S R (1971), Editor. *World vegetation types*. Macmillan.
Goody, R M & Walker, J C G (1972). *Atmospheres*. Prentice-Hall.
Guinness Book of records (published each year).
*Hanwell, J D and Newson, M D (1973). *Techniques in physical geography*. Macmillan.
*Hopkins, B (1965). *Forest and savanna*. Heinemann.
Horsfield, B & Stone, P B (1972). *The great ocean business*. Hodder and Stoughton.
Howe, G M (1972). *Man, environment and disease in Britain*. David and Charles.
Kendrew, W G (1937). *Climates of the continents*, 3rd edition. Oxford.
*Lamb, H H (1966). *The changing climate*. Methuen.
Lane, F W (1966). *The elements rage*. David and Charles; Sphere paperback.

Manley, G (1952). *Climate and the British scene*. Collins.
*Maunder, W J (1970). *The value of the weather*. Methuen.
McBoyle, G (1973), Editor. *Climate in review*. Houghton Mifflin.
*Miles, P M & Miles, H B (1966).
    Seashore ecology. Hulton.
    Freshwater ecology. Hulton.
    Town ecology. Hulton.
    Woodland ecology. Hulton.
    Chalkland and moorland ecology. Hulton.
*Moore, J R (1971), Editor. *Oceanography: readings from Scientific American*. Freeman.
National Academy of Sciences (1969). *Resources and man*. Freeman.
Neuberger, H & Cahir, J (1969). *Principles of climatology: a workbook*. Holt, Rinehart and Winston.
Pearsall, W H (1950). *Mountains and moorland*. Collins.
*Pedgley, D E (1962). *Elementary meteorology*. HMSO.
*Pennington, W (1969). *The history of British vegetation*. English Universities Press.
Perry, R (1972). *The unknown ocean*. David and Charles.
Pilsbury, R K (1969). *Clouds and weather*. Batsford.
*Pirie, R G (1973). *Oceanography: contemporary readings in ocean sciences*. Oxford.
Proctor, R (1972). *Trees of the world*. Hamlyn.
Rumney, G R (1968). *Climatology*. Macmillan.
*SCEP (1970). *Man's impact on the global environment*. Massachusetts Institute of Technology.
Scientific American special issue, September 1970. The biosphere.
Scorer, R & Wexler, H (1967). *Cloud studies in colour*. Pergamon.
Scorer, R (1972). *Clouds of the world*. David and Charles.
*SMIC (1971). *Inadvertent climate modification* (report on Study of Man's Impact on Climate). Massachusetts Institute of Technology.
Stamp, L D (1964). *The geography of life and death*. Fontana.
Steers, J A (1971), Editor. *Applied coastal geomorphology*. Macmillan.
Thomas, W L (1956), Editor. *Man's role in changing the face of the Earth*. University of Chicago.
Vedel, H & Lange, J (1960). *Trees and bushes*. Methuen.
*Wallington, C E (1967). *Your own weather map*. Royal Meteorological Society.
Walter, H (1973). *Vegetation of the Earth*. English Universities Press.
*Wickham, P G (1970). *The practice of weather forecasting*. HMSO.
*Wilson, J T (1972), Editor. *Continents adrift: readings from Scientific American*. Freeman.

Section B

*Barry, R G & Chorley, R J (1968). *Atmosphere, weather and climate*. Methuen.
Billings, W D (1964). *Plants and the ecosystem*. Macmillan.
*Bridges, E M (1970). *World soils*. Cambridge University Press.
Buffaloe, N D (1968). *Animal and plant diversity*. Prentice-Hall.
Cailleux, A (1968). *Anatomy of the Earth*. World University Library.
Chandler, T J (1967). *The air around us*. Aldus.
Day, J A (1966). *The science of weather*. Addison-Wesley.
Flohn, H (1969). *Climate and weather*. World University Library.
Forsdyke, A G (1969). *The weather guide*. Hamlyn.

Hallam, A (1973). *A revolution in the Earth Sciences*. Oxford.
Hare, F K (1953). *The restless atmosphere*. Hutchinson.
Laporte, L F (1968). *Ancient environments*. Prentice-Hall.
McAlester, A (1968). *The history of life*. Prentice-Hall.
Miller, A A (1953). *Climatology*, 8th edition. Methuen.
*Odum, E P (1963). *Ecology*. Holt, Rinehart and Winston.
*Reid, K (1969). *Nature's network*. Aldus.
Riehl, H (1965). *Introduction to the atmosphere*. McGraw-Hill.
Riley, D & Young, A (1966). *World vegetation*. Cambridge University Press.
Selby, M J (1971). *The surface of the Earth*, Volume 2. Cassell.
Strahler, A N (1969). *Physical geography*, 3rd edition. Wiley.
*Strahler, A N (1972). *Planet Earth: its physical systems through geologic time*. Harper Row.
*Strahler, A N & Strahler, A H (1973). *Environmental geoscience*. Wiley.
Sutcliffe, R C (1965). *Weather and climate*. Weidenfeld and Nicholson.
Tait, R V (1968). *Elements of marine ecology*. Butterworths.
Turekian, K K (1968). *Oceans*. Prentice-Hall.
*Whittaker, R H (1970). *Communities and ecosystems*. Macmillan.

Section C

*Barrett, E C & Curtis, L F (1974), Editors. *Environmental remote sensing*. Arnold.
Barrett, E C (1974). *Climatology from satellites*. Methuen.
*Barry, R G & Perry, A H (1973). *Synoptic climatology*. Methuen.
Corby, G A (1970), Editor. *The global circulation of the atmosphere*. Royal Meteorological Society.
Cruickshank, J G (1972). *Soil geography*. David and Charles.
Easton, W H (1960). *Invertebrate palaeontology*. Harper Brothers.
Ekman, S (1953). *Zoogeography of the sea*. Sidgwick and Jackson.
Eyre, S R (1968). *Vegetation and soils*, 2nd edition. Arnold.
*Gross, M G (1972). *Oceanography*. Prentice-Hall.
King, C A M (1962). *Oceanography for geographers*. Arnold.
Lewis, J R (1964). *The ecology of rocky shores*. English Universities Press.
*Lockwood, J G (1974). *World climatology*. Arnold.
Munn, R E (1966). *Descriptive micrometeorology*. Academic Press.
Neill, W T (1969). *The geography of life*. Columbia University Press.
Odum, H T (1971). *Environment, power and society*. Wiley.
Polunin, N (1960). *Introduction to plant geography*. Longmans.
*Richards, P W (1964). *The tropical rain forest*. Cambridge University Press.
Seddon, B (1971). *Introduction to biogeography*. Duckworth.
Sellars, W D (1965). *Physical climatology*. University of Chicago.
Sparks, B W and West, R G (1972). *The Ice Age in Britain*. Methuen.
*Tivy, J (1971). *Biogeography*. Oliver and Boyd.
Trewartha, G T (1961). *The Earth's problem climates*. University of Wisconsin.
*Watts, D (1971). *Principles of biogeography*. McGraw-Hill.
West, R G (1970). *Pleistocene geology and biology*. Longmans.
Weyl, P K (1970). *Oceanography*. Wiley.

Index

Absorption 19-21
Adaptation 269
Adiabatic expansion 31,33
Adiabatic lapse rates 31-33
Aeroplane observation (weather) 131
Airglow 130
Air masses 62-3, 93, 102-3
Albedo 19, 21, 27, 90
Albic zone 139, 193
Alfisol 141, 195
Allerød phase 225-6
Alluvial soils 203
Alpine tundra, mountain vegetation 273-5
Alps 136
Alps — climatic zones 150
Altitude and temperature 23
Altitude and vegetation zones 274-5
Amazon basin 105, 242
Anabatic wind 55
Ancient ecosystems 222-7
Andes 116
Andes — climatic zones 150
Animalia 159, 160
Animals in, arid areas 271
    broad-leaved temperate forests 263
    Mediterranean regions 258
    northern coniferous forests 259
    seasonal tropics 257
    temperate grasslands 266
    tropical forests 249
    tundra 272-3
Anticyclone 69, 70-1, 98, 99, 113, 121-3, 124, 127, 128, 147
Anticyclonic blocking 70, 102
Anti-Locust Research Centre 221-2
Arabia 110
Arctic 170
Arctic brown soil 202
Arctic climate 146-7
Arctic gley soil 202
Argillic horizon 138, 193
Arid area soils 201

Arid biome-types 270-72
Arid climatic areas 108-15, 151
Aridisol 140, 195
Aridity 108-9
Arid region storm 110-111
Atacama desert 112-3, 116
Atlantic phase 226-7
Atmosphere,
    density 12
    gases 12
    heating 7, 12
    movement 54-60
    pollution 89
    vapour 28-30
Australasian realm 232-4
Australasian woodlands 143
Australian desert 113
Autotrophs 166, 218
Azonal soils 196, 203

Bacteria and soil 191
Bahamas 144
Banyan tree 253
Baobab tree 253
Bar graph 95
Bathythermograph 26
Beech wood 142
Bergeron process 44
Biogeography 228
Biomass 181, 217
Biome 239
Biome-types 239-41
Biosphere 159
Boreal forest 258-62
Boreal phase 226
Boundary currents 50-1
British woodlands 158
Broad-leaved temperate evergreen
    forests 263
Broad-leaved temperate forests 262-4
Brown earths 138, 199
Buttress roots 244

Calcic horizon 139
Calcification 194
Calcrete 201
Cambic horizon 139
Cape Town area 121
Carbon 172, 174-5
Carbon dioxide 12, 84
Cardinal points (germination) 182
Carnivore 167-9
Cauliflory 244
Central America, Caribbean 124
Central Brazil 124
Central California 120
Central Chile 120
Chaparral 258
Characteristic groups (animals) 229
Chernozem 200, 265
Chestnut brown soils 201
Chinook 114
Cholera 236-7
Chordates — fossil record 162-3
Cinnamon soils 200
Cirriform clouds 32, 34, 39, 42, 148
City climates 86-90
Clay-colloidal matter 189
Clayskins 137
Climate 85
   and comfort 85
   and soils 190-1
Climates,
   ancient 151-8
   classification 91-5
   cyclic changes 152, 153
   geological evidence 154-8
   historical evidence 151-4
   statistics 95-6
Climatic changes,
   causes 156-8
   implications 158
Climatic climax 214, 220, 264
Climatic diagrams 96
Cloud,
   clusters 71
   height 33
   shape 33
   types 32-4
Cloudiness and altitude 150
   and temperature 23, 24
Clouds,
   cirriform 32, 34, 42, 148
   cumuliform 32-7, 38, 40, 98-9, 117, 125, 132
   stratiform 32-4, 40-2
Coalescence (raindrops) 44
Coal Measures ecosystem 224
Cod 287
Cold area soils 201-2
Cold front 124
Cold water continental shelves 281
Colombia coast 106
Commensalism 206
Community 214
Competition (plants, animals) 206, 245
Condensation,
   nuclei 30
   process 30-1
Conditional stability in atmosphere 35
Conduction 19, 88
Congo basin 107, 242
Consumers 167, 218
Continentality 22
Continental movements 157-8
Contour-ploughing 17, 205
Convection 19, 35, 71
Convergence of air 58, 65-6
Coriolis force 52-3, 57, 105
Cumuliform clouds 36-7, 38, 40, 98-9, 117, 125, 132
Cypress swamp 261

Daily (diurnal) regime 105, 107
Daily Weather Report (U.K.) 76-80
Day and night length 10
Daylight length and plants 179
Deciduous seasonal tropical forest 254
Deciduous trees 251, 254, 256, 262-3
Deep ocean circulation 51
Deep ocean life 283
Density of,
   atmosphere 177
   seawater 177-8
Depressions (midlatitude cyclones) 5, 63-7, 68, 97, 98, 100-4, 119, 126-8, 145, 157-8
Desert vegetation 143
Detritus pathway 168
Dew point 29
Diseases and biogeography 236-8
Dispersal — seeds 207
Disphotic zone 178
Divergence of air 58, 65-6
Dominant species 217

Duripan 139
Dust Bowl 17, 114, 116, 171, 205, 209

Early Tertiary ecosystem 224
Earth environment and life 163-4
Earth,
   axis tilt 10
   orbit 10
   rotation 9, 52-3, 145
   shape 8
East Africa 125-6
Eastern Asia 126-7
Eastern Australia 127
East Indies 107
Ecocline 238-9, 251
Ecology 214
Ecosystem 214, 228, 230
Ecosystems in oceans 278-9
Ecosystems of past 220-7
Ecosystems on world scale 238-41
Ecotone 262
Edaphic factors 220, 256
Ekman spiral 52-3
Electromagnetic spectrum 7
Eluvial horizon 193
Energy,
   and food chain 168
   balance in atmosphere 25
   in biosphere 165-6
   in plant succession 217-9
   pathway 258
   transformations 6, 7
Entisol 140, 195
Environmental gradient analysis 238, 240
Environmental niches 249, 280
Epipedon 195-6
Epiphytes 245, 246-8, 255-6
Equatorial forests 231
Equatorial rainy climates 94, 97, 99, 104-7
Ethiopian realm 232
Eucalyptid woodland 143
Euphotic (photic) zone 280
Eutrophication 176, 183
Evaporation 22, 28, 29, 88, 90
Evapotranspiration 109
Evolution of extratropical forests 263-4
Evolution of life forms 213
Extinction of species 211

Ferralitic soil 201

Ferralization 194
Ferrisol 201
Ferruginous soils 200
Fertiliser 212
Fire and grasslands 210-11
Fisheries 283-7
Fishes 162
Floral kingdoms 228-9
Florida 141
Fog 38, 81, 88-9, 112, 117, 120
Fogs,
   advection 31
   radiation 31-2
   smog 32
Food chain 167, 279
Food web 167, 279-80
Forest structure 143
Formations (plants) 239
Fossils 161
Freezing nuclei 44
Freshwater plants 166
Fronts 62-4, 68, 78-9, 97, 98, 100, 113, 125, 127
   Katafronts 63-4
   anafronts 63-4
   occluded 65-6

Garrigue 258
Geochemical cycles 175
Geostrophic,
   currents 52
   winds 57
Glacial periods 154-5
Gley 138, 197, 202, 203
Gleying 194
Gley-podzol 138
Gradient analysis 238
Grasslands 256, 264-6, 268
Grasslands — southern hemisphere 265-6
Grazing pathway 168
Great Salt Lake, Utah 135
Grey desert soil 201
Ground observer (weather) 131
Growing season 183
Gulf stream 131
Gullying 17
Gyre 50, 53

Hadley cell 60
Hail 45-6, 81

Harmattan 125
Heat budget (balance) 27
'Heat island' effect 86-8
Heat transfer 19-20, 24-5, 60
Heavy metals 84
Hekisotherms 182
Heliophytic plant 178
Helophyte 179
Herbivore 167-9
Herring 285-6
Heterotrophs 166-7
Himalayas and climate 117
Historical records 151-4
Histosol 141, 195
Holocene 226
Humidity,
  absolute 30
  relative 30
Humid temperate soils 198-9
Humid tropical soils 196-8
Humus 159, 191, 200
  moder 193, 199
  mor 193, 198
  mull 193, 199
Hurricane (tropical cyclone) 5, 71-3, 81-2, 123, 124, 134
Hydrological cycle 28, 172
Hydrophyte 179
Hydrosere 216, 249
Hytheragraph 95, 101

Ice crystals 44-8
Illuvial horizon 193
Inceptisol 138, 140, 195
India 117, 121-3
Indo-Malaysian region 242
Industry and weather forecasts 75
Infrared waves 7
Insolation 5, 8
Instability in atmosphere 35
Interglacial periods 154-5
Inter-tropical Convergence Zone (ITCZ) 58, 71-3, 93, 104-7, 124-5
Intrazonal soils 196, 202-3
Inversion 32
Invertebrate (non-chordate) animals 162-3
Ionosphere 12
Irrigation 29
Isanomaly 22-3
Isobars 54-6, 78-9, 110

Isotherms 20-3, 25, 147
Isothermal layer 8

Japan 141
Jet contrails 84
Jet stream 13, 57-8, 60, 66-7, 97, 98, 110-11, 119, 122-3, 126-7, 134
Jupiter 10

Katabatic winds 55
Köppen classification of climate 92-3
Krakatoa 215
Lake forest 259
Land classification 185
Land animals — world distribution 229, 232-4
Land environments 177
Land plants 166
Land plants and environment 206
Land plants — world distribution 228-9
Landsat satellite 26
Land-sea breeze 54-5, 134
Langmuir circulation 52
Lapse rates 35
Latent heat 28-30
Laterisation 209-10
Laterite 197-8
Leaching 190, 193, 197, 198
Leaf shapes — tropical forests 244
Lenticular clouds 41-2
Lianes 244
Life-form classification (Dansereau) 247
Life-form classification (Küchler) 248
Life-form classification (Raunkaier) 246
Life forms in soil 189
Light and plants 178
Lightning 5, 45-8, 81
Light waves 20
Line graph 95
Lithosols 203
Living matter — chemical composition 172
Local winds 119, 120
Locusts 220-2

Mackenzie delta 144
Malaria 237-8
Mallee scrub 258
Man,
  and biosphere 208-13

and animals 211-12
and fire, grazing 210-11
and future of biosphere 212-3
and soil 208-10
Man and ecosystems 220-1, 227
Man and Mediterranean lands 158-9
Man and seasonal tropics 157-8
Mangroves 231, 249
Maquis 258
Marine data buoy 26
Mars 10
Material cycles 11
Mediterranean,
   climate 117, 118-21
   coasts 118-20
   land use 210
   soils 200
   vegetation 141, 257-8
Megatherms 182
Mercury 9
Mesophyte 179
Mesotherms 182
Meteorology 18
Microclimates 85-6, 150
Microphyllous plants 179, 254
Microtherms 182
Middle Jurassic ecosystem 224
Middle latitude,
   weather systems 61-71
   west coast rainy climates 94, 97, 98, 100-4
Migration and dispersal 207-8
Mixed temperate forests — ecotones 262
Mollic epipedon 138
Mollisol 140, 195
Monsoon 5, 93, 107, 117, 121-4
Monsoon forest 250, 254
Moon 27, 130, 170
Moon — heat budget 21
Mountain climates 148-50
Mountain vegetation 273-5
Mountains and climate 93
Moving continents and zoogeography 235-6
Mutualism 206

Natric horizon 138
Nearctic region 232-3
Neotropical region 233-4

Neptune 10
Net production 218
Nitrogen 12, 14, 172, 174-5
North America — interior 145
North American temperate forests 261
Northern coniferous forests 258-9, 262
Nuclear energy 6-7
Nutrients and plants 136, 171, 176
Nutrients in seawater 14, 279
Nutrients in soil 189

Ocean-continent heating contrast 21-23
Oceanic biome-types 280-83
Oceanic deserts 114
Oceanography 18, 26
Oceans,
   and climate 93
   composition 13-4
   currents 50-3, 112-3
   heat store 11, 20, 23-5
   salinity 13-4, 54
   temperature 54
   water density 14, 24, 52
   water masses 53-4
   water movements 50-4
Oceans, animals and plants 234-5
Ocean environments 177
Ocean life zones 180
Oceans — light penetration 178
Oceans with upwelling nutrients 282
Ochric epipedon 138
Oil 84
Old Red Sandstone ecosystem 223
Open oceans, cold 282
Open oceans, warm 283
Optimum growth temperatures 182-3
Orbital period of planets 10
Organic molecules 172
Organic soils 184, 203
Organisms and soil 191
Oriental region 232-3
Origins,
   of atmosphere and oceans 15
   of life 163-4
Oxford Clay (Jurassic) ecosystem 222-3
Oxic horizon 139
Oxisol 141, 195
Oxygen 12, 14, 172-3
Ozonosphere 12

Pacific coast, North America 104
Palaeoecology 222-4
Palearctic region 232-3
Parasitism 206
Parent material (soil) 184, 189-90
Patagonia 114
Peat 138
Pedalfers 190
Pedocals 190
Pedology 186
Pedon 195-6
Peds 189
Permafrost zone 202
Pesticides 84, 212, 220
Phosphorus 172, 174-5
Photic (euphotic) zone 178
Photosynthesis 7, 165-6, 178
Phototrophs 166
pH value 188-9, 191
Phytoplankton 159, 166, 178, 180, 282
Pine-birch-oak association 142
Plagioclimax 216
Plagiosere 216
Plantae 159, 161
Plant forms — tropical evergreen rain-forest 243
Plant nutrients 172
Plant productivity 169
Plants — fossil record 162-3
Plants in oceans 278
Pleistocene changes and vegetation 264
Pleistocene climates 154-5, 157
Pliocene ecosystem 223
Plough pan 204
Pluto 10
Podzolisation 193
Podzols 159, 193, 198-9, 259
Polar climates 146-8
Polar front 60, 63
Polar ice cap climate 147-8
Pollen 151
Pollen diagram 224-6
Pollution 81, 84
Population 214
Potential evapotranspiration 29
Potential land use (world) 212
Prairies 210, 264
Preboreal phase 226
Precipitation 29, 42-4, 91
Pressure,
   and altitude 149

   gradient 55-6
   world distribution 59
Primary producers 166
Primary production 219, 245, 249
Primary productivity,
   arid regions 271
   Mediterranean regions 258
   northern coniferous forests 259
   oceans 279
   temperate grasslands 266
   temperate woodlands 263
   tundra 272
Primary succession 220
Prismatic structure (soil) 184
Pycnocline 14, 15

Quaternary ecosystems in British Isles 224-7

Rabbits 205
Radiation 19, 88, 90
Radiosonde 26
Raindrops 43
Rainfall,
   and altitude 149-50
   tropical 126
   variability 109
   world distribution 43
Rainfall records,
   arid regions 115
   equatorial zone 106
   humid temperate 101
   polar/alpine 146
   wet season climates 128, 145
Rain formation 42-4
Rain-making 81, 83
Rain shadow effect 36, 102, 108, 114, 123, 145
Ranker 203
Raunkiaer spectra 243, 246, 255
Reflection 19-21, 87
Regosols 203
Relative humidity 89, 90
Relict groups (animals) 229
Relief and soils 191-2
Rendzina 200, 202-3, 205
Respiration 218
Rock cycle 172
Rotation rates of planets 10

INDEX 301

Sahara 111-2
Salic horizon 139
Saline soils 205, 208
Salinization 194
Saprophytes 245
Satellites (weather) 131
Saturn 10
Savanna 210-11, 250, 256
Scattering 19
Sciophytic plant 178
Sclreophyllous plant 179
Sclerophyllous scrub vegetation 258
Seasonally wet region soils 199-201
Seasonal weather (British Isles) 103
Seasons 10, 22-3
Seaweed colours 166
Secondary forest 248
Secondary succession 220
Semi-arid regions 114
Semi-evergreen seasonal rain forest 251, 254
Sere 214
Siberia 128
Smog 88-9
Snow line 149
Sodium chloride 12
Soil,
   catena 196
   classification 194-7
   colour 187
   definition 186
   erosion 204-5, 208-9, 258, 260
   formation 189-94
   orders 195
   organisms 189
   profile 189, 190
   series 196
   structure 184
Soils and climate 92
Soils,
   arid areas 201
   azonal 203
   cold areas 201-2
   humid temperate areas 198-9
   humid tropics 196-8
   intrazonal 202-3
   seasonally wet regions 199-201
Solar,
   constant 7-8
   energy 5-10
   flare 16

   radiation 152
   system 5
Solonchak 194, 203
Solonetz 194, 201, 203
Southern oceans (climate) 98, 100
Species 159
Species of conifer 259
Species variety 181, 242
Specific heat capacity 21
Spodic horizon 138, 193
Spodosol 140, 195
Stability in atmosphere 35, 40
Steppes 264-5
Strangler plants 245
Stratification in,
   tropical rainforest 245
   tropical seasonal forest 251, 254
Stratification (layering) of plants 217
Stratiform clouds 39, 40-2, 98-9, 112, 116, 132
Stratosphere 8
Subatlantic phase 227
Subboreal phase 227
Succession of plants 214-6, 230
Sulphur 172-5
Summer wetness index 137
Sun 16
Sunrise 129
Sunset 129
Sunshine (London) 59
Sunspots 152
Supersonic transports (SST) 84
Symbiosis 206
Systems 5

Temperate east coasts, interiors (wet summer climates) 126-8, 145
Temperate forests, grasslands 258, 261, 262-6
Temperate grasslands 264-6
Temperate mountain vegetation zones 274-5
Temperature 91
Temperature and height 149-50
Temperature and living forms 180-3
Temperature at Earth's surface 20-3
Temperature gradient 25
Temperature-humidity index 30

Temperature in cities 87-9
Temperature records,
  arid regions 115
  equatorial zone 107
  humid temperate 101
  polar/alpine 146
  wet seasonal climates 128, 145
Temperature trends 151-2, 155
Territories 207
Thorn forest 254-5
Thorn scrub 255
Thunderstorms 5, 45-7, 48, 89, 123, 124, 135
Time and ecosystem succession 220
Time and soil development 191-2
Tolerance 183
Tornado 47, 49, 83, 133
Toxic chemicals and organisms 173
Trade winds 58-9, 71, 121-5
Transpiration 28
Tree line 274
Tropical cyclone 71
Tropical evergreen rainforest 242-9
Tropical mountain vegetation zones 274
Tropical seasonal forests and savannas 251-7
Tropical weather systems 71-3
Tropical wet summer climates 121-6
Troposphere — winds 8, 13, 55-8
Tundra 160, 269, 272-3
Tundra biome-types 272-3
Tundra soils 202
Typhoon 71

Ultisol 141, 195
Ultraviolet rays 7
Umbric epipedon 138
Underwater habitats 277
United States Soil classification system (7th Approximation) 138-40, 194-5
Unstable atmosphere 98-9
Uranus 10

Valley winds 54-5
Vapour pressure 29, 30

Vegetation and climate 92
Vegetation life-forms 246-8
Vertisol 140, 195, 200, 201
Venus 10

Warm water continental shelves 281
Water and plants 137, 173
Water droplet sizes 42
Water in organisms 172-5
Water phases and changes 29
Water, plants and animals 179
Water vapour 12
Wave disturbances 71-2
Weather forecasts 74-80
Weathering 189-90
Weather maps 76-80
Weather modification 80-1, 83, 84
Weather satellite images 68, 69
Weather systems 61
West Africa 124-5
Westerly winds 58-9
Western Australia 121
Western Europe 101
Western forests (North America) 262
Windbreak 17
Windrows 179-80
Winds 54-60
Winds and altitude 149-50
Winds and organisms 171, 179
Winds — world maps 59
Winter severity index 153
Woodland climates 90

Xerophyte 179
Xerophytic plants 270
Xerosere 216

Zonal soils 196
Zone of eluviation 190
Zone of illuviation 190
Zoogeographical regions 229, 232-4
Zoogeography 236
Zooplankton 159, 180